Universitext

Universitext

Universitext is a series of textbooks that presents material from a wide variety of mathematical disciplines at master's level and beyond. The books, often well class-tested by their author, may have an informal, personal even experimental approach to their subject matter. Some of the most successful and established books in the series have evolved through several editions, always following the evolution of teaching curricula, into very polished texts.

Thus as research topics trickle down into graduate-level teaching, first textbooks written for new, cutting-edge courses may make their way into *Universitext.*

More information about this series at http://www.springer.com/series/223

Marta Lewicka

A Course on Tug-of-War Games with Random Noise

Introduction and Basic Constructions

Springer

Marta Lewicka
Department of Mathematics
University of Pittsburgh
Pittsburgh, PA, USA

ISSN 0172-5939 ISSN 2191-6675 (electronic)
Universitext
ISBN 978-3-030-46208-6 ISBN 978-3-030-46209-3 (eBook)
https://doi.org/10.1007/978-3-030-46209-3

Mathematics Subject Classification (2020): 91A15, 91A24, 31C45, 35B30, 35G30, 35J70

This Springer imprint is published by the registered company Springer Nature Switzerland AG.
The registered company address is: Gewerbestrasse 11, 6330 Cham, Switzerland

Preface

The goal of these Course Notes is to present a systematic overview of the basic constructions and results pertaining to the recently emerged field of Tug-of-War games, as seen from an analyst's perspective. To a large extent, this book represents the author's own study itinerary, aiming at precision and completeness of a classroom text in an upper undergraduate- to graduate-level course.

This book was originally planned as a joint project between Marta Lewicka (University of Pittsburgh) and Yuval Peres (then Microsoft Research). Due to an unforeseen turn of events, neither the collaboration nor the execution of the project in their priorly conceived forms could have been pursued.

The author wishes to dedicate this book to all women in mathematics, with admiration and encouragement. The publishing profit will be donated to the Association for Women in Mathematics.

Pittsburgh, PA, USA
October 2019

Marta Lewicka

Contents

Chapter 1
Introduction

The goal of these Course Notes is to present a systematic overview of the basic constructions pertaining to the recently emerged field of Tug-of-War games with random noise, as seen from an analyst's perspective.

The Linear Motivation The prototypical elliptic equation, arising ubiquitously in Analysis, Function Theory and Mathematical Physics, is the Laplace equation:

$$\Delta u \doteq \operatorname{div}\big(\nabla u\big) = 0.$$

Laplace's equation is linear and its solutions $u : \mathcal{D} \to \mathbb{R}$, defined on a domain $\mathcal{D} \subset \mathbb{R}^N$, are called harmonic functions. They are precisely the critical points (i.e., solutions to the Euler–Lagrange equation) of the quadratic potential energy:

$$\mathcal{I}_2(u) \doteq \int_{\mathcal{D}} |\nabla u(x)|^2 \,\mathrm{d}x.$$

The idea behind the classical interplay of the linear Potential Theory and Probability is that harmonic functions and random walks share a common averaging property. We now briefly recall this relation.

Indeed, on the one hand, for any $u \in C^2(\mathbb{R}^N)$ there holds the expansion:

$$\fint_{B_\epsilon(x)} u(y)\,\mathrm{d}y = u(x) + \frac{\epsilon^2}{2(N+2)}\Delta u(x) + o(\epsilon^2) \qquad \text{as } \epsilon \to 0+.$$

Recall that $\fint_B u \doteq \frac{1}{|B|}\int_B u$ denotes the average of u on a set B. The displayed formula follows by writing the quadratic Taylor expansion of u at x:

M. Lewicka, *A Course on Tug-of-War Games with Random Noise*, Universitext, https://doi.org/10.1007/978-3-030-46209-3_1

$$u(y) = u(x) + \langle \nabla u(x), y - x \rangle + \frac{1}{2} \Big\langle \nabla^2 u(x) : (y-x) \otimes (y-x) \Big\rangle + o(|y-x|^2),$$

and averaging each term separately on the ball $B_\epsilon(x)$. In the right hand side, the zeroth order term $u(x)$ averages to itself, the first order term averages to zero due to the symmetry of the ball, while the second order average consists of the scalar product of two matrices: $\nabla^2 u(x)$ and $\fint_{B_\epsilon(x)} (y-x) \otimes (y-x) \, \mathrm{d}y = \frac{\epsilon^2}{N+2} Id_N$, yielding:

$$\langle \nabla^2 u(x) : Id_N \rangle = \mathrm{Trace} \nabla^2 u(x) = \Delta u(x).$$

We note that the obtained mean value expansion is consistent with the fact that the equivalent condition for $\Delta u = 0$ is the mean value property:

$$\fint_{B_\epsilon(x)} u(y) \, \mathrm{d}y = u(x).$$

Hence, the coefficient $\Delta u(x)$ in the mean value expansion measures the second order (in the averaging radius ϵ) error from the satisfaction of the mean value property and thus from harmonicity of u.

On the other hand, consider the discrete stochastic process where, at each step, the particle is randomly shifted from its current position x to a new position y, uniformly distributed in some ball $B_{\delta(x)}(x)$. Assume that the particle is initially located at $x_0 \in \mathcal{D}$ and that the radius of the random shift $\delta(x)$ at x equals, rather than ϵ as above, the minimum of the parameter ϵ and the particle's current distance from $\partial\mathcal{D}$. We see that this process never reaches the boundary and the position of the token is shifted indefinitely in $\mathcal{D}$; such infinite horizon process is called the ball walk. It is also clear that the current position of the particle equals the mean of its immediate future positions. The same property holds for process' values, defined as follows.

It can be checked that with probability one, the particle's positions accumulate at some limiting point $x_\infty \in \partial\mathcal{D}$. We then set $u^\epsilon(x_0)$ to equal the expected value of the given continuous boundary data F, evaluated at x_∞. It turns out that for each ϵ, all functions u^ϵ are actually one and the same harmonic function u (i.e., they are independent of ϵ), and that they coincide with the so-called Perron's solution of the Laplace–Dirichlet problem on $\mathcal{D}$:

$$\Delta u = 0 \quad \text{in } \mathcal{D}, \qquad u = F \quad \text{on } \partial\mathcal{D}.$$

The Perron solution may actually fail to achieve its defining boundary condition F when $\mathcal{D}$ is irregular. According to what is by now one of the fundamental results in Potential Theory, $u(y)$ coincides with the correct boundary value $F(y)$ precisely at those points $y \in \partial\mathcal{D}$ which satisfy the so-called Wiener regularity condition. This condition is independent of the choice of the continuous data F, and it is worth to mention that it can be derived directly through the ball walk, without appealing to its advanced independent analytical formulation.

We remark that the consecutive positions of the particle in the described process may be characterized as a discrete realization of continuous paths in the Brownian motion on $\mathbb{R}^N$ started at x_0, whereas the limiting position x_∞ is precisely the first exit point from $\mathcal{D}$. The study of these and other relations between linear elliptic PDEs and Stochastic Processes has been instrumental for the interrelated developments of Analysis, PDEs and Probability in the past century.

The Nonlinear Motivation If we replace the quadratic exponent in $\mathcal{I}_2$ by the p-th power, $p \in (1, \infty)$, the resulting p-potential energy functional:

$$\mathcal{I}_p(u) \doteq \int_{\mathcal{D}} |\nabla u(x)|^p \, \mathrm{d}x,$$

has the p-Laplacian as its Euler– Lagrange equation for critical points:

$$\Delta_p u \doteq \mathrm{div}\Big(|\nabla u|^{p-2}\nabla u\Big) = 0.$$

The p-Laplace equation occupies a similar central position for nonlinear phenomena as the Laplacian for the theory of linear PDEs. At points $x \in \mathcal{D}$ where $\nabla u(x) = 0$, this equation is degenerate when $p > 2$, and singular when $p < 2$. Its solutions are called p-harmonic functions; they arise in various contexts of Mathematical Physics and have many applications. We also mention that passing to the limit with $p \to \infty$ leads to (the normalized version of) the celebrated ∞-Laplacian:

$$\Delta_\infty u \doteq \frac{1}{|\nabla u|^2}\Big\langle (\nabla^2 u)\nabla u, \nabla u\Big\rangle,$$

which returns the second derivative of u in the direction of its normalized gradient $\frac{\nabla u}{|\nabla u|}$, whenever defined. The ∞-Laplacian is a fully nonlinear operator arising, for example, in the study of optimal Lipschitz extensions and image processing. It is the most difficult and the least understood among the Laplace-like operators, for which many questions (for instance, related to the regularity or uniqueness of eigenfunctions) remain still open. The p-Laplacian, Laplacian and the ∞-Laplacian are related by the following interpolation identity:

$$|\nabla u|^{2-p}\Delta_p u = \Delta u + (p-2)\Delta_\infty u.$$

We further remark that the operator in the left hand side above, namely:

$$\Delta_p^G u \doteq |\nabla u|^{2-p}\Delta_p u$$

is known as the "game-theoretical" p-Laplacian. Although Δ_p^G is still nonlinear, it is one-homogeneous and less singular than the full p-Laplacian Δ_p.

It is somewhat unexpected that the Probability approach as displayed in the linear case can be implemented with appropriate modifications, also in the setting of the nonlinear Potential Theory. One way of arriving at this observation is as follows. Averaging a function $u \in C^2(\mathbb{R}^N)$ on the ellipsoid, rather than on the ball $B_\epsilon(x)$ which worked well in the linear setting, leads to:

$$\fint_{E(x,\epsilon;\alpha,\nu)} u(y)\, dy = u(x) + \frac{\epsilon^2}{2(N+2)}\Big(\Delta u(x) + (\alpha^2-1)\langle\nabla^2 u(x) : \nu^{\otimes 2}\rangle\Big) + o(\epsilon^2).$$

Here, $E(x,\epsilon;\alpha,\nu) = x + \{y \in \mathbb{R}^N;\ \langle y,\nu\rangle^2 + \alpha^2|y - \langle y,\nu\rangle\nu|^2 < \alpha^2\epsilon^2\}$, where r is the radius, α is the aspect ratio and ν is the unit-length orientation vector. We recall that the scalar product of the two square matrices $\nabla^2 u(x)$ and $\nu^{\otimes 2} = \nu \otimes \nu$ in: $\langle\nabla^2 u(x) : \nu^{\otimes 2}\rangle = \langle\nabla^2 u(x)\nu, \nu\rangle$ returns the second derivative of u at x, in the direction of ν. Thus, for the choice $\alpha = \sqrt{p-1}$ and $\nu = \frac{\nabla u(x)}{|\nabla u(x)|}$, the above displayed mean value expansion becomes:

$$\fint_{E\left(x,\epsilon;\sqrt{p-1},\frac{\nabla u(x)}{|\nabla u(x)|}\right)} u(y)\, dy = u(x) + \frac{\epsilon^2}{2(N+2)}\Delta_p^G u(x) + o(\epsilon^2),$$

with the familiar quantity $\Delta_p^G u(x)$ at the second order in the averaging "radius" ϵ.

To obtain an expansion where the left hand side averaging does not require the knowledge of $\nabla u(x)$ and allows for the identification of a p-harmonic function which is a priori only continuous, one should additionally take the mean over all orientations ν. This idea can indeed be carried out by superposing: the deterministic average "$\frac{1}{2}(\inf + \sup)$", with the stochastic average "$\fint$", leading to the expansion:

$$\begin{aligned}\frac{1}{2}\Big(\inf_{z\in B_\epsilon(x)} + \sup_{z\in B_\epsilon(x)}\Big) & \fint_{E\left(z,\gamma_p\epsilon;\alpha_p\left(\left|\frac{z-x}{\epsilon}\right|\right),\frac{z-x}{|z-x|}\right)} u(y)\, dy \\ &= u(x) + \frac{\gamma_p^2\epsilon^2}{2(N+2)}\Delta_p^G u(x) + o(\epsilon^2).\end{aligned}$$

Above, the constant scaling factor γ_p and the quadratic aspect ratio function α_p satisfy the appropriate compatibility identity depending on N and p. As in the linear case, one can then prove that an equivalent condition for p-harmonicity $\Delta_p u = 0$ is the asymptotic satisfaction of the mean value equation:

$$\frac{1}{2}\Big(\inf_{z\in B_\epsilon(x)} + \sup_{z\in B_\epsilon(x)}\Big)\fint_{E\left(z,\gamma_p\epsilon,\alpha_p,\frac{z-x}{|z-x|}\right)} u(y)\, dy = u(x) + o(\epsilon^2).$$

In order to draw a further parallel, we now describe the discrete stochastic process modelled on the above equation, which is the two-player, zero-sum game, called the Tug-of-War with random noise. In this process, the token is initially placed at $x_0 \in \mathcal{D}$, and at each step it is advanced according to the following rule. First, either of the two players (each acting with probability $\frac{1}{2}$) updates the current position x_n by a chosen vector y of length at most ϵ; second, the token is further randomly shifted to a new position x_{n+1}, uniformly distributed within the ellipsoid $E\big(z, \gamma_p\epsilon, \alpha_p, \frac{y}{|y|}\big)$ that is centred at $z = x_n + y$. The game is stopped whenever the token reaches the ϵ-neighbourhood of $\partial\mathcal{D}$. The game value $u^\epsilon(x_0)$ is then defined as the expectation of the value of the given continuous boundary function F (extended continuously on $\mathbb{R}^N$) at the stopping position x_τ, subject to both players playing optimally.

The applied optimality criterion is based on the rule that Player II pays to Player I the value $F(x_\tau)$. This rule gives Player I the incentive to maximize the gain by pulling the token towards portions of $\partial\mathcal{D}$ with high values of F, whereas Player II will likely try to minimize the loss by pulling towards lower values. Due to the min-max property, the notion of optimality turns out to be well defined, i.e., the order of supremizing the outcome over strategies of the first player and infimizing over strategies of the opponent is immaterial.

We observe that since the random sampling is performed within radius of order ϵ, regardless of the token/particle's position, it is no more true that $\{u^\epsilon\}_{\epsilon\in(0,1)}$ are each, one and the same function, even in the linear case $p = 2$. However, it is expected that the family $\{u^\epsilon\}_{\epsilon\to 0}$ still converges pointwise in $\mathcal{D}$ to the unique Perron solution u of the associated p-Laplace–Dirichlet problem:

$$\Delta_p u = 0 \quad \text{in } \mathcal{D}, \qquad u = F \quad \text{on } \partial\mathcal{D},$$

subject to the continuous boundary data F. In agreement with the linear case, it is also natural to expect that this convergence is uniform for regular boundary, and that in that case $u = F$ on $\partial\mathcal{D}$. While the former result is not yet available at the time when these Course Notes are written (for $p \neq 2$), the latter two assertions hold true and will be precisely the subject of our studies.

The Whys and Wherefores Similarly to the linear case, where the notions and results of the linear Potential Theory find their classical counterparts via Brownian motion, the Game-Theoretical interpretation of the nonlinear Potential Theory brings up a fruitful point of view, allowing to replace involved analytical techniques by relying instead on suitable choices of strategies for the competing players. Applications include: a new proof of Harnack's inequality, a new proof of Hölder's regularity of p-harmonic functions, results on obstacle problems or the study of Tug-of-War games in the Heisenberg group.

Other aspects which are beyond the introductory scope of this textbook, concern the limiting exponent cases $p = \infty$ and $p = 1$. The case $p = \infty$ is related to the absolutely minimizing Lipschitz extension property of the ∞-harmonic functions (which may be defined on arbitrary length spaces), in connection to the pure Tug-of-War games. The case $p = 1$ concerns a variant of Spencer's "pusher-chooser" (deterministic) game and the level-set formulation of motion by mean curvature. Of interest are procedures of passing to the limit with $p \to \infty$ and $p \to 1$.

Parallel problems can be formulated and studied on graphs, and have recently found applications in the graph-based semi-supervised learning problems, yielding new algorithms with theoretical guarantees.

More generally, the approach of:

1. finding an asymptotic expansion in which the second order coefficient matches the prescribed partial differential operator of second order;
2. introducing a related mean value equation by removing higher order error terms in the expansion;
3. interpreting the mean value equation as the dynamic programming principle of a "game" incorporating deterministic and stochastic components;
4. passing to the limit in the radius of sampling in order to recover the continuum solutions from the values of the game process;

is quite flexible and allows to deal with several elliptic and parabolic nonlinear PDEs, including free boundary problems.

The Content of This Book We start with analysing the basic linear case in detail in Chap. 2, where we link the ball walk, the harmonic functions and the mean value property. This chapter presents a more classical material, but it crucially serves as a stepping stone towards gaining familiarity with the more complex nonlinear constructions. In Chap. 3 we are concerned with the case $p \geq 2$, which is somewhat closer to the linear case, by means of the first nonlinear extension of the mean value property (called here the averaging principle) and a resulting Tug-of-War game. Another game process and another asymptotic expansion, valid for $p > 2$, are studied in Chap. 4. Its advantage is that the game values u^ϵ inherit regularity properties of the boundary function F (continuity, Hölder and Lipschitz continuity), thanks to interpolating to the boundary in deterministic and stochastic sampling rules.

The aim of Chap. 5 is to introduce the notion of game-regularity of boundary points, which is essentially equivalent to the local equicontinuity of the family $\{u^\epsilon\}_{\epsilon\to 0}$, and ultimately to its uniform convergence to the unique viscosity solution of the studied Dirichlet problem. This notion extends the walk-regularity introduced in Chap. 2, and both notions may be seen as natural extensions of Doob's regularity for Brownian motion. It is expected that game-regularity is equivalent to the Wiener p-regularity criterion. In the so-far absence of such result, we show its two sufficient conditions: the exterior cone condition and the exponent range $p > N$. In Chap. 6 we derive the ultimate ellipsoid-based averaging principle that serves as the dynamic programming principle for the Tug-of-War game with noise, which is viable for the whole range exponent $p \in (1, \infty)$. We also show that the exterior corkscrew condition is sufficient for game-regularity, and that in dimension $N = 2$, every simply connected domain is game-regular, for any $p \in (1, \infty)$.

The final three chapters gather a background material, which is to be used at the instructor's and students' discretion. Appendix A contains basic definitions and facts in Probability: probability and measurable spaces, conditional expectation, martingales in discrete times, Doob's and Dubins' theorems. Appendix B serves as an introduction to Brownian Motion, where we discuss the Lévy construction, stopping times, Markov properties, the Wiener measure and the Brownian motion harmonic extensions. This material is only used in sections denoted by "$*$", towards comparing the values of different variants of Tug-of-War at $p = 2$ along with their regularity conditions, with the Brownian motion harmonic extension Doob's regularity. Both Appendix B and Sections* are independent of the main material, and may be skipped at first reading.

In Appendix C, we recall the preliminary facts in PDEs: Lebesgue and Sobolev spaces, harmonic functions, p-harmonic functions, viscosity and weak solutions, and also present some aspects of the nonlinear Potential Theory. Finally, Appendix D contains solutions of selected exercises.

Prerequisites The presentation aims to be self-contained, at the level of a classroom text in an upper undergraduate to graduate course. Familiarity with differential and integral calculus in $\mathbb{R}^N$, and with some basic measure theory, is assumed. Familiarity with the concepts of Probability and PDEs at the level of a rigorous core course for Mathematics majors is advised.

At the same time, these Course Notes are equipped with an extensive background and auxiliary material in the three appendix sections, where the range of definitions, facts, proofs and guided exercises is gathered. Students who passed the first courses on Probability and PDEs will be able to go through the main material starting from Chap. 2 with no problems. Students who lack such training should begin by approaching the adequate material from Appendix A and Sects. C.1–C.5 of Appendix C. The usage of the appendices is expected to be gauged to the students' preparation level, determined by the Course instructor.

Notation Having to decide between at times a bit heavier notation, or risking a student wonder about the dependence of quantities on each other, the domains of definiteness of functions, or the structure of the involved spaces, the former has been intentionally chosen. At the same time, the author has strived to make the notation as friendly, balanced and intuitive as possible.

Acknowledgments In preparing these Course Notes, the author has greatly benefited from discussions with Y. Peres, whose seminal work from a decade ago uncovered the deep connections between Nonlinear Potential Theory and Stochastic Processes. The author wishes to thank Y. Peres for advising her studies of Game Theory and Probability, in the oftentimes limiting context of the author's analysis-trained and oriented point of view.

The gratitude extends to J. Manfredi for introducing the author to the topic of this book, for discussions on p-Laplacian and viscosity solutions and for coauthoring joint papers. The author is further grateful to P. Lindqvist for discussions and the continuous kind encouragement.

An acknowledgement is due to Microsoft Research, whose financial support allowed for the author's visits to MSR Redmond in the early stages of this work. As a final word, the author would like to bring the readers' attention to the recent book by Blanc and Rossi (2019), which is concerned with the same topic as these Course Notes, albeit written with different scope and style.

Chapter 2
The Linear Case: Random Walk and Harmonic Functions

In this chapter we present the basic relation between the *linear potential theory* and *random walks*. This fundamental connection, developed by Ito, Doob, Lévy and others, relies on the observation that harmonic functions and martingales share a common cancellation property, expressed via *mean value properties*. It turns out that, with appropriate modifications, a similar observation and approach can be applied also in the nonlinear case, which is of main interest in these Course Notes. Thus, the present chapter serves as a stepping stone towards gaining familiarity with more complex constructions of Chaps. 3–6.

After recalling the equivalent defining properties of *harmonic functions* in Sect. 2.1, in Sect. 2.2 we introduce the *ball walk*. This is an infinite horizon discrete process, in which at each step the particle, initially placed at some point x_0 in the open, bounded domain $\mathcal{D} \subset \mathbb{R}^N$, is randomly advanced to a new position, uniformly distributed within the following open ball: centred at the current placement, and with radius equal to the minimum of the parameter ϵ and the distance from the boundary $\partial\mathcal{D}$. With probability one, such process accumulates on $\partial\mathcal{D}$ and $u^\epsilon(x_0)$ is then defined as the expected value of the given boundary data F at the process limiting position. Each function u^ϵ is harmonic, and we show in Sects. 2.3 and 2.4, that if $\partial\mathcal{D}$ is *regular*, then each u^ϵ coincides with the unique *harmonic extension* of F in $\mathcal{D}$. One sufficient condition for regularity is the *exterior cone condition*, as proved in Sect. 2.5.

Our discussion and proofs are elementary, requiring only a basic knowledge of probabilistic concepts, such as: probability spaces, martingales and Doob's theorem. For convenience of the reader, these are gathered in Appendix A. The slightly more advanced material which may be skipped at first reading, is based on the Potential Theoretic and the Brownian motion arguments from, respectively, Appendix C and Appendix B. Both approaches allow to deduce that functions in the family $\{u^\epsilon\}_{\epsilon\in(0,1)}$ are one and the same function, regardless of the regularity of $\partial\mathcal{D}$. This fact is obtained first in Sect. 2.6* by proving that u^ϵ coincide with the *Perron solution* of the Dirichlet problem for boundary data F. The same follows

M. Lewicka, *A Course on Tug-of-War Games with Random Noise*, Universitext, https://doi.org/10.1007/978-3-030-46209-3_2

in Sect. 2.7* by checking that the ball walk consists of discrete realizations along the Brownian motion trajectories, to the effect that u^ϵ equal the *Brownian motion harmonic extension* of F.

Thus, the three classical approaches to finding the harmonic extension by:

1. evaluating the expectation of the values of the (discrete) ball walk at its limiting infinite horizon boundary position;
2. taking infima/suprema of super- and sub-harmonic functions obeying comparison with the boundary data;
3. evaluating the expectation of the values of the (continuous) Brownian motion at exiting the domain;

are shown to naturally coincide when F is continuous.

2.1 The Laplace Equation and Harmonic Functions

Among the most important of all PDEs is the *Laplace equation.* In this section we briefly recall the relevant definitions and notation; for the proofs and a review of basic properties we refer to Sect. C.3 in Appendix C.

Let $\mathcal{D} \subset \mathbb{R}^N$ be an open, bounded, connected set. The Euler–Lagrange equation for critical points of the following quadratic energy functional:

$$\mathcal{I}_2(u) = \int_{\mathcal{D}} |\nabla u(x)|^2 \, \mathrm{d}x$$

is expressed by the second order partial differential equation:

$$\Delta u \doteq \sum_{i=1}^{N} \frac{\partial^2 u}{(\partial x_i)^2} = 0 \quad \text{in } \mathcal{D},$$

whose solutions are called *harmonic functions*. The operator Δ is defined in the classical sense only for C^2 functions u, however, a remarkable property of harmonicity is that it can be equivalently characterized via *mean value properties* that do not require u to be even continuous. At the same time, harmonic functions are automatically smooth. More precisely, the following conditions are equivalent:

(1) A locally bounded, Borel function $u : \mathcal{D} \to \mathbb{R}$ satisfies the *mean value property on balls*:

$$u(x) = \fint_{B_r(x)} u(y) \, \mathrm{d}y \qquad \text{for all } \bar{B}_r(x) \subset \mathcal{D}.$$

(ii) A locally bounded, Borel function $u : \mathcal{D} \to \mathbb{R}$ satisfies for each $x \in \mathcal{D}$ and almost every $r \in (0, \mathrm{dist}(x, \partial\mathcal{D}))$ the *mean value property on spheres*:

$$u(x) = \fint_{\partial B_r(x)} u(y) \, \mathrm{d}\sigma^{N-1}(y).$$

(iii) The function u is smooth: $u \in C^\infty(\mathcal{D})$, and there holds $\Delta u = 0$ in $\mathcal{D}$.

The proof of the equivalence will be recalled in Sect. C.3. We further remark at this point that, Taylor expanding any function $u \in C^2(\mathcal{D})$ and averaging term by term on $\bar{B}_\epsilon(x) \subset \mathcal{D}$, leads to the *mean value expansion*, also called in what follows the *averaging principle*:

$$\fint_{B_\epsilon(x)} u(y) \, \mathrm{d}y = u(x) + \frac{\epsilon^2}{2(N+2)} \Delta u(x) + o(\epsilon^2) \qquad \text{as } \epsilon \to 0+, \tag{2.1}$$

which in fact is consistent with interpreting Δu as the (second order) error from harmonicity. This point of view is central to developing the probabilistic interpretation of the general p-Laplace equations, which is the goal of these Course Notes. While we will not need (2.1) in order to construct the random walk and derive its connection to the Laplace equation Δ in the linear setting $p = 2$ studied in this chapter, it is beneficial to keep in mind that the mean value property in (i) may be actually "guessed" from the expansion (2.1).

Throughout next chapters, more general averaging principles will be proved (in Sects. 3.2, 4.1 and 6.1), informing the mean value properties that characterize, in the asymptotic sense, zeroes of the nonlinear operators Δ_p at any $p \in (1, \infty)$, and ultimately leading to the Tug-of-War games with random noise.

2.2 The Ball Walk

In this section we construct the discrete stochastic process whose value will be shown to equal the harmonic function with prescribed boundary values.

The probability space of the ball walk process is defined as follows. Consider $(\Omega_1, \mathcal{F}_1, \mathbb{P}_1)$, where Ω_1 is the unit ball $B_1(0) \subset \mathbb{R}^N$, the σ-algebra $\mathcal{F}_1$ consists of Borel subsets of Ω_1, and $\mathbb{P}_1$ is the normalized Lebesgue measure:

$$\mathbb{P}_1(D) = \frac{|D|}{|B_1(0)|} \qquad \text{for all } D \in \mathcal{F}_1,$$

For any $n \in \mathbb{N}$, we denote by $\Omega_n = (\Omega_1)^n$ the Cartesian product of n copies of Ω_1, and by $(\Omega_n, \mathcal{F}_n, \mathbb{P}_n)$ the corresponding product probability space. Further, the

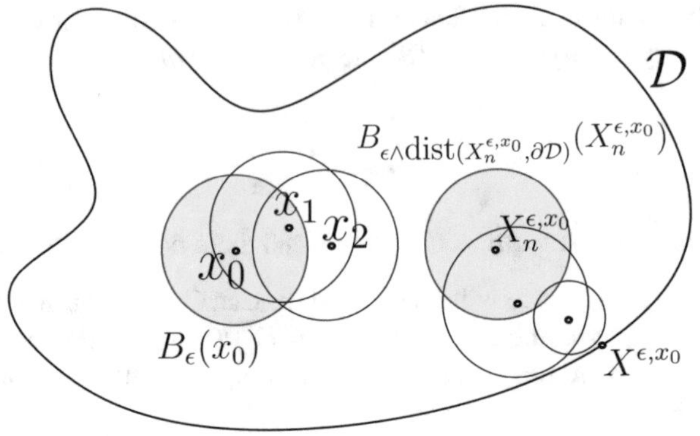

Fig. 2.1 The ball walk and the process $\{X_n^{\epsilon,x_0}\}_{n=0}^\infty$ in (2.2)

countable product $(\Omega,\mathcal{F},\mathbb{P})$ is defined as in Theorem A.12 on:

$$\Omega \doteq (\Omega_1)^{\mathbb{N}} = \prod_{i=1}^\infty \Omega_1 = \Big\{\omega = \{w_i\}_{i=1}^\infty;\ w_i \in B_1(0)\ \text{ for all }\ i\in\mathbb{N}\Big\}.$$

We identify each σ-algebra $\mathcal{F}_n$ with the sub-σ-algebra of $\mathcal{F}$ consisting of sets of the form $F\times\prod_{i=n+1}^\infty \Omega_1$ for all $F\in\mathcal{F}_n$. Note that $\{\mathcal{F}_n\}_{n=0}^\infty$ where $\mathcal{F}_0 = \{\emptyset,\Omega\}$, is a filtration of $\mathcal{F}$ and that $\mathcal{F}$ is the smallest σ-algebra containing $\bigcup_{n=0}^\infty \mathcal{F}_n$.

Definition 2.1 Let $\mathcal{D}\subset\mathbb{R}^N$ be an open, bounded, connected set. The *ball walk* for $\epsilon\in(0,1)$ and $x_0\in\mathcal{D}$, is recursively defined (see Fig. 2.1) as the following sequence of random variables $\{X_n^{\epsilon,x_0}:\Omega\to\mathcal{D}\}_{n=0}^\infty$:

$$\begin{aligned} X_0^{\epsilon,x_0} &\equiv x_0,\\ X_n^{\epsilon,x_0}(w_1,\ldots,w_n) &= X_{n-1}^{\epsilon,x_0}(w_1,\ldots,w_{n-1}) + \big(\epsilon\wedge \mathrm{dist}(X_{n-1}^{\epsilon,x_0},\partial\mathcal{D})\big)\,w_n \qquad (2.2)\\ &\qquad \text{for all }\ n\ge 1\ \text{ and all }\ (w_1,\ldots w_n)\in\Omega_n. \end{aligned}$$

We will often write: $x_n = X_n^{\epsilon,x_0}(w_1,\ldots,w_n)$. Intuitively, $\{x_n\}_{n=0}^\infty$ describe the consecutive positions of a particle initially placed at $x_0\in\mathcal{D}$, along a discrete path consisting of a succession of random steps of magnitude at most ϵ. The size of steps decreases as the particle approaches the boundary $\partial\mathcal{D}$. The position $x_n\in\mathcal{D}$ is obtained from x_{n-1} by sampling uniformly on the open ball $B_{\epsilon\wedge\mathrm{dist}(x_{n-1},\partial\mathcal{D})}(x_{n-1})$. It is clear that each random variable $X_n^{\epsilon,x_0}:\Omega\to\mathbb{R}^N$ is $\mathcal{F}_n$-measurable and that it depends only on the previous position x_{n-1}, its distance from $\partial\mathcal{D}$ and the current random outcome $w_n\in\Omega_1$.

Lemma 2.2 *In the above context, the sequence $\{X_n^{\epsilon,x_0}\}_{n=0}^\infty$ is a martingale relative to the filtration $\{\mathcal{F}_n\}_{n=0}^\infty$, namely:*

$$\mathbb{E}(X_n^{\epsilon,x_0} \mid \mathcal{F}_{n-1}) = X_{n-1}^{\epsilon,x_0} \quad \mathbb{P}-a.s. \qquad \textit{for all } n \geq 1.$$

Moreover, there exists a random variable $X^{\epsilon,x_0} : \Omega \to \partial\mathcal{D}$ such that:

$$\lim_{n\to\infty} X_n^{\epsilon,x_0} = X^{\epsilon,x_0} \qquad \mathbb{P}-a.s. \tag{2.3}$$

Proof

1. Since the sequence $\{X_n^{\epsilon,x_0}\}_{n=0}^\infty$ is bounded in view of boundedness of $\mathcal{D}$, Theorem A.38 will yield convergence in (2.3) provided we check the martingale property. Indeed it follows that (see Lemma A.17):

$$\begin{aligned}\mathbb{E}(X_n^{\epsilon,x_0} \mid \mathcal{F}_{n-1})(w_1,\ldots,w_{n-1}) &= \int_{\Omega_1} X_n^{\epsilon,x_0}(w_1,\ldots,w_n)\,\mathrm{d}\mathbb{P}_1(w_n)\\ &= x_{n-1} + \big(\epsilon \wedge \operatorname{dist}(x_{n-1},\partial\mathcal{D})\big)\int_{\Omega_1} w_n\,\mathrm{d}\mathbb{P}_1(w_n)\\ &= X_{n-1}^{\epsilon,x_0}(w_1,\ldots,w_{n-1}) \quad \text{for } \mathbb{P}_{n-1}\text{-a.e. } (w_1,\ldots,w_{n-1}) \in \Omega_{n-1}.\end{aligned}$$

2. It remains to prove that the limiting random variable $X^{\epsilon,x_0} : \Omega \to \bar{\mathcal{D}}$ satisfies $\mathbb{P}$-a.s. the boundary accumulation property: $X^{\epsilon,x_0} \in \partial\mathcal{D}$. Observe that:

$$\Big\{\lim_{n\to\infty} X_n^{\epsilon,x_0} = X^{\epsilon,x_0}\Big\} \cap \{X^{\epsilon,x_0} \in \mathcal{D}\} \subset \bigcup_{n\in\mathbb{N},\ \delta\in(0,\epsilon)\cap\mathbb{Q}} A(n,\delta), \tag{2.4}$$

where $A(n,\delta) = \big\{\operatorname{dist}(X_i^{\epsilon,x_0},\partial\mathcal{D}) \geq \delta \text{ and } |X_{i+1}^{\epsilon,x_0} - X_i^{\epsilon,x_0}| \leq \frac{\delta}{2} \text{ for all } i \geq n\big\}$. Then:

$$A(n,\delta) \subset \Big\{\omega \in \Omega;\ |w_i| \leq \frac{1}{2} \text{ for all } i > n\Big\}.$$

Indeed, if $\omega = \{w_i\}_{i=1}^\infty \in A(n,\delta)$ with $\delta < \epsilon$, it follows that:

$$\begin{aligned}\frac{\delta}{2} &\geq |X_{i+1}^{\epsilon,x_0}(\omega) - X_i^{\epsilon,x_0}(\omega)| = \big(\epsilon \wedge \operatorname{dist}(X_i^{\epsilon,x_0}(\omega),\partial\mathcal{D})\big)\,|w_{i+1}|\\ &\geq (\epsilon\wedge\delta)|w_{i+1}| = \delta|w_{i+1}|,\end{aligned}$$

which implies that $|w_{i+1}| \leq \frac{1}{2}$ for all $i \geq n$. Concluding:

$$\mathbb{P}\big(A(n,\delta)\big) \leq \lim_{i\to\infty} \mathbb{P}_1(B_{\frac{1}{2}}(0))^{i-n} = 0 \quad \text{for all } n \in \mathbb{N} \text{ and all } \delta \in (0,\epsilon).$$

Hence, the event in the left hand side of (2.4) has probability 0. □

Given now a continuous function $F : \partial\mathcal{D} \to \mathbb{R}$, define:

$$u^\epsilon(x_0) \doteq \mathbb{E}\big[F \circ X^{\epsilon,x_0}\big] = \int_\Omega F \circ X^{\epsilon,x_0}\,\mathrm{d}\mathbb{P}. \tag{2.5}$$

Note that the above construction obeys the *comparison principle*. Namely, if $F, \bar{F} : \partial\mathcal{D} \to \mathbb{R}$ are two continuous functions such that $F \leq \bar{F}$ on $\partial\mathcal{D}$, then the corresponding u^ϵ and $\bar{u}^\epsilon$ satisfy: $u^\epsilon \leq \bar{u}^\epsilon$ in $\mathcal{D}$.

Remark 2.3 It is useful to view the boundary function F as the restriction on $\partial\mathcal{D}$ of some continuous $F : \bar{\mathcal{D}} \to \mathbb{R}$, see Exercise 2.7 (i). Then we may write:

$$u^\epsilon(x_0) = \lim_{n\to\infty} \int_\Omega F \circ X_n^{\epsilon,x_0}\,\mathrm{d}\mathbb{P}. \tag{2.6}$$

Since for each $n \geq 0$ the function $F \circ X_n^{\epsilon,x_0}$ is jointly Borel-regular in the variables $x_0 \in \mathcal{D}$ and $\omega \in \Omega_n$, it follows by Theorem A.11 that $x_0 \mapsto \mathbb{E}[F \circ X_n^{\epsilon,x_0}]$ is Borel-regular. Consequently, $u^\epsilon : \mathcal{D} \to \mathbb{R}$ is also Borel.

In what follows, we will denote the average $\mathcal{A}_\delta u$ of an integrable function $u : \mathcal{D} \to \mathbb{R}$ on a ball $B_\delta(x) \subset \mathcal{D}$ by:

$$\mathcal{A}_\delta u(x) \doteq \fint_{B_\delta(x)} u(y)\,\mathrm{d}y.$$

Directly from Definition 2.1 and (2.5) we conclude the satisfaction of the mean value property for each u^ϵ on the sampling balls from (2.2):

Theorem 2.4 *Let $\mathcal{D} \subset \mathbb{R}^N$ be open, bounded, connected, and let $F : \partial\mathcal{D} \to \mathbb{R}$ be continuous. Then, the function $u^\epsilon : \mathcal{D} \to \mathbb{R}$ defined in (2.5) and equivalently in (2.6), is continuous and satisfies:*

$$u^\epsilon(x) = \mathcal{A}_{\epsilon\wedge dist(x,\partial\mathcal{D})} u^\epsilon(x) \quad \textit{for all } x \in \mathcal{D}.$$

Proof Fix $\epsilon \in (0, 1)$ and $x_0 \in \mathcal{D}$. For each $n \geq 2$ we view $(\Omega_n, \mathcal{F}_n, \mathbb{P}_n)$ as the product of probability spaces $(\Omega_1, \mathcal{F}_1, \mathbb{P}_1)$ and $(\Omega_{n-1}, \mathcal{F}_{n-1}, \mathbb{P}_{n-1})$. Applying Fubini's Theorem (Theorem A.11), we get:

$$\begin{aligned}\mathbb{E}\big[F \circ X_n^{\epsilon,x_0}\big] &= \int_{\Omega_1}\int_{\Omega_{n-1}} \big(F \circ X_n^{\epsilon,x_0}\big)(w_1,\ldots,w_n)\,\mathrm{d}\mathbb{P}_{n-1}(w_2,\ldots,w_n)\,\mathrm{d}\mathbb{P}_1(w_1)\\ &= \int_{\Omega_1} \mathbb{E}\big[F \circ X_{n-1}^{\epsilon,X_1^{\epsilon,x_0}(w_1)}\big]\,\mathrm{d}\mathbb{P}_1(w_1),\end{aligned}$$

where $F : \bar{\mathcal{D}} \to \mathbb{R}$ is some continuous extension of its given values on $\partial\mathcal{D}$, as in (2.6). Passing to the limit with $n \to \infty$ and changing variables, we obtain:

$$
\begin{aligned}
u^\epsilon(x_0) &= \int_{\Omega_1} u^\epsilon\big(X_1^{\epsilon,x_0}(w_1)\big)\,\mathrm{d}\mathbb{P}_1(w_1) \\
&= \int_{\Omega_1} u^\epsilon\big(x_0 + (\epsilon \wedge \mathrm{dist}(x_0, \partial\mathcal{D}))\, w_1\big)\,\mathrm{d}\mathbb{P}_1(w_1) \\
&= \fint_{B_{\epsilon \wedge \mathrm{dist}(x_0,\, \mathcal{D})}(x_0)} u^\epsilon(y)\,\mathrm{d}y.
\end{aligned}
$$

Continuity of u^ϵ follows directly from the averaging formula and we leave it as an exercise (see Exercise 2.7 (ii)). □

The next two statements imply uniqueness of classical solutions to the boundary value problem for the Laplacian. The same property, in the basic analytical setting that we review in Sect. C.3, follows via the maximum principle.

Corollary 2.5 *Given $\epsilon \in (0,1)$ and $x_0 \in \mathcal{D}$, let $\{X_n^{\epsilon,x_0}\}_{n=0}^\infty$ be as defined in (2.2). In the setting of Theorem 2.4, the sequence $\{u^\epsilon \circ X_n^{\epsilon,x_0}\}_{n=0}^\infty$ is then a martingale relative to the filtration $\{\mathcal{F}_n\}_{n=0}^\infty$.*

Proof Indeed, Lemma A.17 yields for all $n \geq 1$:

$$
\begin{aligned}
\mathbb{E}\big(u^\epsilon \circ X_n^{\epsilon,x_0} \mid \mathcal{F}_{n-1}\big)(w_1,\ldots,w_{n-1}) &= \int_{\Omega_1} (u^\epsilon \circ X_n^{\epsilon,x_0})(w_1,\ldots,w_n)\,\mathrm{d}\mathbb{P}_1(w_n) \\
= \int_{\Omega_1} u^\epsilon\Big(X_{n-1}^{\epsilon,x_0}(w_1,\ldots,w_{n-1}) &+ \big(\epsilon \wedge \mathrm{dist}(x_{n-1}, \partial\mathcal{D})\big)\, w_n\Big)\,\mathrm{d}\mathbb{P}_1(w_n) \\
= \fint_{B_{\epsilon \wedge \mathrm{dist}(x_{n-1},\, \mathcal{D})}(x_{n-1})} u^\epsilon(y)\,\mathrm{d}y &= (u^\epsilon \circ X_{n-1}^{x_0})(w_1,\ldots,w_{n-1}),
\end{aligned}
\tag{2.7}
$$

valid for $\mathbb{P}_{n-1}$-a.e. $(w_1,\ldots,w_{n-1}) \in \Omega_{n-1}$. □

Lemma 2.6 *In the setting of Theorem 2.4, assume that $u \in C(\bar{\mathcal{D}})$ solves:*

$$
\Delta u = 0 \quad \text{in } \mathcal{D}, \qquad u = F \quad \text{on } \partial\mathcal{D}. \tag{2.8}
$$

Then $u^\epsilon = u$ for all $\epsilon \in (0,1)$. In particular, (2.8) has at most one solution.

Proof We first claim that given $x_0 \in \mathcal{D}$ and $\epsilon \in (0,1)$, the sequence $\{u \circ X_n^{\epsilon,x_0}\}_{n=0}^\infty$ is a martingale relative to $\{\mathcal{F}_n\}_{n=0}^\infty$. This property follows exactly as in (2.7), where u^ϵ is now replaced by u and where the mean value property for harmonic functions (C.8) is used instead of the single-radius averaging formula of Theorem 2.4. Consequently, we get:

$$u(x_0) = \mathbb{E}[u \circ X_0^{\epsilon,x_0}] = \mathbb{E}[u \circ X_n^{\epsilon,x_0}] \quad \text{for all } n \geq 0.$$

Since the right hand side above converges to $u^\epsilon(x_0)$ with $n \to \infty$, it follows that $u(x_0) = u^\epsilon(x_0)$. To prove the second claim, recall that $u^\epsilon(x_0)$ depends only on the boundary values $u_{|\partial\mathcal{D}} = F$ and not on their extension u on $\bar{\mathcal{D}}$. This yields uniqueness of the harmonic extension in (2.8). □

We finally remark that the mean value property stated in Theorem 2.4 suffices to conclude that each u^ϵ is harmonic (see Sect. C.3). One can also show that all functions in the family $\{u^\epsilon\}_{\epsilon\in(0,1)}$ are the same, even in the absence of the classical harmonic extension u satisfying (2.8). This general result will be given two independent proofs in Sects. 2.6* and 2.7*. In the next section, we provide an elementary proof in domains that are sufficiently regular. An entirely similar strategy, based on showing the uniform convergence of $\{u^\epsilon\}_{\epsilon\to 0}$ in $\bar{\mathcal{D}}$ and analysing its limit, will be adopted in Chaps. 3–6 for the p-harmonic case, $p \in (1, \infty)$, in the context of Tug-of-War with noise.

Exercise 2.7

(i) Let $F : A \to \mathbb{R}$ be a continuous function on a compact set $A \subset \mathbb{R}^N$. Verify that, setting:

$$F(x) \doteq \min_{y\in A} \Big\{F(y) + \frac{|x-y|}{\operatorname{dist}(x, A)} - 1\Big\} \quad \text{for all } x \in \mathbb{R}^N \setminus A,$$

defines a continuous extension of F on $\mathbb{R}^N$. This construction is due to Hausdorff and it provides a proof of the Tietze extension theorem.

(ii) Let $u : \mathbb{R}^N \to \mathbb{R}$ be a bounded, Borel function and let $\epsilon : \mathbb{R}^N \to (0, \infty)$ be continuous. Show that the function: $x \mapsto \mathcal{A}_{\epsilon(x)}u(x)$ is continuous on $\mathbb{R}^N$.

Exercise 2.8 Modify the construction of the ball walk to the *sphere walk* using the outline below.

(i) Let $\Omega_1 = \partial B_1(0) \subset \mathbb{R}^N$ and let $\mathbb{P}_1 = \sigma^{N-1}$ be the normalized spherical measure on the Borel σ-algebra $\mathcal{F}_1$ of subsets of Ω_1 (see Example A.9). Define the induced probability spaces $(\Omega, \mathcal{F}, \mathbb{P})$ and $\{(\Omega_n, \mathcal{F}_n, \mathbb{P}_n)\}_{n=0}^\infty$ as in the case of the ball walk. For every $\epsilon \in (0, 1)$ and $x_0 \in \mathcal{D}$, let $\{X_n^{\epsilon,x_0} : \mathcal{D} \to \mathbb{R}^N\}_{n=0}^\infty$ be the sequence of random variables in:

$$X_0^{\epsilon,x_0} \equiv x_0 \quad \text{and for all } n \geq 1 \text{ and all } (w_1, \ldots, w_{n-1}) \in \Omega_{n-1} :$$

$$X_n^{\epsilon,x_0}(w_1, \ldots, w_n) = x_{n-1} + \Big(\epsilon \wedge \frac{1}{2}\operatorname{dist}(x_{n-1}, \partial\mathcal{D})\Big)\, w_n$$

$$\text{where} \quad x_{n-1} = X_{n-1}^{\epsilon,x_0}(w_1, \ldots, w_{n-1}).$$

Prove that $\{X_n^{\epsilon,x_0}\}_{n=0}^\infty$ is a martingale relative to the filtration $\{\mathcal{F}_n\}_{n=0}^\infty$ and that (2.3) holds for some random variable $X^{\epsilon,x_0} : \Omega \to \partial\mathcal{D}$.

(ii) For a continuous function $F : \partial\mathcal{D} \to \mathbb{R}$, define $u^\epsilon : \mathcal{D} \to \mathbb{R}$ according to (2.5). Show that u^ϵ is Borel-regular and that it satisfies:

$$u^\epsilon(x) = \fint_{\partial B_{\epsilon \wedge \frac{1}{2}\mathrm{dist}(x, \partial\mathcal{D})}(x)} u^\epsilon(y)\, \mathrm{d}\sigma^{N-1}(y) \qquad \text{for all } x \in \mathcal{D}.$$

(iii) Deduce that if F has a harmonic extension u on $\bar{\mathcal{D}}$ as in (2.8), then $u^\epsilon = u$ for all $\epsilon \in (0, 1)$.

2.3 The Ball Walk and Harmonic Functions

The main result of this section states that the uniform limits of values $\{u^\epsilon\}_{\epsilon\to 0}$ of the ball walk that we introduced in Sect. 2.2, are automatically harmonic. The proof relies on checking that each limiting function u satisfies the mean value property on spheres. This is achieved by applying Doob's theorem to u^ϵ evaluated along its own walk process $\{X_n^{\epsilon,x_0}\}_{n=0}^\infty$, and choosing to stop on exiting the ball whose boundary coincides with the given sphere.

Theorem 2.9 *Let $J \subset (0, 1)$ be a sequence decreasing to 0. Assume that $\{u^\epsilon\}_{\epsilon\in J}$ defined in (2.5), converges locally uniformly in $\mathcal{D}$, as $\epsilon \to 0$, $\epsilon \in J$, to some $u \in C(\mathcal{D})$. Then u must be harmonic.*

Proof

1. In virtue of Theorem C.19, it suffices to prove that:

$$u(x_0) = \fint_{\partial B_r(x_0)} u(y)\, \mathrm{d}\sigma^N(y) \qquad \text{for all } B_{2r}(x_0) \subset \mathcal{D}. \tag{2.9}$$

Fix $x_0 \in \mathcal{D}$ and $r \leq \frac{1}{2}\mathrm{dist}(x_0, \partial\mathcal{D})$, and for each $\epsilon \in J$ consider the following random variable $\tau^\epsilon : \Omega \to \mathbb{N} \cup \{+\infty\}$:

$$\tau^\epsilon = \inf\big\{n \geq 1;\ X_n^{\epsilon,x_0} \notin B_r(x_0)\big\},$$

where $\{X_n^{\epsilon,x_0}\}_{n=0}^\infty$ is the usual sequence of the token positions (2.2) in the ϵ-ball walk started at x_0. Clearly, τ^ϵ is finite a.s. in view of convergence to the boundary in (2.3) and it is a stopping time relative to the filtration $\{\mathcal{F}_n\}_{n=0}^\infty$. The corresponding stopping position x_{τ^ϵ} is indicated in Fig. 2.2. By Corollary 2.5, Doob's theorem (Theorem A.31 (ii)) yields:

$$u^\epsilon(x_0) = \mathbb{E}\big[u^\epsilon \circ X_0^{\epsilon,x_0}\big] = \mathbb{E}\big[u^\epsilon \circ X_\tau^{\epsilon,x_0}\big],$$

while by passing to the limit with $\epsilon \to 0$ we obtain, by uniform convergence:

$$u(x_0) = \lim_{\epsilon\to 0,\, \epsilon\in J} \mathbb{E}\big[u^\epsilon \circ X_\tau^{\epsilon,x_0}\big] = \lim_{\epsilon\to 0,\, \epsilon\in J} \int_{B_{r+\epsilon}(x_0)\setminus B_r(x_0)} u(y)\, \mathrm{d}\sigma_\epsilon(y). \tag{2.10}$$

The Borel probability measures $\{\sigma_\epsilon\}_{\epsilon\in(0,r)}$ are here defined on $\bar{B}_{2r}(x_0) \setminus B_r(x_0)$ by the push-forward procedure, as in Exercise A.8:

$$\sigma_\epsilon(A) \doteq \mathbb{P}\big(X_\tau^{\epsilon,x_0} \in A\big).$$

2. We now identify the limit in the right hand side of (2.10). Observe that, by construction, the measures σ_ϵ are rotationally invariant. Further, by Prohorov's theorem (Theorem A.10), each subsequence of $\{\sigma_\epsilon\}_{\epsilon\to 0,\, \epsilon\in J}$ has a further subsequence that converges (weakly-$*$) to a Borel probability measure μ on $\bar{B}_{2r}(x_0) \setminus B_r(x_0)$. Since each σ_ϵ is supported in $B_{r+\epsilon}(x_0) \setminus B_r(x_0)$, the limit μ must be supported on $\partial B_r(x_0)$. Also, μ is rotationally invariant in view of the same property of each σ_ϵ. Consequently, $\mu = \sigma^{N-1}$ must be the uniquely defined, normalized spherical measure on $\partial B_r(x_0)$ (see Exercises 2.10 and 2.11). As the limit does not depend on the chosen subsequence of J, we conclude:

$$\lim_{\epsilon\to 0,\, \epsilon\in J} \int_{\bar{B}_{2r}(x_0)\setminus B_r(x_0)} u(y)\, \mathrm{d}\sigma_\epsilon(y) = \fint_{\partial B_r(x_0)} u(y)\, \mathrm{d}\sigma^{N-1}(y).$$

Together with (2.10), this establishes (2.9) as claimed. □

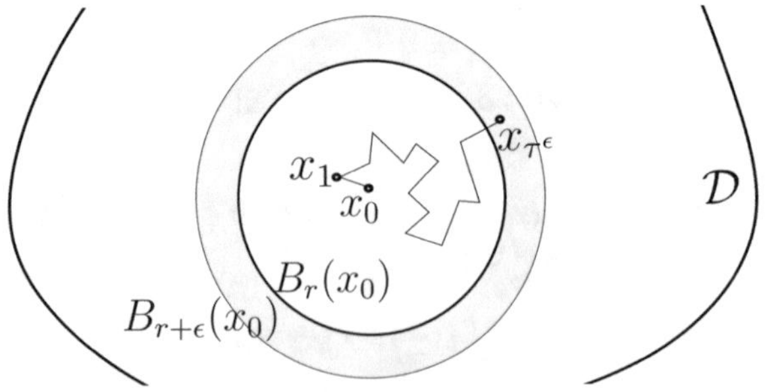

Fig. 2.2 The stopping position x_{τ^ϵ} in the proof of Theorem 2.9

Exercise 2.10 Show that every (weak-$*$) limit point of the family of probability measures $\{\sigma_\epsilon\}_{\epsilon\in(0,r)}$ defined in (2.10) must be rotationally invariant and supported on $\partial B_r(x_0)$.

Exercise 2.11 Using the following outline, prove that the only rotationally invariant, Borel probability measure μ on $\partial B_1(0) \subset \mathbb{R}^N$, is the normalized spherical measure σ^{N-1}.

(i) Fix an open set $U \subset \partial B_1(0)$ and consider the sequence of Borel functions $\Big\{x \mapsto \frac{\mu(U\cap B(x,\frac{1}{n}))}{\mu(B(x,\frac{1}{n}))}\Big\}_{n=1}^{\infty}$, where $B(x,r)$ denotes the $(N-1)$-dimensional curvilinear ball in $\partial B_1(0)$ centred at x and with radius $r \in (0,1)$. Apply Fatou's lemma (Theorem A.6) and Fubini's theorem (Theorem A.11) to the indicated sequence and deduce that:

$$\sigma^{N-1}(U) \leq \Big(\liminf_{n\to\infty} \frac{\sigma^{N-1}(B(x,\frac{1}{n}))}{\mu(B(x,\frac{1}{n}))}\Big)\cdot \mu(U), \tag{2.11}$$

where both quantities $\sigma^{N-1}(B(x,\frac{1}{n}))$ and $\mu(B(x,\frac{1}{n}))$ are independent of $x \in \partial B_1(0)$ because of the rotational invariance.

(ii) Exchange the roles of μ and σ^{N-1} in the above argument and conclude:

$$\lim_{n\to\infty} \frac{\sigma^{N-1}(B(x,\frac{1}{n}))}{\mu(B(x,\frac{1}{n}))} = 1.$$

Thus, $\mu(U) = \sigma^{N-1}(U)$ for all open sets U, so there must be $\mu = \sigma^{N-1}$.

This proof is due to Christensen (1970) and the result is a particular case of Haar's theorem on uniqueness of invariant measures on compact topological groups.

2.4 Convergence at the Boundary and Walk-Regularity

We now investigate conditions assuring the validity of the uniform convergence assumption of Theorem 2.9. It turns out that such condition may be formulated independently of the boundary data F, only in terms of the behaviour of the ball walk (2.2) close to $\partial\mathcal{D}$, which is further guaranteed by a geometrical sufficient condition in the next section. In Theorem 2.14 we will show how the boundary regularity of the process can be translated (via walk coupling) into the interior regularity, resulting in the existence of a harmonic extension u of F on $\bar{\mathcal{D}}$, and ultimately yielding $u^\epsilon = u$ for all $\epsilon \in (0,1)$, in virtue of Lemma 2.6.

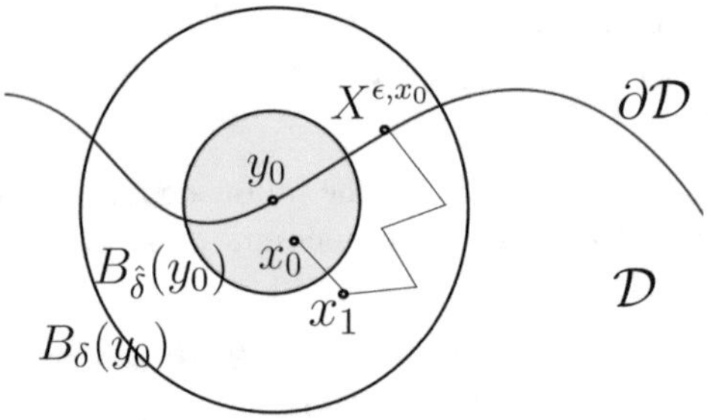

Fig. 2.3 Walk-regularity of a boundary point $y_0 \in \partial\mathcal{D}$

Definition 2.12 Consider the ball walk (2.2) on a domain $\mathcal{D} \subset \mathbb{R}^N$.

(a) We say that a boundary point $y_0 \in \partial\mathcal{D}$ is *walk-regular* if for every $\eta, \delta > 0$ there exists $\hat\delta \in (0, \delta)$ and $\hat\epsilon \in (0, 1)$ such that:

$$\mathbb{P}\big(X^{\epsilon,x_0} \in B_\delta(y_0)\big) \geq 1 - \eta \qquad \text{for all } \ \epsilon \in (0, \hat\epsilon) \ \text{ and all } x_0 \in B_{\hat\delta}(y_0) \cap \mathcal{D},$$

where X^{ϵ,x_0} is the limit in (2.3) of the ϵ-ball walk started at x_0.

(b) We say that $\mathcal{D}$ is *walk-regular* if every $y_0 \in \partial\mathcal{D}$ is walk-regular (Fig. 2.3).

Lemma 2.13 *Assume that the boundary point $y_0 \in \partial\mathcal{D}$ of a given open, bounded, connected domain $\mathcal{D}$ is walk-regular. Then for every continuous $F : \partial\mathcal{D} \to \mathbb{R}$, the family $\{u^\epsilon\}_{\epsilon\to 0}$ defined in (2.5) satisfies the following. For every $\eta > 0$ there is $\hat\delta > 0$ and $\hat\epsilon \in (0, 1)$ such that:*

$$|u^\epsilon(x_0) - F(y_0)| \leq \eta \qquad \textit{for all } \ \epsilon \in (0, \hat\epsilon) \ \textit{ and all } \ x_0 \in B_{\hat\delta}(y_0) \cap \mathcal{D}. \tag{2.12}$$

Proof Given $\eta > 0$, let $\delta > 0$ satisfy:

$$|F(y) - F(y_0)| \leq \frac{\eta}{2} \qquad \text{for all } y \in \partial\mathcal{D} \ \text{ such that } \ |y - y_0| < \delta.$$

By Definition 2.12, we choose $\hat\epsilon$ and $\hat\delta$ corresponding to $\frac{\eta}{4\|F\|_\infty + 1}$ and δ. Then:

$$\begin{aligned}
|u^\epsilon(x_0) - F(y_0)| &\leq \int_\Omega |F \circ X^{\epsilon,x_0} - F(y_0)| \, \mathrm{d}\mathbb{P} \\
&\leq \mathbb{P}\big(X^{\epsilon,x_0} \not\in B_\delta(y_0)\big) \cdot 2\|F\|_\infty + \int_{X^{\epsilon,x_0} \in B_\delta(y_0)} |F \circ X^{\epsilon,x_0} - F(y_0)| \, \mathrm{d}\mathbb{P} \\
&\leq \frac{\eta}{4\|F\|_\infty + 1} \cdot 2\|F\|_\infty + \frac{\eta}{2} \leq \eta,
\end{aligned}$$

for all $x_0 \in B_{\hat\delta}(y_0) \cap \mathcal{D}$ and all $\epsilon \in (0, \hat\epsilon)$. This completes the proof. □

By Lemma 2.13 and Theorem 2.9 we achieve the main result of this chapter:

Theorem 2.14 *Let $\mathcal{D}$ be walk-regular. Then, for every continuous $F : \partial\mathcal{D} \to \mathbb{R}$, the family $\{u^\epsilon\}_{\epsilon\in(0,1)}$ in (2.5) satisfies $u^\epsilon = u$, where $u \in C(\bar{\mathcal{D}})$ is the unique solution of the boundary value problem:*

$$\Delta u = 0 \quad \text{in } \mathcal{D}, \qquad u = F \quad \text{on } \partial\mathcal{D}.$$

Proof

1. Let $F : \partial\mathcal{D} \to \mathbb{R}$ be a given continuous function. We will show that $\{u^\epsilon\}_{\epsilon\to 0}$ is "asymptotically equicontinuous in $\mathcal{D}$", i.e.: for every $\eta > 0$ there exists $\delta > 0$ and $\hat{\epsilon} \in (0, 1)$ such that:

$$\begin{aligned} |u^\epsilon(x_0) - u^\epsilon(y_0)| \leq \eta \qquad & \text{for all } \epsilon \in (0, \hat{\epsilon}) \\ & \text{and all } x_0, y_0 \in \mathcal{D} \text{ with } |x_0 - y_0| \leq \delta. \end{aligned} \tag{2.13}$$

   Since $\{u^\epsilon\}_{\epsilon\to 0}$ is equibounded (by $\|F\|_\infty$), condition (2.13) imply that for every sequence $J \subset (0, 1)$ converging to 0, one can extract a further subsequence of $\{u^\epsilon\}_{\epsilon\in J}$ that converges locally uniformly in $\bar{\mathcal{D}}$. Further, in view of (2.12) it follows that $u \in C(\bar{\mathcal{D}})$ and $u = F$ on $\partial\mathcal{D}$ (see Exercise 2.17). By Theorem 2.9, we get that u is harmonic in $\mathcal{D}$ and the result follows in virtue of Lemma 2.6.
2. To show (2.13), fix $\eta > 0$ and choose $\bar{\delta} > 0$ such that $|F(y) - F(\bar{y})| \leq \frac{\eta}{3}$ for all $y, \bar{y} \in \partial\mathcal{D}$ with $|y - \bar{y}| \leq 3\bar{\delta}$. By (2.12), for each $y_0 \in \partial\mathcal{D}$ there exists $\hat{\delta}(y_0) \in (0, \bar{\delta})$ and $\hat{\epsilon}(y_0) \in (0, 1)$ satisfying:

$$|u^\epsilon(x_0) - F(y_0)| \leq \frac{\eta}{3} \qquad \text{for all } \epsilon \in (0, \hat{\epsilon}(y_0)) \text{ and all } x_0 \in B_{\hat{\delta}(y_0)}(y_0) \cap \mathcal{D}.$$

   The family of balls $\{B_{\hat{\delta}(y)}(y)\}_{y\in\partial\mathcal{D}}$ is then a covering of the compact set $\partial\mathcal{D}$; let $\{B_{\hat{\delta}(y_i)}(y_i)\}_{i=1}^n$ be its finite sub-cover and set $\hat{\epsilon} = \min_{i=1\dots n} \hat{\epsilon}(y_i)$. Clearly:

$$\partial\mathcal{D} + B_{2\delta}(0) \subset \bigcup_{i=1}^n B_{\hat{\delta}(y_i)}(y_i)$$

   for some $\delta > 0$ where we additionally request that $\delta < \bar{\delta}$. This implies:

$$\begin{aligned} |u^\epsilon(x_0) - u^\epsilon(y_0)| \leq \eta \qquad & \text{for all } \epsilon \in (0, \hat{\epsilon}) \\ \text{and all } x_0, y_0 \in \big(\partial\mathcal{D} + B_{2\delta}(0)\big) \cap \mathcal{D} & \text{ with } |x_0 - y_0| \leq \delta. \end{aligned} \tag{2.14}$$

3. To conclude the proof of (2.13), fix $\epsilon \in (0, \hat{\epsilon} \wedge \delta)$ and let $x_0, y_0 \in \mathcal{D}$ satisfy $\mathrm{dist}(x_0, \partial\mathcal{D}) \geq \delta$, $\mathrm{dist}(y_0, \partial\mathcal{D}) \geq \delta$ and $|x_0 - y_0| < \delta$. Define $\tau_\delta : \Omega \to \mathbb{N} \cup \{+\infty\}$:

$$\tau_\delta = \min\big\{n \geq 1;\ \mathrm{dist}(x_n, \partial\mathcal{D}) < \delta \ \text{ or } \ \mathrm{dist}(y_n, \partial\mathcal{D}) < \delta\big\},$$

where $\{x_n = X_n^{\epsilon,x_0}\}_{n=0}^\infty$ and $\{y_n = X_n^{\epsilon,y_0}\}_{n=0}^\infty$ denote the consecutive positions in the process (2.5) started at x_0 and y_0, respectively. It is clear that τ_δ is finite $\mathbb{P}$-a.s. in view of convergence to the boundary in (2.3), and it is a stopping time relative to the filtration $\{\mathcal{F}_n\}_{n=0}^\infty$.
By Corollary 2.5 and Doob's theorem (Theorem A.31 (ii)) it follows that:

$$u^\epsilon(x_0) = \mathbb{E}\big[u^\epsilon \circ X_\tau^{\epsilon,x_0}\big] \quad \text{and} \quad u^\epsilon(y_0) = \mathbb{E}\big[u^\epsilon \circ X_\tau^{\epsilon,y_0}\big].$$

Since $|X_\tau^{\epsilon,x_0} - X_\tau^{\epsilon,y_0}| = |x_0 - y_0| < \delta$ and $X_\tau^{\epsilon,x_0}, X_\tau^{\epsilon,y_0} \in \big(\partial\mathcal{D} + B_{2\delta}(0)\big) \cap \mathcal{D}$ for a.e. $\omega \in \Omega$, we conclude by (2.14) that:

$$|u^\epsilon(x_0) - u^\epsilon(y_0)| \leq \int_\Omega |u^\epsilon \circ X_\tau^{\epsilon,x_0} - u^\epsilon \circ X_\tau^{\epsilon,y_0}|\, \mathrm{d}\mathbb{P} \leq \eta.$$

This ends the proof of (2.13) and of the Theorem. □

Walk-regularity is, in fact, equivalent to convergence of u to the right boundary values. We have the following observation, converse to Lemma 2.13:

Lemma 2.15 *If $y_0 \in \partial\mathcal{D}$ is not walk-regular, then there exists a continuous function $F : \partial\mathcal{D} \to \mathbb{R}$, such that for u^ϵ in (2.5) there holds:*

$$\limsup_{x \to y_0,\ \epsilon \to 0} u^\epsilon(x) \neq F(y_0).$$

Proof Define $F(y) = |y - y_0|$ for all $y \in \partial\mathcal{D}$. By assumption, there exists $\eta, \delta > 0$ and sequences $\{\epsilon_i\}_{i=1}^\infty$, $\{x_j \in \mathcal{D}\}_{j=1}^\infty$ such that:

$$\lim_{j\to\infty} \epsilon_j = 0, \qquad \lim_{j\to\infty} x_j = y_0 \quad \text{and} \quad \mathbb{P}\big(X^{\epsilon_j,x_j} \notin B_\delta(y_0)\big) > \eta \quad \text{for all } \ j \geq 1,$$

where each X^{ϵ_j,x_j} above stands for the limiting random variable in (2.3) corresponding to the ϵ_j- ball walk. By the nonnegativity of F, it follows that:

$$u^{\epsilon_j}(x_j) - F(y_0) = \int_\Omega F \circ X^{\epsilon_j,x_j}\, \mathrm{d}\mathbb{P} \geq \int_{\{X^{\epsilon_j,x_j} \notin B_\delta(y_0)\}} F \circ X^{\epsilon_j,x_j}\, \mathrm{d}\mathbb{P} > \eta\delta > 0,$$

proving the claim. □

Exercise 2.16 Show that if $\mathcal{D}$ is walk-regular then $\hat{\delta}$ and $\hat{\epsilon}$ in Definition 2.12 (a) can be chosen independently of y_0 (i.e., $\hat{\delta}$ and $\hat{\epsilon}$ depend only on the parameters η and δ).

Exercise 2.17 Let $\{u_\epsilon\}_{\epsilon\in J}$ be an equibounded sequence of functions $u_\epsilon : \mathcal{D} \to \mathbb{R}$ defined on an open, bounded set $\mathcal{D} \subset \mathbb{R}^N$, and satisfying (2.12), (2.13) with some continuous $F : \partial\mathcal{D} \to \mathbb{R}$. Prove that $\{u_\epsilon\}_{\epsilon\in J}$ must have a subsequence that converges uniformly, as $\epsilon \to 0$, $\epsilon \in J$, to a continuous function $u : \bar{\mathcal{D}} \to \mathbb{R}$.

2.5 A Sufficient Condition for Walk-Regularity

In this section we state a geometric condition (exterior cone condition) implying the validity of the walk-regularity condition introduced in Definition 2.12. We remark that the exterior cone condition in Theorem 2.19 may be weakened to the so-called exterior corkscrew condition, and that the analysis below is valid not only in the presently studied linear case of $p = 2$, but in the nonlinear setting of an arbitrary exponent $p \in (1, \infty)$ as well. This will be explained in Chap. 6, with proofs conceptually based on what follows.

We begin by observing a useful technical reformulation of the regularity condition in Definition 2.4. Namely, at walk-regular boundary points y_0 not only the limiting position of the ball walk may be guaranteed to stay close to y_0 with high probability, but the same local property may be, in fact, requested for the whole walk trajectory, with uniformly positive probability.

Lemma 2.18 *Let $\mathcal{D} \subset \mathbb{R}^N$ be an open, bounded, connected domain. For a given boundary point $y_0 \in \partial\mathcal{D}$, assume that there exists $\theta_0 < 1$ such that for every $\delta > 0$ there are $\hat{\delta} \in (0, \delta)$ and $\hat{\epsilon} \in (0, 1)$ with the following property. For all $\epsilon \in (0, \hat{\epsilon})$ and all $x_0 \in B_{\hat{\delta}}(y_0) \cap \mathcal{D}$ there holds:*

$$\mathbb{P}\big(\exists n \geq 0 \;\; X_n^{\epsilon,x_0} \notin B_\delta(y_0)\big) \leq \theta_0, \tag{2.15}$$

where $\{X_n^{\epsilon,x_0}\}_{n=0}^\infty$ is the ϵ-ball walk defined in (2.2). Then y_0 is walk-regular.

Proof

1. Fix $\eta, \delta > 0$ and let $m \in \mathbb{N}$ be such that:

$$\theta_0^m \leq \eta.$$

Define the tuples $\{\epsilon_k\}_{k=0}^m$, $\{\hat{\delta}_k\}_{k=0}^{m-1}$ and $\{\delta_k\}_{k=1}^m$ inductively, in:

$$\delta_m = \delta, \quad \epsilon_m = 1,$$

$$\hat{\delta}_{k-1} \in (0, \delta_k), \quad \epsilon_{k-1} \in (0, \epsilon_k) \;\; \text{for all } k = 1, \ldots, m \text{ so that:}$$

$$\mathbb{P}\big(\exists n \geq 0 \;\; X_n^{x_0} \notin B_{\delta_k}(y_0)\big) \leq \theta_0 \tag{2.16}$$

$$\text{for all } x_0 \in B_{\hat{\delta}_{k-1}}(y_0) \cap \mathcal{D} \text{ and all } \epsilon \in (0, \epsilon_{k-1}),$$

$$\delta_{k-1} \in (0, \hat{\delta}_{k-1}) \;\; \text{for all } k = 2, \ldots, m.$$

We finally set:

$$\hat{\epsilon} \doteq \epsilon_0 \wedge \min_{k=1,\dots,m-1} |\hat{\delta}_k - \delta_k| \quad \text{and} \quad \hat{\delta} \doteq \hat{\delta}_0.$$

Fix $x_0 \in B_{\hat{\delta}}(y_0) \cap \mathcal{D}$ and $\epsilon \in (0, \hat{\epsilon})$. We will show that:

$$\begin{aligned} &\mathbb{P}\big(\exists n \geq 0 \;\; X_n^{\epsilon,x_0} \notin B_{\delta_k}(y_0)\big) \\ &\quad \leq \theta_0 \cdot \mathbb{P}\big(\exists n \geq 0 \;\; X_n^{\epsilon,x_0} \notin B_{\delta_{k-1}}(y_0)\big) \quad \text{for all } k = 2, \dots, m. \end{aligned} \tag{2.17}$$

Together with the inequality in (2.16) for $k = 1$, the above bounds will yield:

$$\mathbb{P}\big(X^{\epsilon,x_0} \notin B_{2\delta}(y_0)\big) \leq \mathbb{P}\big(\exists n \geq 0 \;\; X_n^{\epsilon,x_0} \notin B_{\delta}(y_0)\big) \leq \theta_0^m \leq \eta.$$

Since η and δ were arbitrary, the validity of the condition in Definition 2.12 will thus be justified, proving the walk-regularity of y_0.

2. Towards showing (2.17), we denote:

$$\tilde{\Omega} = \{\exists n \geq 0 \;\; X_n^{\epsilon,x_0} \notin B_{\delta_{k-1}}(y_0)\} \subset \Omega.$$

Without loss of generality, we may assume that $\mathbb{P}(\tilde{\Omega}) > 0$, because otherwise $\mathbb{P}\big(\exists n \geq 0 \;\; X_n^{\epsilon,x_0} \notin B_{\delta_k}(y_0)\big) \leq \mathbb{P}\big(\exists n \geq 0 \;\; X_n^{\epsilon,x_0} \notin B_{\delta_{k-1}}(y_0)\big) = 0$ and (2.17) holds then trivially. Consider the probability space $(\tilde{\Omega}, \tilde{\mathcal{F}}, \tilde{\mathbb{P}})$ defined by:

$$\tilde{\mathcal{F}} = \{A \cap \tilde{\Omega};\;\; A \in \mathcal{F}\} \quad \text{and} \quad \tilde{\mathbb{P}}(A) = \frac{\mathbb{P}(A)}{\mathbb{P}(\tilde{\Omega})} \quad \text{for all } A \in \tilde{F}.$$

Also, let the measurable space $(\Omega_{fin}, \mathcal{F}_{fin})$ be given by: $\Omega_{fin} = \bigcup_{n=1}^\infty \Omega_n$ and by taking $\mathcal{F}_{fin}$ to be the smallest σ-algebra containing $\bigcup_{n=1}^\infty \mathcal{F}_n$. Then the following random variable $\tau_k : \tilde{\Omega} \to \mathbb{N}$:

$$\tau_k \doteq \min\big\{n \geq 1;\;\; X_n^{\epsilon,x_0} \notin B_{\delta_{k-1}}(y_0)\big\}$$

is a stopping time on $\tilde{\Omega}$ with respect to the induced filtration $\{\tilde{\mathcal{F}}_n = \{A \cap \tilde{\Omega};\;\; A \in \mathcal{F}_n\}\}_{n=0}^\infty$. We consider two further random variables below:

$$\begin{aligned} &Y_1 : \tilde{\Omega} \to \Omega_{fin} \qquad && Y_1\big(\{w_i\}_{i=1}^\infty\big) \doteq \{w_i\}_{i=1}^{\tau_k} \\ &Y_2 : \tilde{\Omega} \to \Omega \qquad && Y_2\big(\{w_i\}_{i=1}^\infty\big) \doteq \{w_i\}_{i=\tau_k+1}^\infty \end{aligned}$$

and observe that they are independent, namely:

$$\begin{aligned} \tilde{\mathbb{P}}\big(Y_1 \in A_1\big) \cdot \tilde{\mathbb{P}}\big(Y_2 \in A_2\big) = \;& \tilde{\mathbb{P}}\big(\{Y_1 \in A_1\} \cap \{Y_2 \in A_2\}\big) \\ & \text{for all } A_1 \in \mathcal{F}_{fin},\; A_2 \in \mathcal{F}. \end{aligned}$$

We now apply Lemma A.21 to Y_1, Y_2 and to the indicator function:

$$Z\big(\{w_i\}_{i=1}^s, \{w_i\}_{i=s+1}^\infty\big) \doteq \mathbb{1}_{\big\{\exists n\geq 0 \;\; X_n^{\epsilon,x_0}(\{w_i\}_{i=1}^\infty)\not\in B_{\delta_k}(y_0)\big\}}$$

that is a random variable on the measurable space $\Omega_{fin} \times \Omega$, equipped with the product σ-algebra of $\mathcal{F}_{fin}$ and $\mathcal{F}$. It follows that:

$$\mathbb{P}\big(\exists n \geq 0 \;\; X_n^{\epsilon,x_0} \not\in B_{\delta_k}(y_0)\big) = \int_{\tilde{\Omega}} Z \circ (Y_1, Y_2)\, \mathrm{d}\tilde{\mathbb{P}} = \int_{\tilde{\Omega}} f(\omega_1)\, \mathrm{d}\tilde{\mathbb{P}}(\omega_1),$$

where for each $\omega_1 = \{w_i\}_{i=1}^\infty \in \tilde{\Omega}$ we have:

$$\begin{aligned} f(\omega_1) &= \mathbb{P}\Big(\{\bar{w}_i\}_{i=1}^\infty \in \Omega;\;\; \exists n \geq 0 \;\; X_n^{\epsilon,x_0}\big(\{w_i\}_{i=1}^{\tau_k}, \{\bar{w}_i\}_{i=\tau_k+1}^\infty\big) \not\in B_{\delta_k}(y_0)\Big) \\ &= \mathbb{P}(\tilde{\Omega}) \cdot \mathbb{P}\Big(\exists n \geq 0 \;\; X_n^{\epsilon,x_{\tau_k}} \not\in B_{\delta_k}(y_0)\Big) \leq \mathbb{P}(\tilde{\Omega}) \cdot \theta_0, \end{aligned}$$

in view of $x_{\tau_k} \in B_{\hat{\delta}_k}(y_0)$ and the construction assumption (2.16). This ends the proof of (2.17) and of the lemma.

□

The main result of this section is a geometric sufficient condition for walk-regularity. When combined with Theorem 2.14, it implies that every continuous boundary data F admits the unique harmonic extension to any Lipschitz domain $\mathcal{D}$. This extension automatically coincides with all process values u^ϵ, regardless of the choice of the upper bound sampling radius $\epsilon \in (0, 1)$.

Theorem 2.19 *Let $\mathcal{D} \subset \mathbb{R}^N$ be open, bounded, connected and assume that $y_0 \in \partial\mathcal{D}$ satisfies the* exterior cone condition, *i.e., there exists a finite cone $C \subset \mathbb{R}^N \setminus \mathcal{D}$ with the tip at y_0. Then y_0 is walk-regular.*

Proof The exterior cone condition assures the existence of a constant $R > 0$ such that for all sufficiently small $\rho > 0$ there exists $z_0 \in C$ satisfying:

$$|z_0 - y_0| = \rho(1 + R) \quad \text{and} \quad B_{R\rho}(z_0) \subset C \subset \mathbb{R}^N \setminus \mathcal{D}. \tag{2.18}$$

Let $\delta > 0$ be, without loss of generality, sufficiently small and define $z_0 \in \mathbb{R}^N$ as in (2.18) with $\rho = \hat{\delta}$, where we set $\hat{\delta} = \frac{\delta}{4+2R}$ (Fig. 2.4). We will show that condition (2.15) holds for all $\epsilon \in (0, 1)$.

Fix $x_0 \in B_{\hat{\delta}}(y_0) \cap \mathcal{D}$ and consider the profile function $v : (0, \infty) \to \mathbb{R}$ in:

$$v(t) = \begin{cases} \operatorname{sgn}(N-2)\, t^{2-N} & \text{for } N \neq 2, \\ -\log t & \text{for } N = 2. \end{cases}$$

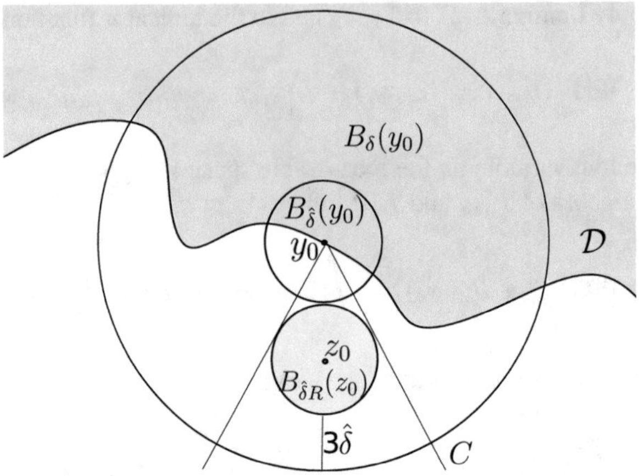

Fig. 2.4 The concentric balls in the proof of Theorem 2.19

By Exercise C.27, the radial function $x \mapsto v(|x - z_0|)$ is harmonic in $\mathbb{R}^N \setminus \{z_0\}$, so in view of Lemma 2.6 the sequence of random variables $\{v \circ |X_n^{\epsilon,x_0} - z_0|\}_{n=0}^\infty$ is a martingale with respect to the filtration $\{\mathcal{F}_n\}_{n=0}^\infty$. Further, define the random variable $\tau : \Omega \to \mathbb{N} \cup \{+\infty\}$ by:

$$\tau \doteq \inf\big\{n \geq 0;\ X_n^{\epsilon,x_0} \notin B_\delta(y_0)\big\},$$

where we suppress the dependence on ϵ in the above notation. Applying Doob's theorem (Theorem A.31 (ii)) we obtain:

$$v\big(|x_0 - z_0|\big) = \mathbb{E}\big[v \circ |X_0^{\epsilon,x_0} - z_0|\big] = \mathbb{E}\big[v \circ |X_{\tau\wedge n}^{\epsilon,x_0} - z_0|\big] \qquad \text{for all } n \geq 0,$$

because for every $n \geq 0$, the a.s. finite random variable $\tau \wedge n$ is a stopping time. Passing to the limit with $n \to \infty$ and recalling the definition (2.3), now yields:

$$v\big(|x_0 - z_0|\big) = \int_{\{\tau<+\infty\}} v\big(|X_\tau^{\epsilon,x_0} - z_0|\big)\,\mathrm{d}\mathbb{P} + \int_{\{\tau=+\infty\}} v\big(|X^{\epsilon,x_0} - z_0|\big)\,\mathrm{d}\mathbb{P}.$$

Since v is a decreasing function, this results in:

$$\begin{aligned} v\big((2+R)\hat\delta\big) \leq v\big(|x_0 - z_0|\big) &\leq \mathbb{P}(\tau < +\infty)\cdot v\big((3+R)\hat\delta\big) + \mathbb{P}(\tau = +\infty)\cdot v(R\hat\delta) \\ &= \mathbb{P}(\tau < +\infty)\cdot\Big(v\big((3+R)\hat\delta\big) - v(R\hat\delta)\Big) + v(R\hat\delta), \end{aligned}$$

in view of the following bounds:

$$
\begin{aligned}
&|x_0 - z_0| \leq |x_0 - y_0| + |y_0 - z_0| < (2+r)\hat{\delta}, \\
&|X_\tau^{\epsilon,x_0} - z_0| \geq |X_\tau^{\epsilon,x_0} - y_0| - |y_0 - z_0| = \delta - (1+R)\hat{\delta} = (3+R)\hat{\delta}, \\
&|X^{\epsilon,x_0} - z_0| \geq R\hat{\delta}.
\end{aligned}
$$

Finally, noting that $v((3+R)\hat{\delta}) - v(R\hat{\delta}) < 0$ we obtain:

$$
\mathbb{P}(\tau < +\infty) \leq \frac{v(R\hat{\delta}) - v((2+R)\hat{\delta})}{v(R\hat{\delta}) - v\big((3+R)\hat{\delta}\big)} = \frac{v(R) - v(2+R)}{v(R) - v(3+R)}.
$$

This establishes (2.15) with the constant $\theta_0 = \frac{v(R)-v(2+R)}{v(R)-v(3+R)} < 1$, that depends only on the dimension N and the cone C. By Lemma 2.18, the proof is done. □

Remark 2.20 An alternative sufficient condition for walk-regularity is the simple-connectedness of $\mathcal{D} \subset \mathbb{R}^2$. The proof follows through the identification of $\{X_n^{\epsilon,x_0}\}_{n=0}^\infty$ as the discrete realization of the Brownian path in Sect. 2.7* and applying the same reasoning as in the proof of Theorem 6.21. Indeed, in Chap. 6 we will give sufficient conditions for the so-called game-regularity, in the context of the Dirichlet problem for p-Laplacian, $p \in (1, \infty)$, encompassing and extending the classical discussion in the present chapter.

2.6* The Ball Walk Values and Perron Solutions

In this section we prove that all functions in the family $\{u^\epsilon\}_{\epsilon\in(0,1)}$ defined in (2.5) for a continuous $F : \partial\mathcal{D} \to \mathbb{R}$, are always one and the same function, coinciding with the so-called Perron solution of the Dirichlet problem:

$$
\Delta u = 0 \quad \text{in } \mathcal{D}, \qquad\qquad u = F \quad \text{on } \partial\mathcal{D}. \tag{2.19}
$$

This material may be skipped at first reading, as it requires familiarity with more advanced PDE notions of Perron's method and Wiener's resolutivity. The related presentation in the general nonlinear case of Δ_p, $p \in (1, \infty)$, can be found in Sect. C.7 of Appendix C. Below we recall this classical approach in the linear setting $p = 2$; for proofs we refer to the textbook by Helms (2014).

Definition 2.21

(i) A function $v \in C(\mathcal{D})$ is called *superharmonic* in $\mathcal{D}$, provided that for every $\bar{B}_r(x) \subset \mathcal{D}$ and every $h \in C(\bar{B}_r(x))$ that is harmonic in $B_r(x)$ and satisfies $h \leq v$ on $\partial B_r(x)$, there holds: $h \leq v$ in $B_r(x)$.
(ii) A function $v \in C(\mathcal{D})$ is *subharmonic* in $\mathcal{D}$, when $(-v)$ is superharmonic.

(iii) Given a continuous boundary data function $F : \partial\mathcal{D} \to \mathbb{R}$, we define the *upper and lower Perron solutions* to (2.19):

$$\bar{h}_F = \inf\Big\{v \in C(\bar{\mathcal{D}}) \text{ superharmonic, such that } F \le v \text{ on } \partial\mathcal{D}\Big\},$$

$$\underline{h}_F = \sup\Big\{v \in C(\bar{\mathcal{D}}) \text{ subharmonic, such that } v \le F \text{ on } \partial\mathcal{D}\Big\}.$$

The usual maximum principle argument implies that if $v_1, v_2 \in C(\bar{\mathcal{D}})$ are, respectively, subharmonic and superharmonic, and if $v_1 \le v_2$ on $\partial\mathcal{D}$, then $v_1 \le v_2$ in $\mathcal{D}$. In this comparison result, the conclusion may be in fact strengthened to: $v_1 < v_2$ or $v_1 \equiv v_2$ in $\mathcal{D}$. It follows that $\bar{h}_F$ and $\underline{h}_F$ are well defined functions, and also: $\underline{h}_F \le \bar{h}_F$. One may further prove, by means of the harmonic lifting, that $\bar{h}_F$ and $\underline{h}_F$ are harmonic in $\mathcal{D}$. The celebrated *Wiener resolutivity theorem* in Wiener (1925) states the uniqueness of this construction:

Theorem 2.22 *Let $\mathcal{D} \subset \mathbb{R}^N$ be open, bounded and connected. Every boundary data $F \in C(\partial\mathcal{D})$ is* resolutive, *i.e., the two functions $\bar{h}_F$ and $\underline{h}_F$ coincide in $\mathcal{D}$. The resulting harmonic function is called the* Perron solution *to (2.19):*

$$h_F = \bar{h}_F = \underline{h}_F. \tag{2.20}$$

We remark that h_F does not have to attain the prescribed boundary value $F(x)$ at each $x \in \partial\mathcal{D}$; it necessarily does so, however, for all points outside of a set whose 2-capacity is zero (see Sect. C.7).

By identifying the super-/subharmonic functions via mean value inequalities and comparing u^ϵ with $\bar{h}_F$ and $\underline{h}_F$, we obtain the main result of this section:

Theorem 2.23 *Let $\mathcal{D} \subset \mathbb{R}^N$ be open, bounded, connected and let $F \in C(\partial\mathcal{D})$. For each $\epsilon \in (0, 1)$, functions u^ϵ in (2.5) satisfy: $u^\epsilon = h_F$ in $\mathcal{D}$.*

Proof Let $v \in C(\bar{\mathcal{D}})$ be superharmonic and satisfy $F \le v$ on $\partial\mathcal{D}$. Observe first that for any ball $\bar{B}_r(x) \subset \mathcal{D}$ we may apply Definition 2.21 to compare v and the harmonic extension u of $v_{|\partial B_r(x)}$ on $B_r(x)$ (see Exercise C.21) and get:

$$\fint_{\partial B_r(x)} v(y)\,\mathrm{d}\sigma^{N-1}(y) = \fint_{\partial B_r(x)} u(y)\,\mathrm{d}\sigma^{N-1}(y) = u(x) \le v(x).$$

Integrating in polar coordinates, as in the proof of Theorem C.19, we obtain:

$$\begin{aligned}\fint_{B_r(x)} v(y)\,\mathrm{d}y &= \frac{1}{|B_r(x)|}\int_0^r\int_{\partial B_s(x)} v(y)\,\mathrm{d}\sigma^{N-1}(y)\,\mathrm{d}s\\ &\leq \frac{1}{|B_r(x)|}\int_0^r\int_{\partial B_s(x)} |\partial B_s(x)|\,\mathrm{d}s\cdot v(x) = v(x).\end{aligned}$$

Fix $\epsilon \in (0, 1)$ and $x_0 \in \mathcal{D}$. The sequence of random variables $\{v \circ X_n^{\epsilon,x_0}\}_{n=0}^{\infty}$ along the ball walk $\{X_n^{\epsilon,x_0}\}_{n=0}^{\infty}$ defined in (2.2), is then a supermartingale with respect to the filtration $\{\mathcal{F}_n\}_{n=0}^{\infty}$, because:

$$\mathbb{E}\big(v \circ X_n^{\epsilon,x_0} \mid \mathcal{F}_{n-1}\big) = \fint_{\epsilon\wedge \mathrm{dist}(X_{n-1},\partial\mathcal{D})} v(y)\,\mathrm{d}y \leq v \circ X_{n-1}^{\epsilon,x_0} \quad a.s.$$

Consequently: $\mathbb{E}[v \circ X_n^{\epsilon,x_0}] \leq \mathbb{E}[v \circ X_0] = v(x_0)$. Passing to the limit with $n \to \infty$ and recalling the boundary comparison assumption, finally yields:

$$v(x_0) \geq \mathbb{E}\big[v \circ X^{\epsilon,x_0}\big] \geq \mathbb{E}\big[F \circ X^{\epsilon,x_0}\big] = u^{\epsilon}(x_0).$$

We conclude that $\bar{h}_F \geq u^{\epsilon}$ by taking the infimum over all v as above. Since by a symmetric argument: $\underline{h}_F \leq u^{\epsilon}$, the result follows in virtue of (2.20). □

2.7* The Ball Walk and Brownian Trajectories

In this section we show that the ball walk, introduced in Sect. 2.2, can be seen as a discrete realization of the Brownian motion. In particular, we will deduce the same result as in Sect. 2.6*, namely that all functions in the family $\{u^{\epsilon}\}_{\epsilon\in(0,1)}$ in (2.5) are always one and the same function. This material may be skipped at first reading; it is slightly more advanced and necessitates familiarity with the construction of Brownian motion in Appendix B.

We start with some elementary technical observations. Denote $(\Omega_{\mathcal{B}}, \mathcal{F}_{\mathcal{B}}, \mathbb{P}_{\mathcal{B}})$ the probability space on which the standard N-dimensional Brownian motion $\{\mathcal{B}_t^N\}_{t\geq 0}$ is defined. We consider the product probability space $(\bar{\Omega}, \bar{\mathcal{F}}, \bar{\mathbb{P}}) = (\Omega_{\mathcal{B}}, \mathcal{F}_{\mathcal{B}}, \mathbb{P}_{\mathcal{B}}) \times (\Omega, \mathcal{F}, \mathbb{P})$ with the space $(\Omega, \mathcal{F}, \mathbb{P})$ in Sect. 2.2, and denote its elements by $(\omega_{\mathcal{B}}, \omega)$ with $\omega = \{w_i\}_{i=1}^{\infty} \in B_1(0)^{\mathbb{N}}$. Clearly, $\{\mathcal{B}_t^N\}_{t\geq 0}$ is also a standard Brownian motion on $(\bar{\Omega}, \bar{\mathcal{F}}, \bar{\mathbb{P}})$. We further denote the product σ-algebras $\bar{\mathcal{F}}_t = \mathcal{F}_t \times \mathcal{F}$, so that $\bar{\mathcal{F}}_s \subset \bar{\mathcal{F}}_t \subset \bar{\mathcal{F}}$ for all $0 \leq s \leq t$; for every $s \in [0, t]$ the random variable $\mathcal{B}_s^N$ is $\bar{\mathcal{F}}_t$-measurable.

We call $\bar{\tau} : \bar{\Omega} \to [0, \infty]$ a stopping time on $(\bar{\Omega}, \bar{\mathcal{F}}, \bar{\mathbb{P}})$ provided that $\{\bar{\tau} \leq t\} \in \bar{\mathcal{F}}_t$ for all $t \geq 0$ and that $\mathbb{P}(\bar{\tau} = +\infty) = 0$. Then, the random variable $\mathcal{B}_{\bar{\tau}}^N$ is $\bar{\mathcal{F}}_{\bar{\tau}}$-

measurable, namely: $\{\mathcal{B}^N_{\bar\tau} \in A\} \cap \{\bar\tau \leq t\} \in \bar{\mathcal{F}}_t$ for all Borel $A \subset \mathbb{R}^N$ and all $t \geq 0$, which can be proved as in Lemma B.21.

Let now $\mathcal{D} \subset \mathbb{R}^N$ be open, bounded, connected and fix $x_0 \in \mathcal{D}$, $\epsilon \in (0,1)$. We inductively define the sequence of random variables $\bar\tau_k : \bar\Omega \to [0,\infty]$ in:

$$\begin{aligned}
&\bar\tau_0 = 0, \\
&\bar\tau_{k+1}\big(\omega_{\mathcal{B}}, \{w_i\}_{i=1}^\infty\big) \\
&\quad = \min\Big\{t \geq \bar\tau_k;\ \big|\mathcal{B}^N_t(\omega_{\mathcal{B}}) - \mathcal{B}^N_{\bar\tau_k(\omega_{\mathcal{B}},\omega)}(\omega_{\mathcal{B}})\big| \\
&\qquad\qquad = \big(\epsilon \wedge \mathrm{dist}(x_0 + \mathcal{B}^N_{\bar\tau_k(\omega_{\mathcal{B}},\omega)}(\omega_{\mathcal{B}}), \partial\mathcal{D})\big)|w_{k+1}|\Big\},
\end{aligned} \tag{2.21}$$

and also:

$$\bar\tau(\omega_{\mathcal{B}},\omega) = \min\big\{t \geq 0;\ x_0 + \mathcal{B}^N_t(\omega_B) \in \partial\mathcal{D}\big\}. \tag{2.22}$$

Lemma 2.24 *Each $\bar\tau_k$ in (2.21) and $\bar\tau$ in (2.22) is a stopping time on $(\bar\Omega, \bar{\mathcal{F}}, \bar{\mathbb{P}})$. Moreover, $\bar\tau_k$ converge to $\bar\tau$ as $k \to \infty$, a.s. in $\bar\Omega$.*

Proof

1. The random variable $\bar\tau$ is a stopping time by Lemma B.9. For $\{\bar\tau_k\}_{k=0}^\infty$, we argue by induction. Assume that $\bar\tau_k$ is a stopping time. Fix $s > 0$ and observe that:

$$\begin{aligned}
&\big\{\bar\tau_{k+1} \leq t\big\} = \{w_{k+1} = 0\} \cup \\
&\cup \bigcap_{m>0} \bigcup_{q\in[0,t]\cap\mathbb{Q}} \Big(\{\bar\tau_k \leq q\} \cap \Big\{|\mathcal{B}^N_q - \mathcal{B}^N_{\bar\tau_k}| \geq \big(\epsilon \wedge \mathrm{dist}(\mathcal{B}^N_{\bar\tau_k}, \partial\mathcal{D} - x_0)\big)|w_{k+1}| - \frac{1}{m}\Big\}\Big).
\end{aligned}$$

Now $\{w_{k+1} = 0\} \in \bar{\mathcal{F}}_t$, whereas each of the countably many sets in the right hand side above may be written as:

$$\{\bar\tau_k \leq q\} \cap \Big\{|\mathcal{B}^N_q - \mathcal{B}^N_{\bar\tau_k} \cdot \mathbb{1}_{\bar\tau_k \leq q}| \geq \big(\epsilon \wedge \mathrm{dist}(\mathcal{B}^N_{\bar\tau_k} \cdot \mathbb{1}_{\bar\tau_k \leq q}, \partial\mathcal{D} - x_0)\big)|w_{k+1}| - \frac{1}{m}\Big\}.$$

Since $\mathcal{B}^N_{\bar\tau_k} \cdot \mathbb{1}_{\bar\tau_k \leq q}$ is $\bar{\mathcal{F}}_q$-measurable, by the inductive assumption it follows that the set above belongs to $\bar{\mathcal{F}}_q \subset \bar{\mathcal{F}}_t$, as claimed. Since $\bar\tau_k \leq \bar\tau$, we further deduce that each $\bar\tau_k$ is a.s. finite, hence a stopping time.
2. Let $\bar\tau_\infty$ be a pointwise limit of the nondecreasing sequence $\{\bar\tau_k\}_{k=0}^\infty$. Clearly $\bar\tau_\infty$ is a stopping time and $\bar\tau_\infty \leq \bar\tau$. To show that $\bar\tau_\infty = \bar\tau$ a.s., consider the event:

$$A = \{\bar\tau_\infty < +\infty\} \cap \Big\{\omega = \{w_k\}_{k=1}^\infty \in \Omega;\ |w_k| \geq \frac{1}{2} \text{ for infinitely many } k = 1\ldots\infty\Big\}.$$

Obviously, $\bar{\mathbb{P}}(A) = 1$, because:

$$\mathbb{P}\Big(\omega \in \Omega;\ |w_k| \geq \frac{1}{2} \text{ for finitely many } k\Big) \leq \sum_{n=1}^{\infty} \mathbb{P}\Big(|w_k| < \frac{1}{2} \text{ for all } k \geq n\Big) = 0.$$

Further, for every $(\omega_{\mathcal{B}}, \omega) \in A$ we get:

$$\lim_{k\to\infty} \mathcal{B}^N_{\bar{\tau}_k} = \mathcal{B}^N_{\bar{\tau}_\infty},$$

and thus: $\lim_{k\to\infty} |\mathcal{B}^N_{\bar{\tau}_{k+1}} - \mathcal{B}^N_{\bar{\tau}_k}| = 0$, while: $|\mathcal{B}^N_{\bar{\tau}_{k+1}} - \mathcal{B}^N_{\bar{\tau}_k}| \geq \frac{1}{2}\big(\epsilon \wedge \mathrm{dist}(x_0 + \mathcal{B}^N_{\bar{\tau}_k}, \partial\mathcal{D})\big)$ whenever $|w_k| \geq \frac{1}{2}$. Consequently:

$$\mathrm{dist}\big(x_0 + \mathcal{B}^N_{\bar{\tau}_\infty}, \partial\mathcal{D}\big) = \lim_{k\to\infty} \mathrm{dist}\big(x_0 + \mathcal{B}^N_{\bar{\tau}_k}, \partial\mathcal{D}\big) = 0,$$

so $x_0 + \mathcal{B}^N_{\bar{\tau}_\infty} \in \partial\mathcal{D}$ and hence $\bar{\tau}_\infty = \bar{\tau}$ on A. □

Given a continuous boundary function $F : \partial\mathcal{D} \to \mathbb{R}$, recall that:

$$u(x_0) = \int_{\bar{\Omega}} F \circ \big(x_0 + \mathcal{B}^N_{\bar{\tau}}\big)\, \mathrm{d}\bar{\mathbb{P}} \tag{2.23}$$

defines a harmonic function $u : \mathcal{D} \to \mathbb{R}$, in virtue of Corollary B.29 that builds on the classical construction and discussion of Brownian motion presented in Appendix B. As in Remark 2.3, we view F as a restriction of some $F \in C(\bar{\mathcal{D}})$. Then, by Lemma 2.24 we also have:

$$u(x_0) = \lim_{k\to\infty} \int_{\bar{\Omega}} F \circ \big(x_0 + \mathcal{B}^N_{\bar{\tau}_k}\big)\, \mathrm{d}\bar{\mathbb{P}}.$$

On the other hand, we recall that in (2.6) we defined:

$$u^\epsilon(x_0) = \int_\Omega F \circ X^{\epsilon,x_0}\, \mathrm{d}\mathbb{P} = \lim_{k\to\infty} \int_\Omega F \circ X_k^{\epsilon,x_0}\, \mathrm{d}\mathbb{P}.$$

We now observe:

Theorem 2.25 *For all $\epsilon \in (0,1)$ and all $x_0 \in \mathcal{D}$ there holds: $u^\epsilon(x_0) = u(x_0)$. In fact, we have:*

$$\mathbb{P}_{\mathcal{B}}\big(x_0 + \mathcal{B}^N_{\bar{\tau}} \in A\big) = \mathbb{P}\big(X^{\epsilon,x_0} \in A\big) \qquad \textit{for all Borel } A \subset \mathbb{R}^N. \tag{2.24}$$

Proof

1. Fix $\epsilon \in (0,1)$ and $x_0 \in \mathcal{D}$. We will prove that for all $f \in C_c(\mathbb{R}^N)$:

$$\mathbb{E}_{\bar{\Omega}}\big[f \circ (x_0 + \mathcal{B}^N_{\bar{\tau}_k})\big] = \mathbb{E}_{\Omega}\big[f \circ X_k^{\epsilon,x_0}\big] \quad \text{for all } k \geq 0. \tag{2.25}$$

We proceed by induction, defining μ_k to be the push-forward of $\bar{\mathbb{P}}$ via the random variable $x_0 + \mathcal{B}^N_{\bar{\tau}_k} : \bar{\Omega} \to \mathcal{D}$, and ν_k to be the push-forward of $\mathbb{P}$ via X_k^{ϵ,x_0}:

$$\mu_k(A) = \bar{\mathbb{P}}\big(x_0 + \mathcal{B}^N_{\bar{\tau}_k} \in A\big), \qquad \nu_k(A) = \mathbb{P}\big(X_k^{\epsilon,x_0} \in A\big) \qquad \text{for all Borel } A \subset \mathcal{D}.$$

For $k = 0$ we have: $\mu_0 = \nu_0 = \delta_{x_0}$. To show that $\mu_{k+1} = \nu_{k+1}$ if $\mu_k = \nu_k$, observe that for $f \in C_c(\mathbb{R}^N)$, there holds:

$$\begin{aligned}\int_{\mathcal{D}} f \,\mathrm{d}\nu_{k+1} &= \int_{\Omega} f \circ X_{k+1}^{\epsilon,x_0} \,\mathrm{d}\mathbb{P} \\ &= \int_{\Omega_k}\int_{\Omega_1} f\big(X_k^{\epsilon,x_0} + (\epsilon \wedge \mathrm{dist}(X_k^{\epsilon,x_0}, \partial\mathcal{D}))w_{k+1}\big) \,\mathrm{d}\mathbb{P}_1(w_{k+1}) \,\mathrm{d}\mathbb{P}_k \\ &= \int_{\Omega_k} \fint_{B_{\epsilon\wedge \mathrm{dist}(X_k,\partial\mathcal{D})}(X_k^{\epsilon,x_0})} f(y) \,\mathrm{d}y \,\mathrm{d}\mathbb{P}_k \\ &= \int_{\mathcal{D}} \fint_{B_{\epsilon\wedge \mathrm{dist}(z,\partial\mathcal{D})}(z)} f(y) \,\mathrm{d}y \,\mathrm{d}\nu_k(z),\end{aligned} \tag{2.26}$$

where we used Fubini's theorem and the definition of ν_k in view of the function $z \mapsto \fint_{B_{\epsilon\wedge \mathrm{dist}(z,\partial\mathcal{D})}(z)} f(y) \,\mathrm{d}y$, being continuous and bounded.

2. On the other hand, we write:

$$\int_{\mathcal{D}} f \,\mathrm{d}\mu_{k+1} = \int_{\bar{\Omega}} f \circ \psi \circ \big(Z_1, Z_2, Z_3\big) \,\mathrm{d}\bar{\mathbb{P}}, \tag{2.27}$$

where $Z_1 = x_0 + \mathcal{B}^N_{\bar{\tau}_k}$ is a $\mathcal{D}$-valued random variable on $(\bar{\Omega}, \bar{\mathcal{F}})$. Further:

$$Z_2(\omega) = \big([0,\infty) \ni t \mapsto \mathcal{B}^N_{\bar{\tau}_k + t} - \mathcal{B}^N_{\bar{\tau}_k} \in \mathbb{R}^N\big)$$

is the E-valued random variable on $(\bar{\Omega}, \bar{\mathcal{F}})$ in the notation of Exercise B.27:

$$E = \Big\{f \in C([0,\infty), \mathbb{R}^N);\ f(0) = 0 \text{ and } |f(t)| > \mathrm{diam}\,\mathcal{D} \text{ for some } t > 0\Big\}.$$

Here, E is a measurable space when equipped with the Borel σ-algebra induced by topology of uniform convergence on compact intervals $[0,T]$, for all $T > 0$.

This σ-algebra is generated by sets of the type:

$$A_{g,T,\delta} = \big\{h \in E;\ \|h - g\|_{L^\infty([0,T])} \leq \delta\big\}$$

for polynomials g with rational coefficients and rational numbers $T, \delta > 0$. Finally, we set $Z_3 = |w_{k+1}|$ to be the $\mathbb{R}$-valued random variable on $(\bar{\Omega}, \bar{\mathcal{F}})$, whereas $\psi : \mathcal{D} \times E \times [0, \operatorname{diam} \mathcal{D}] \to \mathbb{R}^N$ in (2.27) is given by:

$$\psi(z, h, r) = z + h\big(\min\{t \geq 0;\ |h(t)| = \big(\epsilon \wedge \operatorname{dist}(x, \partial\mathcal{D})\big)r\}\big).$$

As in Exercise B.27, ψ is measurable with respect to the product σ-algebra of: Borel subsets of $\mathcal{D}$, the indicated σ-algebra in E and the Borel σ-algebra in $\mathbb{R}$.

We now observe that Z_1, Z_2, Z_3 are independent, by the strong Markov property in Theorem B.24. Indeed, Z_1 is $\bar{\mathcal{F}}_{\bar{\tau}_k}$-measurable, whereas each preimage basis set $Z_2^{-1}(A_{g,T,\delta}) = \bigcap_{q\in[0,T]\cap\mathbb{Q}} \big\{B^N_{\bar{\tau}_k+q} - \mathcal{B}^N_{\bar{\tau}_k} \in \bar{B}_\delta(g(q))\big\}$ belongs to the σ-algebra generated by the Brownian motion $\big\{B^N_{\bar{\tau}_k+t} - \mathcal{B}^N_{\bar{\tau}_k}\}_{t\geq 0}$, which is independent of $\bar{\mathcal{F}}_{\bar{\tau}_k}$. Also, the latter σ-algebra is contained in $\mathcal{F}_{\mathcal{B}} \times \mathcal{F}_k$, as is $\bar{\mathcal{F}}_{\bar{\tau}_k}$.

3. Let $\mu^{(i)}_{k+1}$ denote the push-forward of $\mathbb{P}$ via the corresponding random variable Z_i, $i = 1\dots 3$. By the very definition, we have: $\mu^{(1)}_{k+1} = \mu_k$ and $\mu^{(3)}_{k+1} = dr$, whereas $\mu^{(2)}_{k+1}$ coincides with the Wiener measure μ_W as in Exercise B.17. For each fixed $\bar{r} \in [0, \operatorname{diam}\mathcal{D}]$, consider the stopping time

$$\tau^{\bar{r}} = \min\{t \geq 0;\ |\mathcal{B}^N_t| = \bar{r}\}$$

and observe that the push-forward of $\mu^{(2)}_{k+1}$ on $\partial B_{\bar{r}}(0)$ via the measurable mapping $\mathcal{B}^N_{\tau^{\bar{r}}}$ is rotationally invariant, as in Corollary B.29. Hence, the aforementioned push-forward coincides with the normalized spherical measure σ^{N-1}, by Exercise 2.11.

Fubini's theorem, combined with Exercise A.20, now yield:

$$\begin{aligned}
\int_{\mathcal{D}} f\ \mathrm{d}\mu_{n+1} &= \int_{\mathcal{D}\times E\times[0,\operatorname{diam}\mathcal{D}]} f\circ\psi\ \mathrm{d}\big(\mu^{(1)}_{k+1}\times\mu^{(2)}_{k+1}\times\mu^{(3)}_{k+1}\big) \\
&= \int_{\mathcal{D}}\int_{[0,1]}\int_E \big(f\circ\psi\big)(z,h,r)\ \mathrm{d}\mu^{(2)}_{k+1}(h)\ \mathrm{d}\mu^{(3)}_{k+1}(r)\ \mathrm{d}\mu^{(1)}_{k+1}(z) \\
&= \int_{\mathcal{D}}\int_0^1\int_\Omega f\Big(z + \mathcal{B}^N_{\tau^{(\epsilon\wedge\operatorname{dist}(z,\partial\mathcal{D}))r}}\Big)\ \mathrm{d}\mathbb{P}\ \mathrm{d}r\ \mathrm{d}\mu_k(z).
\end{aligned}$$

Consequently:

$$\begin{aligned}
\int_{\mathcal{D}} f\ \mathrm{d}\mu_{n+1} &= \int_{\mathcal{D}} \fint_{B_{(\epsilon\wedge\operatorname{dist}(z,\partial\mathcal{D}))r}(z)} f(y)\ \mathrm{d}y\ \mathrm{d}\mu_k(z) \\
&= \int_{\mathcal{D}}\int_0^1 \fint_{\partial B_{(\epsilon\wedge\operatorname{dist}(z,\partial\mathcal{D}))r}(z)} f(y)\ \mathrm{d}\sigma^{N-1}(y)\ \mathrm{d}r\ \mathrm{d}\mu_k(z),
\end{aligned} \tag{2.28}$$

and we see that $\mu_k = \nu_k$, (2.26) and (2.28) imply: $\mu_{k+1} = \nu_{k+1}$, together with (2.25). Passing to the limit $k \to \infty$ achieves the proof of (2.24) and implies $u^\epsilon = u$ in $\mathcal{D}$. □

Exercise 2.26 Modify the arguments in this section to the setting of the sphere walk introduced in Exercise 2.8. Follow the outline below:

(i) Given $x_0 \in \mathcal{D}$ and $\epsilon \in (0, 1)$, show that the following are stopping times on $(\Omega_{\mathcal{B}}, \mathcal{F}_{\mathcal{B}}, \mathbb{P}_{\mathcal{B}})$:

$$\tau_0 = 0, \qquad \tau_{k+1} = \min\Big\{t \geq \tau_k;\ |\mathcal{B}_t^N - \mathcal{B}_{\tau_k}^N| = \epsilon \wedge \frac{1}{2}\mathrm{dist}(x_0 + \mathcal{B}_{\tau_k}^N, \partial\mathcal{D})\Big\},$$

that converge a.s. as $k \to \infty$, to the exit time:

$$\tau = \min\{t \geq 0;\ \mathcal{B}_t^N \in \partial\mathcal{D} - x_0\}.$$

(ii) Let $(\Omega, \mathcal{F}, \mathbb{P})$ and $\{X_n^{\epsilon, x_0}\}_{n=0}^\infty$ be as in Exercise 2.8 (i), and define $u^\epsilon : \mathcal{D} \to \mathbb{R}$ according to (2.5) and (2.6). Prove that the push-forward of $\mathbb{P}$ on $\mathcal{D}$ via X_k^{ϵ, x_0}, coincides with the push-forward of $\mathbb{P}_{\mathcal{B}}$ via $x_0 + \mathcal{B}_{\tau_k}^N$, for every $k \geq 0$. Consequently, $u^\epsilon(x_0) = \int_\Omega F \circ (x_0 + \mathcal{B}_\tau^N)\, \mathrm{d}\mathbb{P}$, which is the harmonic extension of a given $F \in C(\partial\mathcal{D})$, independent of $\epsilon \in (0, 1)$.

2.8 Bibliographical Notes

All constructions, statements of results and proofs in this chapter have their continuous random process counterparts through Brownian motion, see Mörters and Peres (2010). The ball walk can be seen as a modification of the sphere walk in Exercise 2.8, which in turn is one of the most commonly used methods for sampling from harmonic measure, proposed in Muller (1956).

The definition of the walk-regularity of a boundary point y_0, which in the context of Sect. 2.7* can be rephrased as:

$$\forall \eta, \delta > 0 \quad \exists \hat{\delta} \in (0, \delta) \quad \forall x_0 \in B_{\hat{\delta}}(y_0) \cap \mathcal{D} \qquad \mathbb{P}\big(x_0 + \mathcal{B}_\tau^N \in B_\delta(y_0)\big) \geq 1 - \eta,$$

is equivalent to the classical definition given in Doob (1984):

$$\mathbb{P}_{\mathcal{B}}\Big(\inf\{t > 0;\ y_0 + \mathcal{B}_t^N \in \mathbb{R}^N \setminus \mathcal{D}\} = 0\Big) = 1;$$

this equivalence will be shown in Sect. 3.7*. The above property is further equivalent to the classical potential theory 2-regularity of y_0 in Definition C.46. Its equivalence with the Wiener regularity criterion, stating that $\mathbb{R}^N \setminus \mathcal{D}$ is 2-thick at y_0 (compare Definition C.47 (ii)) can be proved directly, see Mörters and Peres (2010) for a modern exposition. In working out the proofs of this chapter and the analysis in Sect. 2.7*, the author has largely benefited from the aforementioned book and from personal communications with Y. Peres.

Various averaging principles and related random walks in the Heisenberg group were discussed in Lewicka et al. (2019). In papers by Lewicka and Peres (2019b,a), Laplace's equation augmented by the Robin boundary conditions has been studied from the viewpoint of the related averaging principles in $C^{1,1}$-regular domains. There, the asymptotic Hölder regularity of the values of the ϵ-walk has been proved, for any Hölder exponent $\alpha \in (0, 1)$ and up to the boundary of $\mathcal{D}$, together with the interior asymptotic Lipschitz equicontinuity.

The "ellipsoid walk" linked to the elliptic problem: $\mathrm{Trace}\big(A(x)\nabla^2 u(x)\big) = 0$ has been analysed in Arroyo and Parviainen (2019). For bounded, measurable coefficients matrix A satisfying $\det A = 1$, and uniformly elliptic with the elliptic distortion ratio that is close to 1 in $\mathcal{D}$, this leads to proving the local asymptotic uniform Hölder continuity of the associated process values u^ϵ.

Chapter 3
Tug-of-War with Noise: Case $p \in [2, \infty)$

Many properties of the ordinary Laplacian $\Delta = \Delta_2$ studied in Chap. 2 also hold for the p-Laplacian Δ_p. While the classical potential theory deals with harmonic functions, the discussed here case of the *p-harmonic functions* or solutions to more general nonlinear equations of divergence type requires more refined techniques. These were developed in the framework of the *nonlinear potential theory*, combining ideas and notions in partial differential equations, calculus of variations and mathematical physics.

The purpose of this and the following chapters is to present one of such remarkable connections, namely the connection to probability. The nonlinear counterpart of the arguments in Chap. 2, relying on the so-called Tug-of-War games, has been discovered and studied only recently by Peres et al. (2009); Peres and Sheffield (2008); Manfredi et al. (2012b). We presently start with treating the case $p \geq 2$, which is somewhat closer to the linear case. The remaining singular case $p \in (1, 2)$ will be discussed in Chap. 6, in the unified manner to $p > 1$.

In Sect. 3.1 below, we briefly recall the definition and notation for Δ_p. Section 3.2 develops the *mean value expansions* with Δ_p as the second order term, in the spirit of the familiar expansion $\fint_{B_\epsilon(x)} u(y)\,\mathrm{d}y = u(x) + \frac{\epsilon^2}{2(N+2)}\Delta u(x) + o(\epsilon^2)$ and consistent with the mean value property $\fint_{B_\epsilon(x)} u(y)\,\mathrm{d}y = u(x)$ of harmonic functions. In Sect. 3.3 we focus on the first of the introduced expansions and show that the boundary value problem for the *mean value equation* resulting by neglecting the $o(\epsilon^2)$ error term has a unique solution u_ϵ on $\mathcal{D} \subset \mathbb{R}^N$. In Sects. 3.4 and 3.5, we prove that each u_ϵ coincides with the value u^ϵ of the *Tug-of-War game with noise*, which can be seen as a deterministic modification of the finite-horizon counterpart to the ball walk in Chap. 2. In this process, a token that is initially placed at $x_0 \in \mathcal{D}$, at each step of the game is advanced by one of the players (both acting with probability $\frac{\beta}{2}$) or by a random shift (activated with probability α), within the ball centred at the current position and with radius ϵ. The probabilities in: $\alpha + \beta = 1$ depend on p and the dimension N. The process is stopped when the token reaches the

M. Lewicka, *A Course on Tug-of-War Games with Random Noise*, Universitext, https://doi.org/10.1007/978-3-030-46209-3_3

ϵ-neighbourhood of the boundary $\partial\mathcal{D}$, and $u^\epsilon(x_0)$ is then defined as the expected value of the boundary data F (extended continuously on the neighbourhood of $\partial\mathcal{D}$) at the stopping position x_τ, subject to both players playing optimally. The optimality criterion is based on the rule that Player II pays to Player I the value $F(x_\tau)$, thus giving Player I the incentive to maximize the gain by pulling towards the portions of $\partial\mathcal{D}$ with high values of F, whereas Player II will likely try to minimize the loss by pulling, locally, towards the low values of F.

In Sect. 3.6*, we identify the Tug-of-War game process corresponding to $p = 2$, with the discrete realization of the Brownian motion along its continuous trajectories, similarly as in Sect. 2.7*. Since the sampling takes place on balls of radius ϵ regardless of the position of the particle, it is no more true that $\{u^\epsilon\}_{\epsilon\in(0,1)}$ are each, one and the same function. However, we show that $\{u^\epsilon\}_{\epsilon\to 0}$ converge pointwise to a harmonic function u that is the *Brownian motion harmonic extension* of F introduced in Appendix B. When $\partial\mathcal{D}$ is regular, then $u_{|\partial\mathcal{D}} = F$ so that u is the classical harmonic extension of F. Various regularity conditions are compared in Sect. 3.7*; these last two sections require the basic familiarity with Brownian motion and may be skipped at first reading.

3.1 The p-Harmonic Functions and the p-Laplacian

We briefly recall the definition and notation for the p-Laplacian. An expanded overview of the nonlinear Potential Theory can be found in Appendix C.

Let $\mathcal{D} \subset \mathbb{R}^N$ be an open, bounded, connected set. Consider the integral:

$$\mathcal{I}_p(u) = \int_{\mathcal{D}} |\nabla u(x)|^p \,\mathrm{d}x \qquad \text{for all } u \in W^{1,p}(\mathcal{D}),$$

where, in this chapter, the exponent p belongs to the range:

$$p \in [2, \infty),$$

replacing the harmonic exponent $p = 2$ from Chap. 2. We want to minimize the energy $\mathcal{I}_p$ among all functions u subject to given boundary data. The condition for vanishing of the first variation of $\mathcal{I}_p$ (see Lemma C.28) takes form:

$$\int_{\mathcal{D}} \Big\langle |\nabla u|^{p-2}\nabla u, \nabla\eta \Big\rangle \,\mathrm{d}x = 0 \qquad \text{for all } \eta \in C_c^\infty(\mathcal{D}).$$

Assuming sufficient regularity of u, the divergence theorem then yields:

$$\int_{\mathcal{D}} \eta \,\mathrm{div}\big(|\nabla u|^{p-2}\nabla u\big) \,\mathrm{d}x = 0 \qquad \text{for all } \eta \in C_c^\infty(\mathcal{D}),$$

which, by the fundamental theorem of Calculus of Variations, becomes:

$$\Delta_p u \doteq \operatorname{div}\left(|\nabla u|^{p-2}\nabla u\right) = 0 \quad \text{in } \mathcal{D}. \tag{3.1}$$

Definition 3.1 The second order differential operator in (3.1) is called the *p-Laplacian*, the partial differential equation (3.1) is called the *p-harmonic equation* and its solution u is a *p-harmonic function*.

To examine the differential expression in (3.1) more closely, compute:

$$\nabla\big(|\nabla u|^{p-2}\big) = \nabla\Big(\big(\sum_{i=1}^N |\partial_i u|^2\big)^{\frac{p-2}{2}}\Big) = (p-2)\big(\sum_{i=1}^N |\partial_i u|^2\big)^{\frac{p-4}{2}} (\nabla^2 u)\nabla u,$$

which implies:

$$\begin{aligned}\big\langle\nabla\big(|\nabla u|^{p-2}\big), \nabla u\big\rangle &= (p-2)|\nabla u|^{p-2}\Big\langle\frac{(\nabla^2 u)\nabla u}{|\nabla u|}, \frac{\nabla u}{|\nabla u|}\Big\rangle \\ &= (p-2)\Big\langle\nabla^2 u : \frac{\nabla u}{|\nabla u|}\otimes\frac{\nabla u}{|\nabla u|}\Big\rangle.\end{aligned}$$

Consequently:

$$\begin{aligned}\Delta_p u &= |\nabla u|^{p-2}\Delta u + \Big\langle\nabla\big(|\nabla u|^{p-2}\big), \nabla u\Big\rangle \\ &= |\nabla u|^{p-2}\Big(\Delta u + (p-2)\Big\langle\nabla^2 u : \frac{\nabla u}{|\nabla u|}\otimes\frac{\nabla u}{|\nabla u|}\Big\rangle\Big),\end{aligned} \tag{3.2}$$

and we see that when $p \to \infty$, the second term in parentheses above prevails, which motivates the following definition of the *∞-Laplacian*:

$$\Delta_\infty u = \Big\langle\nabla^2 u : \frac{\nabla u}{|\nabla u|}\otimes\frac{\nabla u}{|\nabla u|}\Big\rangle. \tag{3.3}$$

The equation $\Delta_\infty u = 0$ is called the *∞-harmonic equation* and its solution is a *∞-harmonic function*. Some authors, e.g., Lindqvist (2019), prefer to write $\Delta_\infty u = \langle\nabla^2 u : \nabla u \otimes \nabla u\rangle = \frac{1}{2}\langle\nabla|\nabla u|^2, \nabla u\rangle$, which results in the interpolation: $\Delta_p u = |\nabla u|^{p-4}\left(|\nabla u|^2\Delta u + (p-2)\Delta_\infty\right)$, and call Δ_∞ in (3.3) the "game-theoretic" ∞-Laplacian. Since this book features connections of Δ_p to game theory, we indeed use this definition.

We conclude by pointing out the following useful decomposition which will be at the centre of our attention in the next Section:

Lemma 3.2 *Let $\mathcal{D} \subset \mathbb{R}^N$ be an open set, and let $u \in C^2(\mathcal{D})$. If $\nabla u(x) \neq 0$, then:*

$$\Delta_p u(x) = |\nabla u|^{p-2}\Big(\Delta u + (p-2)\Delta_\infty u\Big)(x), \tag{3.4}$$

$$\Delta_p u(x) = |\nabla u|^{p-2}\Big(|\nabla u|\Delta_1 u + (p-1)\Delta_\infty u\Big)(x). \tag{3.5}$$

Proof The formula (3.4) follows directly from (3.2). In particular, from the same expansion expression it also follows that $\Delta u = |\nabla u|\Delta_1 u + \Delta_\infty u$, which together with (3.4) results in (3.5). □

Exercise 3.3 Let $1 < q < p < r < \infty$. Prove the identity:

$$(r-q)|\nabla u|^{2-p}\Delta_p u = (r-p)|\nabla u|^{2-q}\Delta_q u + (p-q)|\nabla u|^{2-r}\Delta_r u.$$

3.2 The Averaging Principles

The purpose of this section is to derive the *averaging principles* for the operator Δ_p. The ultimate goal is to find the approximations of solutions to (3.1), analogous to the construction in Chap. 2 where the mean value property of harmonic functions was precisely the averaging principle responsible for the correspondence between the Laplace operator $\Delta = \Delta_2$ and the ball walk. The expansion formulas (3.13) and (3.14) below will, in the same spirit, serve as the dynamic programming principles for Tug-of-War games with noise whose values yield p-harmonic functions.

Recall that given a function $u : \mathcal{D} \to \mathbb{R}$, its average on $B_\epsilon(x) \subset \mathcal{D}$ is denoted:

$$\mathcal{A}_\epsilon u(x) = \fint_{B_\epsilon(x)} u(y)\,\mathrm{d}y.$$

We start with proving the basic averaging expansions for: the linear operator Δ corresponding to $p = 2$, and the fully nonlinear Δ_∞ corresponding to $p = \infty$:

Theorem 3.4 *Let $\mathcal{D} \subset \mathbb{R}^N$ be an open set and assume that $u \in C^2(\mathcal{D})$. Then, for every $x \in \mathcal{D}$ and $\epsilon > 0$ such that $\bar{B}_\epsilon(x) \subset \mathcal{D}$, we have:*

(i)
$$\mathcal{A}_\epsilon u(x) = u(x) + \frac{\epsilon^2}{2(N+2)}\Delta u(x) + o(\epsilon^2). \tag{3.6}$$

(ii) If $\nabla u(x) \neq 0$, then:

$$\frac{1}{2}\Big(\inf_{y \in B_\epsilon(x)} u(y) + \sup_{y \in B_\epsilon(x)} u(y)\Big) = u(x) + \frac{\epsilon^2}{2}\Delta_\infty u(x) + o(\epsilon^2). \tag{3.7}$$

A precise bound on the $o(\epsilon^2)$ error terms above is given in Exercise 3.7.

Proof

1. Fix $\bar{B}_\epsilon(x) \subset \mathcal{D}$ and observe that by subtracting constants from both sides of (3.6) and (3.7), we may without loss of generality assume that $u(x) = 0$. Consider an approximation v of u, given by its quadratic Taylor polynomial:

$$v(y) = \langle a, y - x\rangle + \frac{1}{2}\Big\langle A : (y - x) \otimes (y - x)\Big\rangle \qquad \text{for all } y \in \bar{B}_\epsilon(x), \tag{3.8}$$

where we set:

$$a = \nabla u(x) = \nabla v(x), \qquad A = \nabla^2 u(x) = \nabla^2 v(x),$$

so that, in particular:

$$\Delta u(x) = \Delta v(x) = \text{trace}\, A \quad \text{and} \quad \Delta_\infty u(x) = \Delta_\infty v(x) = \Big\langle A : \frac{a}{|a|} \otimes \frac{a}{|a|}\Big\rangle.$$

Since $\|u - v\|_{C^0(B_\epsilon(x))} = o(\epsilon^2)$, it also follows that:

$$\begin{aligned} &|\mathcal{A}_\epsilon u(x) - \mathcal{A}_\epsilon v(x)| = o(\epsilon^2), \\ &\Big| \inf_{B_\epsilon(x)} u - \inf_{B_\epsilon(x)} v\Big| + \Big| \sup_{B_\epsilon(x)} u - \sup_{B_\epsilon(x)} v\Big| = o(\epsilon^2). \end{aligned}$$

Thus, proving (3.6) and (3.7) for v will automatically imply the validity of the same asymptotic expressions for u.

2. In order to show (3.6), we integrate (3.8) over the ball $B_\epsilon(x)$. Observe that: $\fint_{B_\epsilon(x)} (y - x)\,\mathrm{d}y = \fint_{B_\epsilon(0)} y\,\mathrm{d}y = 0$ by the symmetry of the ball, which results in: $\fint_{B_\epsilon(x)} \langle a, y - x\rangle\,\mathrm{d}y = 0$. Further, the entries of the matrix $\fint_{B_\epsilon(x)} (y - x) \otimes (y - x)\,\mathrm{d}y$ are given by: $\fint_{B_\epsilon(0)} y_i y_j\,\mathrm{d}y$, equalling 0 for $i \neq j$, while for any $i = j$ we get:

$$\mathcal{A}_\epsilon |y_i|^2(0) = \mathcal{A}_\epsilon |y_1|^2(0) = \fint_{B_\epsilon(0)} |y_1|^2\,\mathrm{d}y = \frac{\epsilon^2}{N+2}. \tag{3.9}$$

The above simple calculation is left as Exercise 3.7 (i). Consequently:

$$\begin{aligned} \mathcal{A}_\epsilon v(x) &= \frac{1}{2}\Big\langle A : \fint_{B_\epsilon(x)} (y - x) \otimes (y - x)\,\mathrm{d}y\Big\rangle \\ &= \frac{1}{2}\Big\langle A : \Big(\mathcal{A}_\epsilon |y_1|^2(0)\Big)\,\mathrm{Id}_N\Big\rangle \\ &= \frac{\epsilon^2}{2(N+2)}\,\text{trace}\, A = \frac{\epsilon^2}{2(N+2)} \Delta v(x), \end{aligned} \tag{3.10}$$

proving (3.6) for v and hence for u.

3. Assume that $a = \nabla u(x) \neq 0$. In order to show (3.7) for v, we write:

$$v(y) = \epsilon |a| \psi\Big(\frac{y-x}{\epsilon}\Big),$$

where the function $\psi : \bar{B}_1(0) \to \mathbb{R}$ is the according rescaling of v, of the form:

$$\psi(z) = \langle b, z\rangle + \epsilon \langle B : z \otimes z\rangle \quad \text{with } b = \frac{a}{|a|}, \quad B = \frac{1}{2|a|}A.$$

We will prove that:

$$\frac{1}{2}\Big(\inf_{z\in B_1(0)} \psi(z) + \sup_{z\in B_1(0)} \psi(z)\Big) = \epsilon \langle B : b \otimes b\rangle + o(\epsilon), \tag{3.11}$$

which after substituting the definition of ψ and changing variable, directly translates into the claimed bound (3.7) for the function v:

$$\frac{1}{2}\Big(\inf_{y\in B_\epsilon(x)} v(y) + \sup_{y\in B_\epsilon(x)} v(y)\Big) = \frac{\epsilon^2}{2}\Delta_\infty v(x) + o(\epsilon^2).$$

To prove (3.11), let $z_{max} \in \bar{B}_1(0)$ be some maximizer of ψ, so that: $\psi(z_{max}) = \max_{z\in \bar{B}_1(0)} \psi(z)$. Recalling that $|b| = 1$ and writing $\psi(z_{max}) \geq \psi(b)$, we obtain:

$$\begin{aligned}\langle z_{max}, b\rangle &\geq \langle b, b\rangle + \epsilon\big\langle B : b\otimes b - z_{max} \otimes z_{max}\big\rangle \\ &\geq 1 - \epsilon |B| \big|b \otimes b - z_{max} \otimes z_{max}\big| \geq 1 - 2\epsilon |B||z_{max} - b|.\end{aligned}$$

Observe now that $\nabla \psi(z) \neq 0$ in $B_1(0)$, for every ϵ satisfying $\epsilon|B| < 1$. In that case, z_{max} must belong to $\partial B_1(0)$ and $|z_{max}| = 1$. Consequently:

$$\begin{aligned}|z_{max} - b|^2 &= |z_{max}|^2 + |b|^2 - 2\langle z_{max}, b\rangle \leq 2 - \Big(2 - 4\epsilon |B||z_{max} - b|\Big) \\ &= 4\epsilon |B||z_{max} - b|,\end{aligned}$$

which results in:

$$|z_{max} - b| \leq 4\epsilon |B|.$$

Since $\langle b, z_{max}\rangle \leq 1 = \langle b, b\rangle$, we conclude that:

$$\begin{aligned}0 \leq \psi(z_{max}) - \psi(b) &= \langle b, z_{max} - b\rangle + \epsilon\big\langle B : z_{max} \otimes z_{max} - b \otimes b\big\rangle \\ &\leq \epsilon\big\langle B : z_{max} \otimes z_{max} - b \otimes b\big\rangle \leq 2\epsilon |B||z_{max} - b| \leq 8\epsilon^2 |B|^2.\end{aligned}$$

Likewise, denoting a minimizer of ψ by z_{min}, so that $\psi(z_{min}) = \min_{z\in\bar{B}_1(0)} \psi(z)$, we obtain that $|z_{min}| = 1$ and:

$$0 \geq \psi(z_{min}) - \psi(-b) \geq -8\epsilon^2|B|^2.$$

Combining the last two displayed inequalities, we arrive at:

$$\begin{aligned}
&\Big|\frac{1}{2}\big(\psi(z_{max})+\psi(z_{min})\big) - \epsilon\langle B : b\otimes b\rangle\Big| \\
&\qquad = \frac{1}{2}\Big|\big(\psi(z_{max}) + \psi(z_{min})\big) - \big(\psi(b) + \psi(-b)\big)\Big| \\
&\qquad \leq \frac{1}{2}\Big(\big|\psi(z_{max}) - \psi(b)\big| + \big|\psi(z_{min}) - \psi(-b)\big|\Big) \leq 8\epsilon^2|B|^2.
\end{aligned}$$

This ends the proof of (3.11) and of (ii). □

In the next result we interpolate the two averaging expansions in Theorem 3.4, motivated by the formula (3.4) in which Δ_p is shown to be an interpolation between Δ_2 and Δ_∞. Given $p \in [1, \infty)$, we define the interpolation coefficients:

$$\alpha_{N,p} = \frac{N+2}{N+p}, \qquad \beta_{N,p} = 1 - \alpha_{N,p} = \frac{p-2}{N+p}, \tag{3.12}$$

Theorem 3.5 *Let $\mathcal{D} \subset \mathbb{R}^N$ be an open set, $u \in C^2(\mathcal{D})$ and let $x \in \mathcal{D}$ be such that $\nabla u(x) \neq 0$ and $\Delta_p u(x) = 0$. Then:*

(i) For every $\epsilon > 0$ such that $\bar{B}_\epsilon(x) \subset \mathcal{D}$, we have:

$$u(x) = \alpha_{N,p}\,\mathcal{A}_\epsilon u(x) + \frac{\beta_{N,p}}{2}\Big(\inf_{y\in B_\epsilon(x)} u(y) + \sup_{y\in B_\epsilon(x)} u(y)\Big) + o(\epsilon^2). \tag{3.13}$$

(ii) Let $p > 2$. Fix $\alpha \in [0, 1)$ and $\beta = 1 - \alpha$, and define:

$$r = \sqrt{\frac{1}{\beta}\cdot\frac{\beta_{N,p}}{\alpha_{N,p}}} = \sqrt{\frac{1}{\beta}\cdot\frac{p-2}{N+2}}.$$

Then for every $\epsilon > 0$ small enough, we have:

$$u(x) = \alpha\,\mathcal{A}_\epsilon u(x) + \frac{\beta}{2}\Big(\inf_{y\in B_{r\epsilon}(x)} \mathcal{A}_\epsilon u(y) + \sup_{y\in B_{r\epsilon}(x)} \mathcal{A}_\epsilon u(y)\Big) + o(\epsilon^2). \tag{3.14}$$

A precise bound on the $o(\epsilon^2)$ error terms above is given in Exercise 3.7.

Proof

1. Summing the expansions in Theorem 3.4, weighted with coefficients $\alpha_{N,p}$ and $\beta_{N,p}$ in (3.12), we arrive at:

$$\begin{aligned}\alpha_{N,p}\mathcal{A}_\epsilon u(x) + \frac{\beta_{N,p}}{2}\Big(\inf_{y\in B_\epsilon(x)} u(y) + \sup_{y\in B_\epsilon(x)} u(y)\Big) \\ &= u(x) + \frac{1}{2}\frac{\epsilon^2}{N+p}\Big(\Delta u(x) + (p-2)\Delta_\infty u(x)\Big) + o(\epsilon^2) \\ &= u(x) + \frac{\epsilon^2}{2(N+p)}\frac{1}{|\nabla u(x)|^{p-2}}\Delta_p u(x) + o(\epsilon^2).\end{aligned} \tag{3.15}$$

This proves (3.13) under the indicated assumptions.

2. To show (3.14), consider the function $v(x) = \mathcal{A}_\epsilon u(x)$. Clearly, for ϵ small enough, we have $\bar{B}_{\epsilon+r\epsilon}(x) \subset \mathcal{D}$ and also $\nabla v(x) \neq 0$ in view of $\nabla u(x) \neq 0$. We may thus apply the expansion (3.7) to v on the ball $B_{r\epsilon}(x)$, and obtain:

$$\frac{1}{2}\Big(\inf_{y\in B_{r\epsilon}(x)} \mathcal{A}_\epsilon u(y) + \sup_{y\in B_{r\epsilon}(x)} \mathcal{A}_\epsilon u(y)\Big) = \mathcal{A}_\epsilon u(x) + \frac{r^2\epsilon^2}{2}\Delta_\infty v(x) + o(\epsilon^2).$$

Further, a straightforward error analysis in which we replace $\nabla^2 v$ and ∇v in the expression for $\Delta_\infty v$ by $\mathcal{A}_\epsilon \nabla^2 u$ and $\mathcal{A}_\epsilon \nabla u$ leads to:

$$\Delta_\infty v(x) = \Big\langle \mathcal{A}_\epsilon \nabla^2 u(x) : \frac{\mathcal{A}_\epsilon \nabla u(x)}{|\mathcal{A}_\epsilon \nabla u(x)|} \otimes \frac{\mathcal{A}_\epsilon \nabla u(x)}{|\mathcal{A}_\epsilon \nabla u(x)|}\Big\rangle = \Delta_\infty u(x) + o(1).$$

Thus, in virtue of (3.6) and (3.4), we obtain:

$$\begin{aligned}\alpha\, \mathcal{A}_\epsilon u(x) + \frac{\beta}{2}\Big(\inf_{y\in B_{r\epsilon}(x)} \mathcal{A}_\epsilon u(y) + \sup_{y\in B_{r\epsilon}(x)} \mathcal{A}_\epsilon u(y)\Big) \\ &= \mathcal{A}_\epsilon u(x) + \frac{\beta r^2\epsilon^2}{2}\Delta_\infty u(x) + o(\epsilon^2) \\ &= u(x) + \frac{\epsilon^2}{2(N+2)}\Big(\Delta u(x) + \beta r^2(N+2)\Delta_\infty u(x)\Big) + o(\epsilon^2) \\ &= u(x) + \frac{\epsilon^2}{2(N+2)}\Big(\Delta u(x) + (p-2)\Delta_\infty u(x)\Big) + o(\epsilon^2) \\ &= u(x) + \frac{\epsilon^2}{2(N+2)}\frac{1}{|\nabla u(x)|^{p-2}}\Delta_p u(x) + o(\epsilon^2).\end{aligned} \tag{3.16}$$

This completes the proof of (3.14). □

Remark 3.6

(i) The formula in (3.13), although true for any $p \in [1, \infty)$, is most useful for $p \geq 2$ when $\alpha_{N,p}, \beta_{N,p} \geq 0$. Since $\alpha_{N,p} + \beta_{N,p} = 1$, these coefficients can then be interpreted as proportions of the linear part (integral average) and the fully nonlinear (arithmetic mean) part in the operator Δ_p.

(ii) When $p = 2$, then (3.14) can be interpreted as: $u(x) = \mathcal{A}_\epsilon u(x) + o(\epsilon^2)$, which also coincides with (3.13), since there we have $\beta_{N,2} = 0$. Indeed, from (3.6) we see that the average of u on a small ball $B_\epsilon(x)$ differs from the value $u(x)$ at the centre of the ball by order of square of its radius, and the leading coefficient in this expansion is given precisely by $\Delta u(x)$. These observations are in agreement with our discussion in Chap. 2, and reflect the mean value property of harmonic functions, stating that $u(x) = \mathcal{A}_\epsilon u(x)$ for every $B_r(x)$ contained in the domain where $\Delta u = 0$.

(iii) Taking $p \to \infty$, the formula in (3.13) asymptotically becomes: $u(x) = \frac{1}{2}\big(\inf_{y \in B_\epsilon(x)} u(y) + \sup_{y\in B_\epsilon(x)} u(y)\big)$. The same is implied by (3.14), where we accordingly set: $\alpha = 0$ and replace $r\epsilon$ with ϵ, and $\mathcal{A}_\epsilon u(y)$ with $u(y)$. On the other hand, we see by (3.7) that the arithmetic mean of the extreme values of u on a small ball $B_\epsilon(x)$ differs from the value $u(x)$ at the centre of the ball by, as before, order of square of the radius, whereas the leading order coefficient is specified by $\Delta_\infty u(x)$. This observation is in agreement with the Absolutely Minimizing Lipschitz Extension (AMLE) property of ∞-harmonic functions, which states that for every open subset $V \subset \mathcal{D}$, the restriction $u_{|V}$ has the smallest Lipschitz constant among all the extensions of $u_{|\partial V}$, see Jensen (1993).

Exercise 3.7

(i) Compute the integral in the formula (3.9).

(ii) Work out another proof of (3.7) using the following outline. Let y_ϵ^{min} be a minimizer of u on $\bar{B}_\epsilon(x)$. Summing the Taylor expansion (3.8) where $y = y_\epsilon^{min}$, with the estimate: $\sup_{y\in B_\epsilon(x)} u(y) \geq u(x) - \langle \nabla u(x), y_\epsilon^{min} - x\rangle + \frac{1}{2}\langle \nabla^2 u(x) : (y_\epsilon^{min} - x) \otimes (y_\epsilon^{min} - x)\rangle + o(\epsilon^2)$, we get:

$$\frac{1}{2}\Big(\inf_{y\in B_\epsilon(x)} u(y) + \sup_{y\in B_\epsilon(x)} u(y)\Big) \geq u(x) + \frac{1}{2}\Big\langle \nabla^2 u(x) : (y_\epsilon^{min} - x) \otimes (y_\epsilon^{min} - x)\Big\rangle + o(\epsilon^2).$$

Note that:

$$\lim_{\epsilon\to 0} \frac{y_\epsilon^{min} - x}{\epsilon} = -\frac{\nabla u(x)}{|\nabla u(x)|}.$$

This follows by a simple blow-up argument, observing that the maps $u_\epsilon(z) = \frac{1}{\epsilon}(u(x + \epsilon z) - u(x))$ converge uniformly on $\bar{B}_1(0)$ to the linear function

$z \mapsto \langle \nabla u(x), z\rangle$, so that the limit of any converging subsequence of their minimizers must be a minimizer of $\langle \nabla u(x), z\rangle$. Conclude that:

$$\frac{1}{2}\Big(\inf_{y\in B_\epsilon(x)} u(y) + \sup_{y\in B_\epsilon(x)} u(y)\Big) \geq u(x) + \frac{\epsilon^2}{2}\Delta_\infty u(x) + o(\epsilon^2).$$

The same reasoning applied to the maximizers y_ϵ^{max} instead of the minimizers gives the reversed inequality in the above formula.

(iii) Revisit the proof of Theorem 3.4 to quantify the bounds in (3.6) and (3.7). Namely, let $\mathcal{D} \subset \mathbb{R}^N$ be an open, bounded set and denote $\mathcal{D}^\diamond = \bar{\mathcal{D}} + \bar{B}_{\epsilon_0}(0)$, for some $\epsilon_0 > 0$. Let $u \in C^2(\operatorname{int}\mathcal{D}^\diamond)$. Prove that for all $\epsilon \in (0, \epsilon_0)$ one has:

$$\Big|\mathcal{A}_\epsilon u(x) - \Big(u(x) + \frac{\epsilon^2}{2(N+2)}\Delta u(x)\Big)\Big| \leq C\epsilon^2\omega_{\nabla^2 u}(B_\epsilon(x)) \quad \text{for all } x \in \mathcal{D},$$

where $\omega_{\nabla^2 u}$ stands for the modulus of continuity of the function $\nabla^2 u$, namely: $\omega_{\nabla^2 u}(B_\epsilon(x)) = \sup_{z,w\in B_\epsilon(x)} |\nabla^2 u(z) - \nabla^2 u(w)|$, and C is a universal constant. If additionally $\nabla u(x) \neq 0$ for all $x \in \mathcal{D}^\diamond$, then prove the following estimate, valid uniformly for all $x \in \mathcal{D}$ and all $\epsilon \in (0, \hat{\epsilon})$, with $\hat{\epsilon}$ depending on u:

$$\begin{aligned}\Big|\frac{1}{2}\Big(\inf_{y\in B_\epsilon(x)} u(y) + \sup_{y\in B_\epsilon(x)} u(y)\Big) - \Big(u(x) + \frac{\epsilon^2}{2}\Delta_\infty u(x)\Big)\Big| \\ \leq C\epsilon^2\Big(\omega_{\nabla^2 u}(B_\epsilon(x)) + \epsilon\frac{|\nabla^2 u(x)|^2}{|\nabla u(x)|}\Big).\end{aligned}$$

Conclude then that:

$$\begin{aligned}\Big|\alpha_{N,p}\mathcal{A}_\epsilon u(x) + \frac{\beta_{N,p}}{2}\Big(\inf_{y\in B_\epsilon(x)} u(y) + \sup_{y\in B_\epsilon(x)} u(y)\Big) \\ - \Big(u(x) + \frac{\epsilon^2}{2(N+p)}\frac{1}{|\nabla u(x)|^{p-2}}\Delta_p u(x)\Big)\Big| \\ \leq C\epsilon^2\Big(\omega_{\nabla^2 u}(B_\epsilon(x)) + \beta_{N,p}\epsilon\frac{|\nabla^2 u(x)|^2}{|\nabla u(x)|}\Big).\end{aligned}$$

(iv) Revisit the proof of Theorem 3.5 to deduce the following bound. Let $\mathcal{D}$, $\mathcal{D}^\diamond$ be as in Exercise (iii) above and let $p > 2$, $\alpha \in [0, 1)$, $\beta = 1 - \alpha$ and r be as in Theorem 3.5 (ii). Assume that $u \in C^2(\operatorname{int}\mathcal{D}^\diamond)$ satisfies $\nabla u(x) \neq 0$ for every

$x \in \bar{\mathcal{D}}$. Then, for all $\epsilon \in (0, \hat{\epsilon}(u))$ and all $x \in \bar{\mathcal{D}}$, there holds:

$$\begin{aligned}
&\Big|\alpha \mathcal{A}_\epsilon u(x) + \frac{\beta}{2}\Big(\inf_{y\in B_{r\epsilon}(x)} \mathcal{A}_\epsilon u(y) + \sup_{y\in B_{r\epsilon}(x)} \mathcal{A}_\epsilon u(y)\Big) \\
&\qquad - \Big(u(x) + \frac{\epsilon^2}{2(N+2)}\frac{1}{|\nabla u(x)|^{p-2}}\Delta_p u(x)\Big)\Big| \\
&\leq C\epsilon^2\bigg(\omega_{\nabla^2 u}(B_{(1+r)\epsilon}(x)) + \epsilon\frac{\|\nabla^2 u\|^2_{C^0(B_\epsilon(x))} + \omega_{\nabla^2 u}(B_\epsilon(x))^2}{|\nabla u(x)|}\bigg),
\end{aligned}$$

where the constant C depends on N and p but not on u.

3.3 The First Averaging Principle

The asymptotic mean value expansions in Theorem 3.4 suggest to seek p-harmonic functions as limits, when $\epsilon \to 0$, of solutions to the exact ϵ-averaging formulas. This will be implemented in the following chapters where we prove the uniform convergence of such solutions, interpreted as values of a Tug-of-War game, corresponding to (3.14) with, in particular, $\alpha = \frac{1}{3}$, $\beta = \frac{2}{3}$.

Below, we first discuss the variant (3.13) that is quite simple and leads to the Tug-of-War game described in Sect. 3.4. The more involved variant (3.14), modified in order to ensure continuity (Lipschitz continuity) of the approximate solutions for continuous (Lipschitz) boundary data, will be implemented in Sect. 4.1, together with the game interpretation in Sect. 4.3. The reader interested in the main convergence result of Theorem 5.2 may directly move to Sect. 4.1.

Definition 3.8 Let $\mathcal{D} \subset \mathbb{R}^N$ be open, bounded and connected. For a parameter $\epsilon \in (0, 1)$, define the thickened inner boundary Γ_ϵ of $\mathcal{D}$, the outer boundary Γ_{out} and the closed domain $\mathcal{D}^\diamond$ (Fig. 3.1):

$$\begin{aligned}
\Gamma_\epsilon &= \{x \in \mathcal{D};\ \mathrm{dist}(x, \partial\mathcal{D}) \leq \epsilon\}, \quad \mathcal{D}^\diamond = \mathcal{D} \cup \Gamma_{out}, \\
\Gamma_{out} &= \{x \in \mathbb{R}^N \setminus \mathcal{D};\ \mathrm{dist}(x, \partial\mathcal{D}) \leq 1\}.
\end{aligned} \tag{3.17}$$

Theorem 3.9 *Fix $\epsilon \in (0, 1)$ and let $\mathcal{D}, \Gamma_\epsilon$ be as in Definition 3.8. Let $\alpha \in (0, 1]$, $\beta = 1 - \alpha$. Given a bounded, Borel function $F : \Gamma_\epsilon \to \mathbb{R}$, there exists a unique bounded, Borel function $u_\epsilon : \mathcal{D} \to \mathbb{R}$, such that:*

$$u_\epsilon(x) = \begin{cases} \alpha\, \mathcal{A}_\epsilon u_\epsilon(x) + \\ \qquad + \dfrac{\beta}{2}\Big(\inf\limits_{y\in B_\epsilon(x)} u_\epsilon(y) + \sup\limits_{y\in B_\epsilon(x)} u_\epsilon(y)\Big) & \textit{if } x \in \mathcal{D}\setminus\Gamma_\epsilon \\ F(x) & \textit{if } x \in \Gamma_\epsilon. \end{cases} \tag{3.18}$$

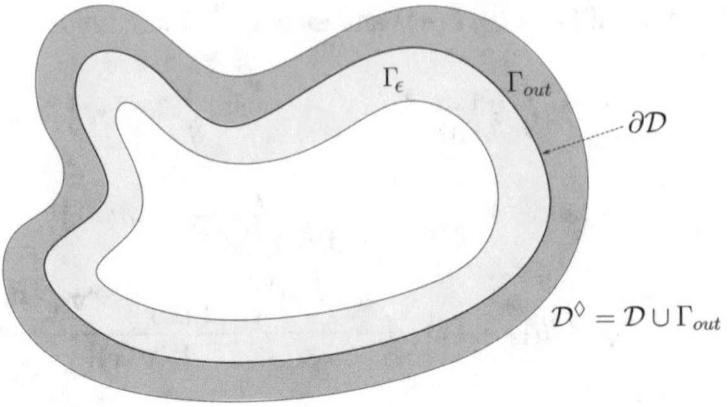

Fig. 3.1 The thickened boundaries Γ_ϵ, Γ_{out} and the closed domain $\mathcal{D}^\diamond$ in Definition 3.8

Proof

1. To ease the notation, we drop the subscript ϵ and write u, Γ instead of $u_\epsilon, \Gamma_\epsilon$. Towards proving existence of solutions, we define the operator T, which to a bounded, Borel $v : \mathcal{D} \to \mathbb{R}$ associates $Tv : \mathcal{D} \to \mathbb{R}$ given by:

$$(Tv)(x) = \begin{cases} \alpha\, \mathcal{A}_\epsilon v(x) + \frac{\beta}{2}\Big(\inf_{y \in B_\epsilon(x)} v(y) + \sup_{y \in B_\epsilon(x)} v(y)\Big) & \text{if } x \in \mathcal{D} \setminus \Gamma \\ F(x) & \text{if } x \in \Gamma. \end{cases} \tag{3.19}$$

The first easy observation is that the function Tv is Borel, as a sum of: a continuous function $x \mapsto \mathcal{A}_\epsilon v(x)$ and the lower-semicontinuous and upper-semicontinuous functions $x \mapsto \sup_{B_\epsilon(x)} v(x)$ and $x \mapsto \inf_{B_\epsilon(x)} v(x)$ (see Exercise 3.12 (i)). It is also easy to observe that T is monotone: $Tv \leq T\bar{v}$ if $v \leq \bar{v}$. Further, for any two bounded, Borel functions $v, \bar{v} : \mathcal{D} \to \mathbb{R}$ and any $x \in \mathcal{D} \setminus \Gamma$ there holds:

$$\begin{aligned} &|(Tv)(x) - (T\bar{v})(x)| \\ &\quad \leq \alpha\, |\mathcal{A}_\epsilon(v - \bar{v})(x)| \\ &\qquad + \frac{\beta}{2}\Big(\Big|\inf_{y \in B_\epsilon(x)} v(y) - \inf_{y \in B_\epsilon(x)} \bar{v}(y)\Big| + \Big|\sup_{y \in B_\epsilon(x)} v(y) - \sup_{y \in B_\epsilon(x)} \bar{v}(y)\Big|\Big) \\ &\quad \leq \alpha\, \mathcal{A}_\epsilon |v - \bar{v}|(x) + \beta \sup_{y \in B_\epsilon(x)} |v(y) - \bar{v}(y)|. \end{aligned} \tag{3.20}$$

2. The solution u of (3.18) is obtained as the limit of iterations $u_{n+1} = Tu_n$, where we set u_0 to be a constant function, satisfying:

$$u_0 \equiv \text{const} \leq \inf F.$$

An easy direct calculation shows that $u_1 = Tu_0 \geq u_0$ in $\mathcal{D}$. By monotonicity of T, the sequence $\{u_n\}_{n=1}^{\infty}$ is nondecreasing and also it is bounded, because:

$$u_0 \leq u_n(x) \leq \sup F \qquad \text{for all } x \in \mathcal{D}, \quad n \geq 1.$$

Hence, $\{u_n\}_{n=1}^{\infty}$ converges pointwise to a bounded, Borel function $u : \mathcal{D} \to \mathbb{R}$. We now prove that the above convergence is actually uniform. For every $m, n \geq 1$ the estimate (3.20) yields the following bound:

$$\sup_{\mathcal{D}} |Tu_m - Tu_n| \leq \frac{\alpha}{|B_\epsilon(0)|} \|u_m - u_n\|_{L^1(\mathcal{D})} + \beta \sup_{\mathcal{D}} |u_m - u_n|.$$

Passing to the limit with $m \to \infty$ implies, in view of the boundedness of $\{u_n\}_{n=1}^{\infty}$ and its pointwise convergence to u, that:

$$\sup_{\mathcal{D}} |u - Tu_n| \leq \frac{\alpha}{|B_\epsilon(0)|} \|u - u_n\|_{L^1(\mathcal{D})} + \beta \sup_{\mathcal{D}} |u - u_n| \qquad \text{for all } n \geq 1.$$

Passing now to the limit with $n \to \infty$, gives:

$$\lim_{n\to\infty} \sup_{\mathcal{D}} |u - u_n| \leq \beta \lim_{n\to\infty} \sup_{\mathcal{D}} |u - u_n|,$$

so there must be:

$$\lim_{n\to\infty} \sup_{\mathcal{D}} |u - u_n| = 0,$$

because $\beta < 1$. The uniform convergence follows. Consequently, the limit function u is a fixed point of T and thus a solution to (3.18):

$$u = \lim_{n\to\infty} Tu_n = T \lim_{n\to\infty} u_n = Tu.$$

3. To show uniqueness, assume that u and $\bar{u}$ satisfy (3.18) and call:

$$M = \sup_{x\in\mathcal{D}} |u(x) - \bar{u}(x)|.$$

By (3.20) we get: $M \leq \alpha \sup_{x\in\mathcal{D}\setminus\Gamma} \mathcal{A}_\epsilon |u - \bar{u}|(x) + \beta M$. Subtracting βM from both sides and dividing by $1 - \beta = \alpha > 0$, we obtain the first inequality in the following bound, whereas the second one is obvious:

$$M \leq \sup_{x\in\mathcal{D}\setminus\Gamma} \mathcal{A}_\epsilon |u - \bar{u}|(x) \leq M.$$

Since the function $x \mapsto \mathcal{A}_\epsilon|u - \bar{u}|(x)$ is well defined and continuous in $\overline{\mathcal{D} \setminus \Gamma}$, we now conclude that the following set must be nonempty:

$$D_M = \left\{x \in \overline{\mathcal{D} \setminus \Gamma};\ \mathcal{A}_\epsilon|u - \bar{u}|(x) = M\right\}.$$

Clearly, D_M is relatively closed in $\overline{\mathcal{D} \setminus \Gamma}$, as a level set of a continuous function. To prove that it is relatively open, take $x \in D_M$ and note that almost every point $y \in B_\epsilon(x) \cap \overline{\mathcal{D} \setminus \Gamma}$ has the property that $|u(y) - \bar{u}(y)| = M$. The same argument as before, based on (3.20), yields that $y \in D_M$. Consequently: $B_\epsilon(x) \cap \overline{\mathcal{D} \setminus \Gamma} \subset D_M$, proving the openness of D_M in $\overline{\mathcal{D} \setminus \Gamma}$ and that $D_M = \overline{\mathcal{D} \setminus \Gamma}$.

Let now $x \in \partial\mathcal{D}$ be a maximizer of the function $x \mapsto |x|$ on $\overline{\mathcal{D} \setminus \Gamma}$. Then, $|B_\epsilon(x) \cap \Gamma| > 0$ and since $u = \bar{u}$ on Γ and $x \in D_M$, we get that $M = 0$. □

Directly from the proof of Theorem 3.9 and the monotonicity of the operator T, we obtain that: $\inf F \le u_\epsilon(x) \le \sup F$ for every $x \in \mathcal{D}$. More generally:

Corollary 3.10 *In the setting of Theorem 3.9, let u_ϵ and $\bar{u}_\epsilon$ be the unique solutions to (3.18) with the respective Borel, bounded boundary data F and $\bar{F}$. If $F \le \bar{F}$ in Γ, then $u_\epsilon \le \bar{u}_\epsilon$ in $\mathcal{D}$.*

Corollary 3.11 *In the setting of Theorem 3.9, let u_0 be any bounded, Borel function on $\mathcal{D}$. Then the sequence $\{u_n\}_{n=1}^\infty$, defined recursively by: $u_{n+1} = Tu_n$ where T is as in (3.19), converges uniformly to the unique solution u_ϵ to (3.18).* .

Proof We first observe that by the same argument as in the proof of Theorem 3.9, the new sequence:

$$\bar{u}_n = T^n \bar{u}_0 \quad \text{where } \bar{u}_0 \equiv \text{const} \ge \sup F,$$

converges uniformly to the same unique solution of (3.18) on $\mathcal{D}$.

Given u_0 as in the statement, call $\underline{u}_n = T^n \underline{u}_0$ and $\bar{u}_n = T^n \bar{u}_0$, where $\underline{u}_0 = \inf u_0$ and $\bar{u}_0 = \sup u_0$. By the monotonicity of the operator T, it follows that:

$$\underline{u}_n \le u_n \le \bar{u}_n \quad \text{in } \mathcal{D},$$

which yields the uniform convergence of $\{u_n\}_{n=1}^\infty$ to u_ϵ. □

Exercise 3.12

(i) Let v be a bounded, Borel function on $\mathbb{R}^N$. Show that $x \mapsto \inf_{B_\epsilon(x)} v$ is upper-semicontinuous and $x \mapsto \sup_{B_\epsilon(x)} v$ is lower-semicontinuous. In particular, both functions are Borel and bounded.

(ii) Let $p \ge 2$ and let $F : \Gamma_{out} \cup \Gamma_\epsilon \to \mathbb{R}$ be a given bounded, Borel function. For each $\epsilon \in (0, 1)$ denote $d_\epsilon(x) = \frac{1}{\epsilon}\min\{\epsilon, \text{dist}(x, \mathbb{R}^N \setminus \mathcal{D})\}$. Modify the proof of Theorem 3.9 to construct an iterated sequence $\{u_n\}_{n=1}^{\infty}$ of approximate solutions to the following problem:

$$u_\epsilon(x) = d_\epsilon(x)\left(\alpha_{N,p}\,\mathcal{A}_\epsilon u_\epsilon(x) + \frac{\beta_{N,p}}{2}\Big(\inf_{y\in B_\epsilon(x)} u_\epsilon(y) + \sup_{y\in B_\epsilon(x)} u_\epsilon(y)\Big)\right) + (1 - d_\epsilon(x))F(x) \qquad \text{for all } x \in \mathcal{D}^{\diamond}, \tag{3.21}$$

and prove their uniform convergence to the unique solution of (3.21).

(iii) Modify the proof of Corollary 3.11 to the setting of (3.21) and show that every initial (bounded, Borel) function u_0 results in an iterated sequence $\{u_n\}_{n=1}^{\infty}$ that converges uniformly to the unique solution of (3.21).

(iv) Deduce that if F is continuous, then u_ϵ solving (3.21) is also continuous.

3.4 Tug-of-War with Noise: A Basic Construction

In this and the next sections we relate solutions of the ϵ-averaging formulas to the probabilistic setting and interpret them as the values of an appropriate *Tug-of-War game with noise*. We start with the averaging principle (3.18).

The game play is as follows: a token is initially placed at a position x_0 on the board game $\mathcal{D}$. At a n-th step of the game, a biased coin is tossed: if the outcome is heads (with probability $\alpha > 0$), then the token is moved randomly in the ball $B_\epsilon(x_{n-1})$ around the current position x_n. If the outcome is tails (with probability $\beta = 1 - \alpha$), then another fair coin is tossed and the player who wins the toss may move the token to any x_n with $|x_n - x_{n-1}| < \epsilon$, according to their strategy (Fig. 3.2). The game ends the first time when $x_n \in \Gamma_\epsilon$ and Player I payoff (equal to Player II loss) is the value of the boundary data $F(x_n)$.

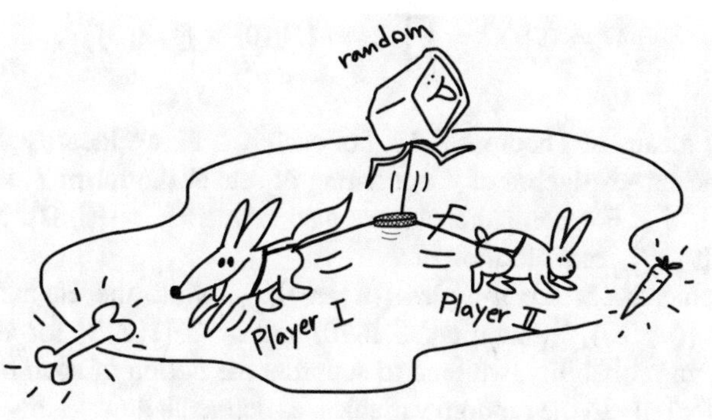

Fig. 3.2 Player I and Player II compete in a Tug-of-War with random noise

Note that since $\mathcal{D}$ is bounded, the game ends almost surely for any choice of strategies (Lemma 3.13). This is due to the fact that for a large $n \geq 1$ satisfying $n\epsilon > 2 \operatorname{diam} \mathcal{D}$, almost surely there will be a block of consecutive n random moves, advancing the token by the distance at least $\frac{\epsilon}{2}$ in the x_1 direction.

Since Player I seeks to maximize the payoff, and since the game is zero-sum so that Player II seeks to minimize the loss, then naturally Player I will try to "tug" the token towards portions of the boundary where F is maximized, while Player II will tug away from his opponent's target towards the set where F is minimized. Below we define this setting in detail.

1. Fix parameters $\alpha \in (0, 1]$, $\beta = 1-\alpha$ and let $\Omega_1 = B_1(0) \times \{1, 2, 3\}$. We consider the product probability space $(\Omega_1, \mathcal{F}_1, \mathbb{P}_1)$, where $\mathcal{F}_1$ is the smallest σ-algebra of subsets of Ω_1, containing all the sets of the form $D \times A$ with $D \subset B_1(0) \subset \mathbb{R}^N$ Borel and $A \subset \{1, 2, 3\}$. The probability measure $\mathbb{P}_1$ is uniquely defined by requiring that $\mathbb{P}_1(D \times A)$ is given as the product:

$$\mathbb{P}_1(D \times A) = \frac{|D|}{|B_1(0)|} \cdot |A|,$$

of the normalized Lebesgue measure of D and the normalized discrete measure of A, where (see Chap. A and Example A.2):

$$|\{1\}| = |\{2\}| = \frac{\beta}{2}, \qquad |\{3\}| = \alpha$$

.

For any number $n \in \mathbb{N}$, let $\Omega_n = (\Omega_1)^n$ be the Cartesian product of n copies of Ω_1, and $(\Omega_n, \mathcal{F}_n, \mathbb{P}_n)$ be the corresponding product probability space. Finally:

$$(\Omega, \mathcal{F}, \mathbb{P}) = (\Omega_1, \mathcal{F}_1, \mathbb{P}_1)^{\mathbb{N}}$$

will denote the probability space on the countable product:

$$\Omega = (\Omega_1)^{\mathbb{N}} = \prod_{i=1}^{\infty} \Omega_1 = \big(B_1(0) \times \{1, 2, 3\}\big)^{\mathbb{N}},$$

defined by means of Theorem A.12. For each $n \in \mathbb{N}$, we identify the σ-algebra $\mathcal{F}_n$ with the sub-σ-algebra of $\mathcal{F}$ consisting of sets of the form: $F \times \prod_{i=n+1}^{\infty} \Omega_1$ for all $F \in \mathcal{F}_n$. For completeness, we also define $\mathcal{F}_0 = \{\emptyset, \Omega\}$. Note that the sequence $\{\mathcal{F}_n\}_{n=0}^{\infty}$ is a filtration of $\mathcal{F}$.

The elements Ω_n are n-tuples $\{(w_i, a_i)\}_{i=1}^{n}$, while the elements of Ω are sequences $\{(w_i, a_i)\}_{i=1}^{\infty}$ with $w_i \in B_\epsilon(0)$ and $a_i \in \{1, 2, 3\}$ for all $i \in \mathbb{N}$. As customary in probability, we tend to suppress the notion of such outcomes and instead refer only to the random variables, as defined below.

2. We now introduce the strategies $\sigma_I = \{\sigma_I^n\}_{n=0}^\infty$ and $\sigma_{II} = \{\sigma_{II}^n\}_{n=0}^\infty$ of Players I and II, respectively. For every $n \geq 0$, these are:

$$\sigma_I^n, \sigma_{II}^n : H_n \to B_1(0) \subset \mathbb{R}^N$$

the vector-valued random variables on the spaces of "finite histories": $H_n = \mathbb{R}^N \times (\mathbb{R}^N \times \Omega_1)^n$, endowed with the product σ-algebra, where the σ-algebra of subsets of $\mathbb{R}^N$ is, as usual, taken to be Borel.
3. Let $\epsilon \in (0, 1)$ and let $\mathcal{D}$, Γ_ϵ be as in Definition 3.8. Fix an initial point (the position of the token) $x_0 \in \mathcal{D}^\diamond$, and the strategies σ_I and σ_{II} as above. We now recursively define a sequence of vector-valued random variables:

$$X_n^{x_0,\sigma_I,\sigma_{II}} : \Omega \to \mathcal{D} \qquad \text{for } n = 0, 1, \ldots.$$

For simplicity of notation, we momentarily suppress the superscripts $x_0, \sigma_I, \sigma_{II}$ and write X_n instead of $X_n^{x_0,\sigma_I,\sigma_{II}}$. We begin by setting X_0 to be constant:

$$X_0 \equiv x_0 \qquad \text{in } \Omega.$$

The sequence $\{X_n\}_{n=0}^\infty$ will be adapted to the filtration $\{\mathcal{F}_n\}_{n=0}^\infty$ of $\mathcal{F}$, and thus each X_n for $n \geq 1$ is effectively defined on Ω_n. We set, for all $(w_1, a_1) \in \Omega_1$:

$$X_1(w_1, a_1) = \begin{cases} x_0 + \epsilon\sigma_I^0(x_0) & \text{for } a_1 = 1 \text{ and } x_0 \in \mathcal{D} \setminus \Gamma_\epsilon \\ x_0 + \epsilon\sigma_{II}^0(x_0) & \text{for } a_1 = 2 \text{ and } x_0 \in \mathcal{D} \setminus \Gamma_\epsilon \\ x_0 + \epsilon w_1 & \text{for } a_1 = 3 \text{ and } x_0 \in \mathcal{D} \setminus \Gamma_\epsilon \\ x_0 & \text{for } x_0 \in \Gamma_\epsilon. \end{cases}$$

Note that, calling $x_1 = X_1(w_1, a_1)$, we have: $h_1 = (x_0, (x_1, w_1, a_1)) \in H_1$ and that h_1 is a Borel function of the argument $(w_1, a_1) \in \Omega_1$.

We now proceed by setting, for $n \geq 2$:

$$X_n\big((w_1, a_1), \ldots, (w_n, a_n)\big) = \begin{cases} x_{n-1} + \epsilon\sigma_I^{n-1}(h_{n-1}) & \text{for } a_n = 1 \text{ and } x_{n-1} \in \mathcal{D} \setminus \Gamma_\epsilon \\ x_{n-1} + \epsilon\sigma_{II}^{n-1}(h_{n-1}) & \text{for } a_n = 2 \text{ and } x_{n-1} \in \mathcal{D} \setminus \Gamma_\epsilon \\ x_{n-1} + \epsilon w_n & \text{for } a_n = 3 \text{ and } x_{n-1} \in \mathcal{D} \setminus \Gamma_\epsilon \\ x_{n-1} & \text{for } x_{n-1} \in \Gamma_\epsilon, \end{cases} \tag{3.22}$$

together with the n-th position of the token:

$$x_n = X_n\big((w_1, a_1), \ldots, (w_n, a_n)\big)$$

and the n-th augmented history h_n that, as before, can be seen as a measurable function of the argument $\big((w_1, a_1), \ldots, (w_n, a_n)\big)$:

$$h_n = \big(x_0, (x_1, w_1, a_1), \ldots, (x_n, w_n, a_n)\big) \in H_n.$$

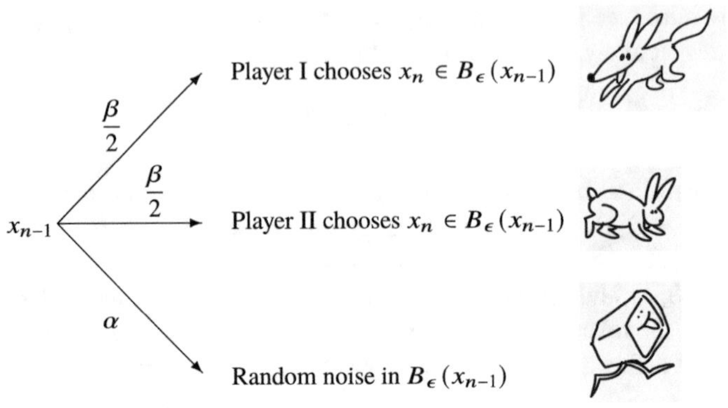

Fig. 3.3 Player I, Player II and random noise with their probabilities

It is clear that each X_n is a $\mathcal{F}_n$-measurable random variable on Ω, taking values in $\mathcal{D}$. It represents the token position x_n, which was initially placed at x_0 and has then been advanced (as long as in $\mathcal{D} \setminus \Gamma$) by:

(i) the random shifts ϵw_n of length at most ϵ, and:
(ii) the ϵ-scaled outputs of the deterministic strategies σ_I and σ_{II}.

These are activated according to the results of the biased "3-sided dice" tosses a_n, where $a_n = 1$ corresponds to activating σ_I and $a_n = 2$ to σ_{II}, whereas $a_n = 3$ results in not activating any of them (Fig. 3.3). The strategies depend on the partial histories h_n of the game that record: the positions x_i of the token, the random shifts w_i, and the toss outcomes a_i, for all $i \leq n$. When the token reaches some (enlarged) boundary position $x_n \in \Gamma_\epsilon$ then it is stopped, i.e., $x_k = x_n$ for every $k \geq n$.

4. We now define the random variable $\tau^{x_0,\sigma_I,\sigma_{II}} : \Omega \to \{0, 1, 2, \ldots, +\infty\}$:

$$\tau^{x_0,\sigma_I,\sigma_{II}}\big((w_1, a_1), (w_2, a_2), \ldots\big) = \min\{n \geq 0;\ x_n \in \Gamma_\epsilon\},$$

where as before $x_n = X_n\big((w_1, a_1), \ldots, (w_n, a_n)\big)$. We drop the superscript $x_0, \sigma_I, \sigma_{II}$ and write τ instead of $\tau^{x_0,\sigma_I,\sigma_{II}}$ if no ambiguity arises. Clearly, τ is $\mathcal{F}$-measurable and, in fact, it is a stopping time relative to the filtration $\{\mathcal{F}_n\}_{n=0}^\infty$, as:

Lemma 3.13 *In the above setting,* $\mathbb{P}(\tau < +\infty) = 1$.

Proof Consider the following set D_{adv} of "advancing" random outcomes (Fig. 3.4):

$$D_{adv} = \Big\{w \in B_1(0);\ \ \langle w, e_1\rangle > \frac{1}{2}\Big\}.$$

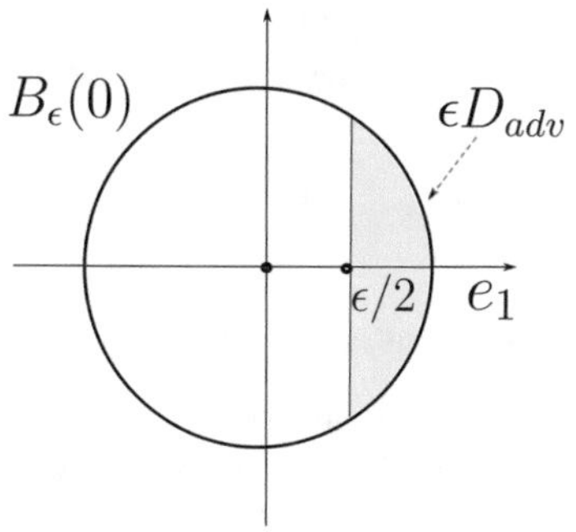

Fig. 3.4 The set D_{adv} of "advancing" random noise outcomes in the proof of (3.23)

By the boundedness of $\mathcal{D}$, it follows that there exists $n \geq 1$ such that:

$$x + \epsilon \sum_{i=1}^{n} w_i \notin \mathcal{D} \setminus \Gamma_\epsilon \quad \text{for all } x \in \mathcal{D} \setminus \Gamma_\epsilon \text{ and } w_i \in D_{adv}, \quad i = 1 \ldots n.$$

Defining $\delta = \Big(\frac{|D_{adv}|}{|B_1(0)|}\alpha\Big)^n > 0$, we thus obtain the first easy bound:

$$\mathbb{P}\big(\tau \leq n\big) \geq \mathbb{P}\Big((D_{adv} \times \{3\})^n \times \prod_{i=n+1}^{\infty} \Omega_1\Big) = \Big(\mathbb{P}_1\big(D_{adv} \times \{3\}\big)\Big)^n = \delta. \tag{3.23}$$

We now prove by induction that:

$$\mathbb{P}\big(\tau > kn\big) \leq (1-\delta)^k \qquad \text{for all } k \in \mathbb{N}. \tag{3.24}$$

For $k = 1$, the claim follows directly from (3.23). On the other hand:

$$\begin{aligned} \mathbb{P}\Big(\tau > (k+1)n\Big) &= \mathbb{E}\big[\mathbb{1}_{\tau > (k+1)n}\big] = \mathbb{E}\Big[\mathbb{E}\big(\mathbb{1}_{\tau > (k+1)n} \mid \mathcal{F}_{kn}\big)\Big] \\ &\leq \mathbb{E}\big[(1-\delta)\mathbb{1}_{\tau > kn}\big] = (1-\delta)\,\mathbb{E}\big[\mathbb{1}_{\tau > kn}\big] \leq (1-\delta)^{k+1}, \end{aligned}$$

where we used the inductive assumption in addition to Lemma A.17 (i) and (3.23), to deduce that:

$$\begin{aligned} \mathbb{E}\big(\mathbb{1}_{\tau > (k+1)n} \mid \mathcal{F}_{kn}\big) &= \int_{\big(\prod_{i=kn+1}^{\infty}\Omega_1\big)} \mathbb{1}_{\tau > (k+1)n} \,\mathrm{d}\Big(\prod_{i=kn+1}^{\infty} \mathbb{P}_1(w_i, a_i)\Big) \\ &\leq (1-\delta)\mathbb{1}_{\tau > kn} \quad \text{a.s.} \end{aligned}$$

and conclude (3.24). Finally, since $\{\tau > kn\}_{k=1}^{\infty}$ is a sequence of decreasing subsets of Ω, the bound in (3.24) yields:

$$\mathbb{P}(\tau = +\infty) = \mathbb{P}\Big(\bigcap_{k=1}^{\infty}\{\tau > kn\}\Big) = \lim_{k\to\infty} \mathbb{P}(\tau > kn) \leq \lim_{k\to\infty}(1-\delta)^k = 0,$$

This completes the proof of Lemma 3.13. □

5. Consequently, given a starting point $x_0 \in \mathcal{D}$ and two strategies σ_I and σ_{II}, one can define the vector-valued random variable $\big(X^{x_0,\sigma_I,\sigma_{II}}\big)_{\tau^{x_0,\sigma_I,\sigma_{II}}} : \Omega \to \mathcal{D}$:

$$\big(X^{x_0,\sigma_I,\sigma_{II}}\big)_{\tau^{x_0,\sigma_I,\sigma_{II}}}(\omega) = X^{x_0,\sigma_I,\sigma_{II}}_{\tau^{x_0,\sigma_I,\sigma_{II}}(\omega)}(\omega) \qquad \mathbb{P}-\text{a.s. in } \Omega.$$

For a bounded, Borel function $F : \Gamma_1 \to \mathbb{R}$, we now set:

$$\begin{aligned} u_I^\epsilon(x_0) &= \sup_{\sigma_I}\inf_{\sigma_{II}} \mathbb{E}\Big[F \circ \big(X^{x_0,\sigma_I,\sigma_{II}}\big)_{\tau^{x_0,\sigma_I,\sigma_{II}}}\Big] \\ u_{II}^\epsilon(x_0) &= \inf_{\sigma_{II}}\sup_{\sigma_I} \mathbb{E}\Big[F \circ \big(X^{x_0,\sigma_I,\sigma_{II}}\big)_{\tau^{x_0,\sigma_I,\sigma_{II}}}\Big], \end{aligned} \tag{3.25}$$

where sup and inf are taken over all strategies as above and the expectation $\mathbb{E}$ is with respect to the probability measure $\mathbb{P}$ on $(\Omega, \mathcal{F}, \mathbb{P})$. These can be understood as the values of the game for the respective Players I and II: the minimum gain for Player I if he plays optimally, choosing the strategy σ_I that maximizes the expectation $\mathbb{E}[F \circ X_\tau]$ under the assumption that Player II always responds with his best strategy; and the minimum loss for Player II if he plays optimally (and assumes that Player I is the perfect player as well).

It turns out that the game values u_I^ϵ and u_{II}^ϵ coincide (i.e., the Tug-of-War game described at the beginning of this Section has a value) and that they equal the solution of the averaging principle (3.18). We now state the main theorem in this setting that will be proved in the next section:

Theorem 3.14 *In the setting of Theorem 3.9, let $F : \Gamma_1 \to \mathbb{R}$ be a bounded, Borel function. Let u_ϵ be the solution to (3.18) and let u_I^ϵ, u_{II}^ϵ be the two game values as in (3.25). Then:*

$$u_I^\epsilon = u_\epsilon = u_{II}^\epsilon \qquad \text{in } \mathcal{D}. \tag{3.26}$$

We remark that, in the context of game theory, the equality $u_I^\epsilon = u_{II}^\epsilon$ in (3.26) is not automatic. Consider the matrix $A = [a_{ij}]_{i,j=1,2} = \text{diag}(1,2) \in \mathbb{R}^{2\times 2}$ and the corresponding game where Player I selects a row i, Player II selects a column j, and then Player II pays Player I the value a_{ij}. Then:

$$u_I = \max_i \min_j a_{ij} = 0 < 1 = \min_j \max_i a_{ij} = u_{II}.$$

3.5 The First Averaging as the Dynamic Programming Principle

We will use the notation of Definition 3.8 and Theorem 3.14. An important role in the proof of (3.26) will be played by the almost-optimal selection lemma:

Lemma 3.15 *Let $u : \mathcal{D} \to \mathbb{R}$ be a bounded, Borel function. Fix $\epsilon \in (0, 1)$ and $\delta > 0$. There exist Borel functions $\sigma_{sup}, \sigma_{inf} : \mathcal{D} \setminus \Gamma_\epsilon \to \mathcal{D}$ such that:*

$$\sigma_{sup}(x),\ \sigma_{inf}(x) \in B_\epsilon(x) \qquad \textit{for all } x \in \mathcal{D} \setminus \Gamma_\epsilon \tag{3.27}$$

and:

$$\begin{aligned} &u(\sigma_{sup}(x)) \geq \sup_{y\in B_\epsilon(x)} u(y) - \delta, \\ &u(\sigma_{inf}(x)) \leq \inf_{y\in B_\epsilon(x)} u(y) + \delta \qquad \textit{for all } x \in \mathcal{D} \setminus \Gamma_\epsilon. \end{aligned} \tag{3.28}$$

Proof

1. We will prove existence of σ_{sup}, while existence of σ_{inf} follows in a similar manner. By adding a constant, we may assume that u is nonnegative.

   Firstly, let $u = \mathbb{1}_A$ for some Borel set $A \subset \mathcal{D}$. We write $A + B_\epsilon(0) = \bigcup_{i=1}^\infty B_\epsilon(x_i)$ as the union of countably many open balls centred at points $x_i \in A$. Define, for all $x \in \mathcal{D} \setminus \Gamma_\epsilon$:

   $$\sigma_{sup}(x) = \begin{cases} x & \text{if } x \notin A + B_\epsilon(0) \\ x_i & \text{if } x \in B_\epsilon(x_i) \setminus \bigcup_{j=1}^{i-1} B_\epsilon(x_j). \end{cases}$$

   Clearly, σ_{sup} above is Borel, it satisfies (3.27) and also we have:

   $$u(\sigma_{sup}(x)) = \sup_{B_\epsilon(x)} u \qquad \text{for all } x \in \mathcal{D} \setminus \Gamma_\epsilon. \tag{3.29}$$

2. Let $u = \sum_{k=1}^n \alpha_k \mathbb{1}_{A_k}$ be a simple function, given by n disjoint Borel sets $\{A_k \subset \mathcal{D}\}_{k=1}^n$ satisfying $\bigcup_{k=1}^n A_k = \mathcal{D}$, and scalars $\alpha_1 > \ldots > \alpha_n$. We now write, as before: $A_k + B_\epsilon(0) = \bigcup_{i=1}^\infty B_\epsilon(x_i^k)$, with $x_i^k \in A_k$ for all i and k. Similarly as before, (3.27) and (3.29) hold by setting for all $x \in \mathcal{D} \setminus \Gamma_\epsilon$:

   $$\sigma_{sup}(x) = \begin{cases} x_i^k & \text{if } \ x \in B_\epsilon(x_i^k) \setminus \Big(\bigcup_{j=1}^{i-1} B_\epsilon(x_j^k) \cup \bigcup_{j<k} (A_j + B_\epsilon(0))\Big) \\ x & \text{otherwise.} \end{cases}$$

3. In the general case consider a simple function v such that $\sup_{\mathcal{D}} |u - v| \leq \frac{\delta}{2}$. Let $\sigma_{sup} : \mathcal{D} \setminus \Gamma_\epsilon \to \mathcal{D}$ be defined as above. Then we have:

$$\begin{aligned} u(\sigma_{sup}(x)) &\geq v(\sigma_{sup}(x)) - \frac{\delta}{2} = \sup_{y \in B_\epsilon(x)} v(y) - \frac{\delta}{3} \\ &\geq \sup_{y \in B_\epsilon(x)} u(y) - \delta \qquad \text{for all } x \in \mathcal{D} \setminus \Gamma_\epsilon, \end{aligned}$$

and therefore σ_{sup} is also the required sup-selection for the function u. □

Proof of Theorem 3.14

1. We drop the sub/superscript ϵ to ease the notation and write u, u_I, u_{II} instead of $u_\epsilon, u_I^\epsilon, u_{II}^\epsilon$. We will show that $u_{II} \leq u$ and $u \leq u_I$. Since $u_I \leq u_{II}$ (see Exercise 3.16), this will establish (3.26).

   We first prove that $u_{II} \leq u$ in $\mathcal{D}$. Fix $x_0 \in \mathcal{D}$ and let $\eta > 0$. By Lemma 3.15, there exists a strategy $\sigma_{0,II}$ for Player II (as defined in Sect. 3.4), such that $\sigma_{0,II}^n(h_n) = \sigma_{0,II}^n(x_n)$ and that, for every $h_n \in H_n$ there holds:

$$\begin{aligned} & u(x_n + \epsilon\sigma_{0,II}^n(x_n)) \leq \inf_{y \in B_\epsilon(x_n)} u(y) + \frac{\eta}{2^{n+1}} \qquad \text{if } x_n \in \mathcal{D} \setminus \Gamma_\epsilon \\ & \sigma_{0,II}^n(x_n) = x_n \qquad \text{if } x_n \in \Gamma_\epsilon. \end{aligned} \tag{3.30}$$

   Clearly then, we have:

$$u_{II}(x_0) = \sup_{\sigma_I} \inf_{\sigma_{II}} \mathbb{E}\big[F \circ \big(X^{x_0,\sigma_I,\sigma_{II}}\big)_{\tau^{x_0,\sigma_I,\sigma_{II}}}\big] \leq \sup_{\sigma_I} \mathbb{E}\big[F \circ \big(X^{x_0,\sigma_I,\sigma_{0,II}}\big)_{\tau^{x_0,\sigma_I,\sigma_{II}}}\big].$$

   We will prove that, for every σ_I:

$$\mathbb{E}\big[F \circ \big(X^{x_0,\sigma_I,\sigma_{0,II}}\big)_{\tau^{x_0,\sigma_I,\sigma_{II}}}\big] \leq u(x_0) + \eta, \tag{3.31}$$

   which, in view of arbitrariness of $\eta > 0$ will yield $u_{II}(x_0) \leq u(x_0)$, as claimed.
2. Let σ_I be any strategy of Player I, and consider the sequence of random variables $\{M_n\}_{n=0}^\infty$ on Ω, given by:

$$M_n = u \circ X_n^{x_0,\sigma_I,\sigma_{0,II}} + \frac{\eta}{2^n}.$$

   As usual, we drop the superscripts in X_n and simply write: $M_n = u \circ X_n + \frac{\eta}{2^n}$. We claim that $\{M_n\}_{n=0}^\infty$ is a supermartingale with respect to the filtration $\{\mathcal{F}_n\}_{n=0}^\infty$:

$$\begin{aligned} \mathbb{E}(M_n \mid \mathcal{F}_{n-1}) &= \mathbb{E}(u \circ X_n \mid \mathcal{F}_{n-1}) + \frac{\eta}{2^n} \\ &= u \circ X_{n-1} + \frac{\eta}{2^n} + \frac{\eta}{2^n} = M_{n-1} \quad \text{a.s.} \end{aligned} \tag{3.32}$$

To prove (3.32), fix $n \geq 1$ and compute:

$$
\begin{aligned}
&\mathbb{E}\big(u \circ X_n \mid \mathcal{F}_{n-1}\big)\big((w_1, a_1), \dots (w_{n-1}, a_{n-1})\big) \\
&= \int_{\Omega_1} (u \circ X_n)\big((w_1, a_1), \dots (w_{n-1}, a_{n-1}), (w_n, a_n)\big)\, \mathrm{d}\mathbb{P}_1(w_n, a_n) \\
&= \begin{cases} \alpha \fint_{B_1(0)} u(x_{n-1} + \epsilon w_n)\, \mathrm{d}w_n & \\ \quad + \frac{\beta}{2} u(x_{n-1} + \epsilon \sigma_I^{n-1}(h_{n-1})) + \frac{\beta}{2} u(x_{n-1} + \epsilon \sigma_{0,II}^{n-1}(h_{n-1})) & \text{if } x_{n-1} \notin \Gamma_\epsilon \\ u(x_{n-1}) & \text{if } x_{n-1} \in \Gamma_\epsilon \end{cases} \\
&\leq \begin{cases} \alpha\, \mathcal{A}_\epsilon u(x_{n-1}) + \frac{\beta}{2}\Big(\sup_{y \in B_\epsilon(x_{n-1})} u(y) + \inf_{y \in B_\epsilon(x_{n-1})} u(y) + \frac{\eta}{2^n}\Big) & \text{if } x_{n-1} \notin \Gamma_\epsilon \\ u(x_{n-1}) & \text{if } x_{n-1} \in \Gamma_\epsilon \end{cases} \\
&\leq u(x_{n-1}) + \frac{\eta}{2^n} \quad \text{a.s.}
\end{aligned}
$$

where we used Lemma A.17 (i) in the first equality, the Fubini–Tonelli theorem (Theorem A.11) in the second equality, the bound (3.30) in the successive inequality and the averaging principle (3.18) on $\mathcal{D}$ in the final bound.

The supermartingale property (3.32) thus follows. Applying Doob's theorem (Theorem A.34 (ii)) in view of the uniform boundedness of $\{M_n\}_{n=0}^\infty$, yields:

$$\mathbb{E}\big[M_\tau\big] \leq \mathbb{E}[M_0] = u(x_0) + \eta.$$

Since $F \circ X_\tau = u \circ X_\tau$, we easily conclude (3.31):

$$\mathbb{E}\big[F \circ X_\tau\big] \leq \mathbb{E}\big[M_\tau\big] \leq u(x_0) + \eta.$$

3. To prove that $u \leq u_I$ in $\mathcal{D}$, we argue exactly as above, choosing an almost-optimal strategy $\sigma_{0,I}$ for Player I, so that $\sigma_{0,I}^n(h_n) = \sigma_{0,I}^n(x_n)$ and:

$$
\begin{aligned}
& u(x_n + \epsilon \sigma_{0,I}^n(x_n)) \geq \sup_{y \in B_\epsilon(x_n)} u(y) - \frac{\eta}{2^{n+1}} && \text{if } x_n \in \mathcal{D} \setminus \Gamma_\epsilon \\
& \sigma_{0,I}(x_n) = x_n \qquad \text{if } x_n \in \Gamma_\epsilon &&
\end{aligned}
$$

for all $h_n \in H_n$. It follows that for every strategy σ_{II} of Player II, the sequence of random variables $\{M_n\}_{n=0}^\infty$, defined by:

$$M_n = u \circ X_n^{x_0, \sigma_{0,I}, \sigma_{II}} - \frac{\eta}{2^n} \tag{3.33}$$

is a submartingale adapted to the filtration $\{\mathcal{F}_n\}_{n=0}^\infty$. This allows to conclude:

$$u_I(x_0) \geq \inf_{\sigma_{II}} \mathbb{E}\big[F \circ \big(X^{x_0,\sigma_{0,I},\sigma_{II}}\big)_{\tau^{x_0,\sigma_I,\sigma_{II}}}\big]$$
$$\geq \mathbb{E}\big[M_{\tau^{x_0,\sigma_I,\sigma_{II}}}\big] \geq \mathbb{E}\big[M_0\big] = u(x_0) - \eta,$$

proving $u \leq u_I$, since $\eta > 0$ was arbitrary.

We remark that one can alternatively deduce $u \leq u_I$ from the already established inequality $u_{II} \leq u$, by replacing u with $-u$ and switching the roles of the players. □

Exercise 3.16

(i) In the setting of Theorem 3.14, show that $u_I^\epsilon \leq u_{II}^\epsilon$.
(ii) Prove that the sequence defined in (3.33) is a submartingale with respect to the filtration $\{\mathcal{F}_n\}_{n=0}^\infty$. Deduce that $u^\epsilon(x_0) \leq u_I^\epsilon(x_0)$.

Exercise 3.17 If we replace the open balls $B_\epsilon(x)$ in the requirement (3.27) by the closed ones, then the Borel selection in Lemma 3.15 may not exist. Let $\epsilon = 1$, $\delta = \frac{1}{3}$ and let $u = \mathbb{1}_A$ where $A \subset \mathbb{R}^3$ is a bounded Borel set with the property that $A + \bar{B}_1(0)$ is not Borel. Show that (3.28) does not hold.

The existence of A is nontrivial (see Luiro et al. 2014) and relies on constructing a Borel set in $\mathbb{R}^2$ whose projection on the x_1 axis is not Borel. This extends the famous example in Erdos and Stone (1970) of a compact (Cantor) set A and a G_δ set B such that $A + B$ is not Borel.

3.6* Case $p = 2$ and Brownian Trajectories

In this section we compare the Tug-of-War game process corresponding to $p = 2$, with the appropriate discrete realization of the Brownian motion trajectories. Similarly as in the discussion of Sect. 2.7* for the ball walk studied in Chap. 2, familiarity with the material in Appendix B will be assumed.

We will use notation of Sect. 2.7*, for $(\bar{\Omega}, \bar{\mathcal{F}}, \bar{\mathbb{P}}) = (\Omega_{\mathcal{B}}, \mathcal{F}_{\mathcal{B}}, \mathbb{P}_{\mathcal{B}}) \times (\Omega, \mathcal{F}, \mathbb{P})$ being the product probability space where $(\Omega, \mathcal{F}, \mathbb{P})$ denotes the space in Sect. 3.4, and for the filtration $\{\bar{\mathcal{F}}_t\}_{t\geq 0}$. We also refer to the standard Brownian motion $\{\mathcal{B}_t^N\}_{t\geq 0}$ discussed in Appendix B.

For $\mathcal{D} \subset \mathbb{R}^N$ open, bounded, connected, and given a starting position $x_0 \in \mathcal{D}$, recall the definition of the exit time:

$$\bar{\tau}(\omega_{\mathcal{B}}) = \min\big\{t \geq 0;\ x_0 + \mathcal{B}_t^N(\omega_{\mathcal{B}}) \in \partial\mathcal{D}\big\}. \tag{3.34}$$

Fix $\epsilon \in (0,1)$. Since the discrete process (3.22) utilizes random sampling on balls of constant radii ϵ, in distinction from the shrinking radii construction in the ball walk (2.2), we modify definition (2.21) to:

$$\begin{aligned} &\bar{\tau}_0 = 0, \\ &\bar{\tau}_{n+1}\big(\omega_{\mathcal{B}}, \{w_i\}_{i=1}^{\infty}\big) \\ &\qquad = \min\Big\{t \geq \bar{\tau}_n;\ \big|\mathcal{B}_t^N(\omega_{\mathcal{B}}) - \mathcal{B}^N_{\bar{\tau}_n(\omega_{\mathcal{B}},\omega)}(\omega_{\mathcal{B}})\big| = \epsilon|w_{n+1}|\Big\}. \end{aligned} \tag{3.35}$$

As in Lemma 2.24 and Theorem 2.25, one can deduce the following properties of $\bar{\tau}_0$ and the relation to the discrete random walk:

Exercise 3.18 In the above context, prove that:

(i) Each $\bar{\tau}_k : \bar{\Omega} \to [0, \infty]$ in (3.35) is a stopping time on $(\bar{\Omega}, \bar{\mathcal{F}}, \bar{\mathbb{P}})$.
(ii) Consider the random variables $X_n^{\epsilon,x_0} : \Omega \to \mathbb{R}^N$ in:

$$X_0^{\epsilon,x_0} = x_0, \quad X_{n+1}^{\epsilon,x_0}\big(w_1, \ldots, w_{n+1}\big) = X_n^{\epsilon,x_0} + \epsilon w_{n+1}.$$

Then for all $n \geq 0$ and all Borel $A \subset \mathbb{R}^N$ we have:

$$\bar{\mathbb{P}}\big(x_0 + \mathcal{B}^N_{\bar{\tau}_n} \in A\big) = \mathbb{P}_\Omega\big(X_n^{\epsilon,x_0} \in A\big).$$

In what follows, we will consider the same stopping time as in Sect. 3.4:

$$\tau_1^{\epsilon,x_0} = \min\big\{n \geq 0;\ \operatorname{dist}(X_n^{\epsilon,x_0}, \partial\mathcal{D}) \leq \epsilon\big\}.$$

For a given $F \in C_c(\mathbb{R}^N)$, the formulas (3.25) applied with $p = 2$ reduce to:

$$u^\epsilon(x_0) = \int_\Omega F \circ X^{\epsilon,x_0}_{\tau_1}\, \mathrm{d}\mathbb{P}. \tag{3.36}$$

We now prove the main statement of this Section, which is:

Theorem 3.19 *In the above context, $\{u^\epsilon\}_{\epsilon\to 0}$ in (3.36) converge pointwise on $\mathcal{D}$, to the harmonic extension of $F_{|\partial\mathcal{D}}$ in:*

$$u(x_0) = \int_{\Omega_{\mathcal{B}}} F \circ \big(x_0 + \mathcal{B}_\tau^N\big)\, \mathrm{d}\mathbb{P}_{\mathcal{B}}. \tag{3.37}$$

When $\mathcal{D}$ is regular, in the sense that for every $y_0 \in \partial\mathcal{D}$ there holds:

$$\begin{aligned} \forall \eta, \delta > 0 \quad \exists \hat{\delta} \in (0, \delta),\ \hat{\epsilon} \in (0, 1) \quad \forall \epsilon \in (0, \hat{\epsilon}),\ x_0 \in B_{\hat{\delta}}(y_0) \cap \mathcal{D} \\ \mathbb{P}\big(X^{\epsilon,x_0}_{\tau_1} \in B_\delta(y_0)\big) \geq 1 - \eta, \end{aligned} \tag{3.38}$$

then convergence of $\{u^\epsilon\}_{\epsilon\to 0}$ to u is uniform.

Proof

1. Fix $x_0 \in \mathcal{D}$, $\epsilon \in (0, 1)$ and in addition to τ_1^{ϵ,x_0} define the corresponding stopping time for $\{\mathcal{B}_t^N\}_{t\geq 0}$ on $\bar{\Omega}$:

$$T_1^{\epsilon,x_0} = \min\big\{\bar{\tau}_n \geq 0;\ \mathrm{dist}\big(x_0 + \mathcal{B}^N_{\bar{\tau}_n}, \partial\mathcal{D}\big) \leq \epsilon\big\}.$$

The fact that T_1^{ϵ,x_0} is a.s. well defined, follows as in the proof of Lemma 2.24, where we note that on the full measure set:

$$A = \{\bar{\tau} < +\infty\} \cap \Big\{\omega = \{w_i\}_{i=1}^\infty;\ |w_i| \geq \frac{1}{2} \text{ for infinitely many } i = 1, \ldots, \infty\Big\} \subset \bar{\Omega},$$

it is not possible to have: $\mathrm{dist}(x_0 + \mathcal{B}^N_{\bar{\tau}_n}, \partial\mathcal{D}) > \epsilon$ for all $n \geq 0$. Indeed, otherwise, the bounded sequence $\{x_0 + \mathcal{B}^N_{\bar{\tau}_n}(\bar{\omega})\}_{n=0}^\infty$ along some $\bar{\omega} = (\omega_{\mathcal{B}}, \omega) \in A$ would have a converging subsequence, and since $\lim_{n\to\infty} \bar{\tau}_n(\bar{\omega}) < +\infty$ in view of $\bar{\tau}(\omega_{\mathcal{B}}) < +\infty$, then: $\lim_{n\to\infty} \mathcal{B}^N_{\bar{\tau}_n}(\bar{\omega}) = \mathcal{B}^N_{\lim_{n\to\infty}\bar{\tau}_n(\bar{\omega})}(\omega_{\mathcal{B}})$. However, the sequence $\{\mathcal{B}^N_{\bar{\tau}_n}(\bar{\omega})\}_{n=0}^\infty$ being Cauchy stands in contradiction with having infinitely many $|w_i| \geq \frac{1}{2}$ in ω.

It is also straightforward to check (using Lemma B.21) that T_1^{ϵ,x_0} is a stopping time. We now argue that for every $f \in C_c(\mathbb{R}^N)$ there holds:

$$\int_{\bar{\Omega}} f \circ \big(x_0 + \mathcal{B}^N_{T_1}\big)\, \mathrm{d}\bar{\mathbb{P}} = \int_{\Omega} f \circ X^{\epsilon,x_0}_{\tau_1}\, \mathrm{d}\mathbb{P}. \tag{3.39}$$

The argument follows along the lines of proof of Theorem 2.25 and Exercise 3.18 (ii), by inductively checking that for all Borel $A \subset \mathbb{R}^N$:

$$\bar{\mathbb{P}}\Big(\big\{x_0 + \mathcal{B}^N_{\bar{\tau}_n} \in A\big\} \cap \bigcap_{k=0}^{n-1} \big\{\mathrm{dist}(x_0 + \mathcal{B}^N_{\bar{\tau}_k}, \partial\mathcal{D}) > \epsilon\big\}\Big)$$
$$= \mathbb{P}\Big(\big\{X_n^{\epsilon,x_0} \in A\big\} \cap \bigcap_{k=0}^{n-1} \big\{\mathrm{dist}(X_k^{\epsilon,x_0}, \partial\mathcal{D}) > \epsilon\big\}\Big) \quad \text{for all } n \geq 0.$$

Applying the above assertion to the Borel subsets $A \subset \{x \in \mathcal{D};\ \mathrm{dist}(x, \partial\mathcal{D}) \leq \epsilon\}$, we obtain:

$$\bar{\mathbb{P}}\Big(\big\{T_1 = \bar{\tau}_n\big\} \cap \big\{x_0 + \mathcal{B}^N_{\bar{\tau}_n} \in A\big\}\Big) = \mathbb{P}\Big(\big\{\tau_1 = n\big\} \cap \big\{X_n^{\epsilon,x_0} \in A\big\}\Big) \quad \text{for all } n \geq 0.$$

Consequently, the claimed identity (3.39) follows through:

$$\sum_{n=0}^\infty \int_{\bar{\Omega}\cap\{T_1 = \bar{\tau}_n\}} f \circ \big(x_0 + \mathcal{B}^N_{\bar{\tau}_n}\big)\, \mathrm{d}\bar{\mathbb{P}} = \sum_{n=0}^\infty \int_{\Omega\cap\{\tau_1 = n\}} f \circ X_n^{\epsilon,x_0}\, \mathrm{d}\mathbb{P}.$$

2. We now observe that: $\lim_{\epsilon\to 0} T_1^{\epsilon,x_0} = \bar{\tau}$ a.s. in $\bar{\Omega}$. Indeed, we have: $T_1^{\epsilon,x_0} < \bar{\tau}$, so on the event $\{\bar{\tau} < +\infty\}$, any decreasing to 0 sequence $J \subset (0,1)$ must have a subsequence along which $\{T_1^{\epsilon,x_0}\}_{\epsilon\to 0}$ converges to some $T \leq \bar{\tau}$, whereas $\big\{x_0 + \mathcal{B}^N_{T_1^{\epsilon,x_0}}\big\}_{\epsilon\to 0}$ converges to $x_0 + \mathcal{B}^N_T \in \partial\mathcal{D}$, implying $T = \bar{\tau}$. We now apply (3.39) with F, to the effect of:

$$\lim_{\epsilon\to 0} u^\epsilon(x_0) = \lim_{\epsilon\to 0}\int_{\bar{\Omega}} F\circ\big(x_0 + \mathcal{B}^N_{T_1^{\epsilon,x_0}}\big)\,\mathrm{d}\bar{\mathbb{P}} = \int_{\bar{\Omega}} F\big(x_0 + \mathcal{B}^N_{\bar{\tau}}\big)\,\mathrm{d}\bar{\mathbb{P}} = u(x_0),$$

which proves the first part of Theorem. Towards the second part, we begin by the following useful implication (which in fact is an equivalence, see Lemma 3.21):

Lemma 3.20 *Given $y_0 \in \partial\mathcal{D}$, condition (3.38) implies the following regularity:*

$$\begin{aligned}\forall \eta,\delta > 0\quad \exists\hat{\delta}\in(0,\delta)\quad &\forall x_0 \in B_{\hat{\delta}}(y_0)\cap\mathcal{D}\\ &\mathbb{P}_{\mathcal{B}}\big(x_0 + \mathcal{B}^N_{\bar{\tau}} \in B_\delta(y_0)\big) \geq 1-\eta.\end{aligned}\tag{3.40}$$

Proof Clearly, (3.39) allows for replacing $X^{\epsilon,x_0}_{\tau_1}$ by $x_0 + \mathcal{B}^N_{T_1^{\epsilon,x_0}}$ in (3.38), namely:

$$\begin{aligned}\forall \eta,\delta > 0\quad \exists\hat{\delta}\in(0,\delta),\ \hat{\epsilon}\in(0,1)\quad &\forall\epsilon\in(0,\hat{\epsilon}),\ x_0 \in B_{\hat{\delta}}(y_0)\cap\mathcal{D}\\ &\bar{\mathbb{P}}\big(x_0 + \mathcal{B}^N_{T_1^{\epsilon,x_0}} \in B_\delta(y_0)\big) \geq 1-\eta.\end{aligned}\tag{3.41}$$

Fix $\eta,\delta > 0$ and find $\hat{\delta},\hat{\epsilon} > 0$ such that: $\bar{\mathbb{P}}(x_0 + \mathcal{B}^N_{T_1^{1/n,x_0}} \in B_{\frac{\delta}{2}}(y_0)) \geq 1-\eta$ for all $x_0 \in B_{\hat{\delta}}(y_0)\cap\mathcal{D}$ and $n > \frac{1}{\hat{\epsilon}}$. Since:

$$\big\{x_0 + \mathcal{B}^N_{\bar{\tau}} \in B_\delta(y_0)\big\} \supset \bigcap_{k\geq 1}\bigcup_{n>k}\big\{x_0 + \mathcal{B}^N_{T_1^{1/n,x_0}} \in B_{\frac{\delta}{2}}(y_0)\big\}$$

holds on the full measure event $\{\lim_{\epsilon\to 0} T_1^{\epsilon,x_0} = \bar{\tau}\}$, if follows that:

$$\mathbb{P}\big(x_0 + \mathcal{B}^N_{\bar{\tau}} \in B_\delta(y_0)\big) \geq 1-\eta$$

for all $x_0 \in B_{\hat{\delta}}(y_0)\cap\mathcal{D}$, as claimed in (3.40). □

1. We are now ready to complete the proof of Theorem 3.19. By Lemma 3.20, condition (3.40) holds for all $y_0 \in \partial\mathcal{D}$. Similarly as in Lemma 5.6, observe that $\hat{\delta}$ may be chosen uniformly in y_0, due to compactness of $\partial\mathcal{D}$:

$$\begin{aligned}\forall \eta,\delta > 0\quad \exists\hat{\delta}\in(0,\delta)\quad &\forall y_0\in\partial\mathcal{D},\quad x_0 \in B_{\hat{\delta}}(y_0)\cap\mathcal{D}\\ &\mathbb{P}_{\mathcal{B}}\big(x_0 + \mathcal{B}^N_{\bar{\tau}} \in B_\delta(y_0)\big) \geq 1-\eta.\end{aligned}\tag{3.42}$$

Fix $\delta_0 \in (0, 1)$. By continuity of F on $\bar{\mathcal{D}}$, there exists $\delta > 0$ such that:

$$\forall x, y \in \bar{\mathcal{D}} \quad |x - y| < \delta \Rightarrow |F(x) - F(y)| \leq \frac{\delta_0}{2}.$$

We take $\epsilon \in (0, \hat{\delta})$, where $\hat{\delta} < \delta/2$ is obtained by applying (3.42) to $\delta/2$ in place of δ, and to $\eta = \frac{\delta_0}{4\|F\|_\infty + 1}$. Firstly, there holds:

$$\begin{aligned}
&\int_{\bar{\Omega}} \left|F \circ \left(x_0 + \mathcal{B}^N_{T_1^{\epsilon,x_0}}\right) - F \circ \left(x_0 + \mathcal{B}^N_{\tau}\right)\right| \, \mathrm{d}\bar{\mathbb{P}} \\
&= \int_{\bar{\Omega}} \mathbb{E}\left(\left|F \circ \left(x_0 + \mathcal{B}^N_{T_1^{\epsilon,x_0}}\right) - F \circ \left(x_0 + \mathcal{B}^N_{\tau}\right)\right| \mid \mathcal{F}_{T_1^{\epsilon,x_0}}\right) \mathrm{d}\bar{\mathbb{P}} \\
&= \int_{\bar{\Omega}} \int_{\Omega_{\mathcal{B}}} \Big|F \circ \left(x_0 + \mathcal{B}^N_{T_1^{\epsilon,x_0}}(\bar{\omega})\right) \\
&\qquad - F \circ \left(x_0 + \mathcal{B}^N_{T_1^{\epsilon,x_0}}(\bar{\omega}) + \mathcal{B}^N_{\tau_{x_0 + \mathcal{B}^N_{T_1^{\epsilon,x_0}}(\bar{\omega})}}(\omega)\right)\Big| \, \mathrm{d}\mathbb{P}_{\mathcal{B}}(\omega) \, \mathrm{d}\bar{\mathbb{P}}(\bar{\omega}),
\end{aligned} \tag{3.43}$$

where we used the strong Markov property, reasoning as in the proof of Theorem B.28. To estimate the internal integral above, for a given $x \in \mathcal{D}$ with $\mathrm{dist}(x, \partial\mathcal{D}) < \epsilon$ let $y_0 \in \partial\mathcal{D}$ be such that $|x - y_0| < \hat{\delta}$. Then, (3.42) yields:

$$\begin{aligned}
&\int_{\Omega_{\mathcal{B}}} \left|F(x) - F \circ \left(x + \mathcal{B}^N_{\tau}\right)\right| \, \mathrm{d}\mathbb{P}_{\mathcal{B}} \\
&\qquad = \int_{\{x + \mathcal{B}^N_{\tau} \in B_\delta(y_0)\}} \left|F(x) - F \circ \left(x + \mathcal{B}^N_{\tau}\right)\right| \, \mathrm{d}\mathbb{P}_{\mathcal{B}} \\
&\qquad\qquad + \int_{\{x + \mathcal{B}^N_{\tau} \notin B_\delta(y_0)\}} \left|F(x) - F \circ \left(x + \mathcal{B}^N_{\tau}\right)\right| \, \mathrm{d}\mathbb{P}_{\mathcal{B}} \\
&\qquad \leq \frac{\delta_0}{2} + 2\|F\|_\infty \eta \leq \delta_0.
\end{aligned}$$

Consequently, (3.43) becomes:

$$\int_{\bar{\Omega}} \left|F \circ \left(x_0 + \mathcal{B}^N_{T_1^{\epsilon,x_0}}\right) - F \circ \left(x_0 + \mathcal{B}^N_{\tau}\right)\right| \, \mathrm{d}\bar{\mathbb{P}} \leq \delta_0 \qquad \text{for all } \epsilon < \hat{\delta},$$

achieving the proof of Theorem 3.19. □

3.7* Equivalence of Regularity Conditions

In this section, we note equivalence of all the boundary regularity conditions used so far, with the classical Doob's regularity condition (3.44) below.

Theorem 3.21 *Given $y_0 \in \partial\mathcal{D}$, conditions (3.38) and (3.40) are equivalent. They are further equivalent to:*

$$\mathbb{P}_{\mathcal{B}}\Big(\inf\big\{t > 0;\ y_0 + \mathcal{B}_t^N \notin \mathcal{D}\big\} = 0\Big) = 1, \tag{3.44}$$

and also to the strengthened version of (3.40), namely:

$$\begin{aligned} &\forall \eta, \delta > 0 \quad \exists \hat{\delta} \in (0, \delta) \quad \forall x_0 \in B_{\hat{\delta}}(y_0) \cap \mathcal{D} \\ &\qquad \mathbb{P}_{\mathcal{B}}\big(\bar{\tau} < \min\{t \geq 0;\ x_0 + \mathcal{B}_t^N \notin B_\delta(y_0)\}\big) \geq 1 - \eta. \end{aligned} \tag{3.45}$$

Proof

1. In Lemma 3.20 we already showed that (3.38)⇒(3.40). Also, the implication (3.45)⇒(3.38) is elementary, as $T_1^{\epsilon,x_0} < \bar{\tau}$ and hence:

$$\big\{\bar{\tau} < \min\{t \geq 0;\ x_0 + \mathcal{B}_t^N \notin B_\delta(y_0)\big\} \subset \big\{x_0 + \mathcal{B}^N_{T_1^{\epsilon,x_0}} \in B_\delta(y_0)\big\}.$$

We now prove (3.40)⇒(3.44). Denote $T = \inf\big\{t > 0;\ y_0 + \mathcal{B}_t^N \notin \mathcal{D}\big\}$ and for each $r > 0$, define the stopping time $\tau_r = \min\big\{t > 0;\ |\mathcal{B}_t^N| \geq r\big\}$. It follows that:

$$\begin{aligned} \mathbb{P}\big(\inf_{r>0} \tau_r > 0\big) &= \mathbb{P}\Big(\bigcup_{n\geq 1}\big\{\inf_{r>0} \tau_r > \frac{1}{n}\big\}\Big) = \lim_{n\to\infty} \mathbb{P}\Big(\inf_{r>0} \tau_r > \frac{1}{n}\Big) \\ &\leq \lim_{n\to\infty} \mathbb{P}\big(\mathcal{B}^N_{1/n} = 0\big) = 0. \end{aligned}$$

Thus, to show (3.44) it suffices to check that:

$$\mathbb{P}_{\mathcal{B}}\Big(\exists t \in (0, \tau_r) \quad y_0 + \mathcal{B}_t^N \notin \mathcal{D}\Big) = \mathbb{P}_{\mathcal{B}}\big(T < \tau_r\big) = 1 \qquad \text{for all } r > 0. \tag{3.46}$$

Fix $\eta > 0$ and for each $\delta \in (0, r)$ consider the events:

$$A_r(\delta) = \big\{T > \tau_r\big\} \cap \big\{|\mathcal{B}_T^N| < \delta\big\},$$

which consist of $\omega \in \Omega_{\mathcal{B}}$ such that the trajectory $t \mapsto y_0 + \mathcal{B}_t^N(\omega)$ exists $B_r(y_0)$ while remaining in $\mathcal{D}$ and further returns to $B_\delta(y_0)$ before exiting $\mathcal{D}$. Since $\mathbb{P}_{\mathcal{B}}\big(\bigcap_{n=1}^\infty A_r(\frac{1}{n})\big) = \mathbb{P}_{\mathcal{B}}\big(\{T > \tau_r\}\cap\{\mathcal{B}_T^N = 0\}\big) = 0$, there exists $\delta > 0$ satisfying:

$$\mathbb{P}_{\mathcal{B}}\big(A_r(\delta)\big) < \eta.$$

For the chosen $\delta, \eta > 0$ we assign $\hat\delta \in (0,\delta)$ according to (3.40) and write:

$$\begin{aligned}
&\mathbb{P}_{\mathcal{B}}\Big(\exists t \in (0,\tau_r) \quad y_0 + \mathcal{B}_t^N \not\in \mathcal{D}\Big) \\
&\geq \mathbb{P}_{\mathcal{B}}\Big(\exists t \in (0,\tau_r) \quad y_0 + \mathcal{B}_t^N \in B_\delta(y_0)\setminus\mathcal{D}\Big) \\
&= \mathbb{P}_{\mathcal{B}}\Big(\exists t \in (0,\tau_{\hat\delta/2}] \quad y_0 + \mathcal{B}_t^N \not\in \mathcal{D}\Big) \\
&\quad + \mathbb{P}_{\mathcal{B}}\Big(\big\{\forall t \in (0,\tau_{\hat\delta/2}] \quad y_0 + \mathcal{B}_t^N \in \mathcal{D}\big\} \\
&\qquad\qquad \cap \big\{\exists t > \tau_{\hat\delta/2} \quad y_0 + \mathcal{B}_t^N \in B_\delta(y_0)\setminus\mathcal{D}\big\} \cap \big(\Omega_{\mathcal{B}}\setminus A_r(\delta)\big)\Big),
\end{aligned} \tag{3.47}$$

and estimate the second probability in the right hand side above by:

$$\begin{aligned}
&\mathbb{P}_{\mathcal{B}}\Big(\mathbb{P}_{\mathcal{B}}\Big(\exists t > \tau_{\hat\delta/2} \quad y_0 + \mathcal{B}_t^N \in B_\delta(y_0)\setminus\mathcal{D} \mid \forall t \in (0,\tau_{\hat\delta/2}] \quad y_0 + \mathcal{B}_t^N \in \mathcal{D}\Big)\Big) \\
&\qquad - \mathbb{P}_{\mathcal{B}}\big(A_r(\delta)\big) \\
&\qquad \geq \mathbb{P}_{\mathcal{B}}\Big(\forall t \in (0,\tau_{\hat\delta/2}] \quad y_0 + \mathcal{B}_t^N \in \mathcal{D}\Big)\cdot(1-\eta) - \eta,
\end{aligned}$$

since the indicated conditional probability is a.s. bounded from below by $1-\eta$, in view of the assumed (3.40) and the strong Markov property. In conclusion, (3.47) yields:

$$\mathbb{P}_{\mathcal{B}}\Big(\exists t \in (0,\tau_r) \quad y_0 + \mathcal{B}_t^N \not\in \mathcal{D}\Big) \geq 1 - 2\eta,$$

implying (3.46) and (3.44) in view of $\eta > 0$ being arbitrarily small.

2. In this Step, we check the remaining implication (3.44)⇒(3.45). For a fixed $\delta, \eta > 0$, by (3.44) it follows that $\mathbb{P}\big(\exists t \in (0,\tau_{\delta/2}) \quad y_0 + \mathcal{B}_t^N \not\in \mathcal{D}\big) = 1$, hence for some large $n > 0$ we get:

$$\mathbb{P}\Big(\exists t \in \big(\frac{1}{n}, \tau_{\delta/2}\big) \quad y_0 + \mathcal{B}_t^N \not\in \mathcal{D}\Big) \geq 1 - \frac{\eta}{2}. \tag{3.48}$$

We will show that taking $\hat{\delta} \leq \frac{\eta}{2\sqrt{n}}$ results in:

$$\mathbb{P}\Big(\exists t \in \big(\frac{1}{n}, \tau_{\delta/2}\big) \quad x_0 + \mathcal{B}_t^N \notin \mathcal{D}\Big) \geq 1 - \frac{\eta}{2} \quad \text{for all } x_0 \in B_{\hat{\delta}}(y_0) \cap \mathcal{D}, \tag{3.49}$$

as the probabilities in the left hand sides of (3.48) and (3.49) differ at most by $\frac{\eta}{2}$. The argument uses the "reflection coupling" as follows.

Fix $x_0 \in B_{\hat{\delta}}(y_0) \cap \mathcal{D}$ and let H denote the hyperplane bisecting the segment $[x_0, y_0]$. By the usual application of Exercise B.10, the random variable $T_H = \min\{t > 0;\ y_0 + \mathcal{B}_t^H \in H\}$ is a stopping time for $\{\mathcal{B}_t^N\}_{t \geq 0}$. We will employ the fact that (see Exercise 3.22):

$$\mathbb{P}\big(T_H > \frac{1}{n}\big) \leq |x_0 - y_0|\sqrt{n}. \tag{3.50}$$

We denote $\{\mathcal{B}_t^{N,H}\}_{t \geq 0}$ the Brownian motion reflected across the $(N-1)$-dimensional subspace $H - \frac{x_0 + y_0}{2} \subset \mathbb{R}^N$. Then the newly defined process:

$$\bar{\mathcal{B}}_t = \begin{cases} \mathcal{B}_t^{N,H} & \text{for } t \leq T_H \\ \mathcal{B}_t^N + \mathcal{B}_{T_H}^{N,H} - \mathcal{B}_{T_H}^N & \text{for } t > T_H, \end{cases}$$

is again a Brownian motion (see Exercise B.25). The uniqueness of the Wiener measure in Theorem B.15 and Exercise B.17, imply that:

$$\mathbb{P}\Big(\exists t \in \big(\frac{1}{n}, \tau_{\delta/2}\big) \quad x_0 + \mathcal{B}_t^N \notin \mathcal{D}\Big) = \mathbb{P}\Big(\exists t \in \big(\frac{1}{n}, \tau_{\delta/2}\big) \quad x_0 + \bar{\mathcal{B}}_t \notin \mathcal{D}\Big).$$

At the same time, for $t \geq T_H$ there holds: $x_0 + \bar{\mathcal{B}}_t = y_0 + \mathcal{B}_t^N$ because:

$$\mathcal{B}_{T_H}^{N,H} - \mathcal{B}_{T_H}^N = y_0 - x_0.$$

Consequently:

$$\begin{aligned} &\Big\{\exists t \in \big(\frac{1}{n}, \tau_{\delta/2}\big) \quad x_0 + \bar{\mathcal{B}}_t \notin \mathcal{D}\Big\} \cap \big\{T_H \leq \frac{1}{n}\big\} \\ &\qquad = \Big\{\exists t \in \big(\frac{1}{n}, \tau_{\delta/2}\big) \quad y_0 + \mathcal{B}_t^N \notin \mathcal{D}\Big\} \cap \big\{T_H \leq \frac{1}{n}\big\}, \end{aligned}$$

which, in view of (3.50), results in:

$$\begin{aligned} &\Big|\mathbb{P}\Big(\exists t \in \big(\frac{1}{n}, \tau_{\delta/2}\big) \quad x_0 + \bar{\mathcal{B}}_t \notin \mathcal{D}\Big) - \mathbb{P}\Big(\exists t \in \big(\frac{1}{n}, \tau_{\delta/2}\big) \quad y_0 + \mathcal{B}_t^N \notin \mathcal{D}\Big)\Big| \\ &\qquad \leq \mathbb{P}\Big(T_H > \frac{1}{n}\Big) \leq \hat{\delta}\sqrt{n} \leq \frac{\eta}{2}, \end{aligned}$$

and in (3.49). Requesting further that $|x_0 - y_0| < \frac{\delta}{2}$, we obtain (3.45). □

Exercise 3.22 Use the following outline to show that for every $\theta, \zeta > 0$, we have the following identity on the standard 1-dimensional Brownian motion:

$$\mathbb{P}\Big(\min\big\{t \geq 0;\ \mathcal{B}_t^1 = \theta\big\} > \zeta\Big) = \mathbb{P}\Big(|\mathcal{B}_1^1| < \frac{\theta}{\sqrt{\zeta}}\Big).$$

(i) Consider the stopping time: $\tau_\theta = \min\big\{t \geq 0;\ \mathcal{B}_t^1 = \theta\big\}$ and define a reflected process by:

$$\bar{\mathcal{B}}_t = \begin{cases} \mathcal{B}_t^1 & \text{for } t \leq \tau_\theta \\ -\mathcal{B}_t^1 + 2\theta & \text{for } t > \tau_\theta. \end{cases}$$

Deduce that $\{\bar{\mathcal{B}}_t\}_{t \geq 0}$ is a standard 1-dimensional Brownian motion.

(ii) We write: $\{\tau_\theta > \zeta\} = \{\mathcal{B}_\xi^1 < \theta\} \setminus \big(\{\tau_\theta < \zeta\} \cap \{\mathcal{B}_\zeta^1 < \theta\}\big)$. Observe that:

$$\mathbb{P}\Big(\{\tau_\theta < \zeta\} \cap \{\mathcal{B}_\zeta^1 < \theta\}\Big) = \mathbb{P}\Big(\{\tau_\theta < \zeta\} \cap \{\bar{\mathcal{B}}_\zeta < \theta\}\Big) = \mathbb{P}\big(\mathcal{B}_\zeta^1 > \theta\big)$$

and conclude, towards completing the proof:

$$\begin{aligned} \mathbb{P}\big(\tau_\theta > \zeta\}\big) &= \mathbb{P}\big(\mathcal{B}_\zeta^1 < \theta\big) - \mathbb{P}\big(\mathcal{B}_\zeta^1 > \theta\big) = \mathbb{P}\big(\mathcal{B}_\zeta^1 < \theta\big) - \mathbb{P}\big(\mathcal{B}_\zeta^1 < -\theta\big) \\ &= \mathbb{P}\big(|\mathcal{B}_\zeta^1| < \theta\big) = \mathbb{P}\Big(|\mathcal{B}_1^1| < \frac{\theta}{\sqrt{\zeta}}\Big). \end{aligned}$$

(iii) Deduce further that: $\mathbb{P}\Big(\min\big\{t \geq 0;\ \mathcal{B}_t^1 = \theta\big\} > \zeta\Big) \leq \sqrt{\frac{2}{\pi}} \cdot \frac{\theta}{\sqrt{\zeta}}$.

3.8 Bibliographical Notes

The asymptotic mean value property (3.13) (called here "averaging principle") for p-harmonic functions with exponent $p \in (1, \infty]$, has been first put forward in Juutinen et al. (2010), where the authors proved that $u \in C(\mathcal{D})$ is a viscosity solution to $\Delta_p u = 0$ in $\mathcal{D}$ if and only if (3.13) holds in the viscosity sense as $\epsilon \to 0$, for all $x \in \mathcal{D}$. Paper by Kawohl et al. (2012) concerns further properties of solutions to nonlinear PDEs in the sense of averages, including a discussion of case $p = 1$. It is also worth mentioning that the interpolation (3.5) has been used in Kawohl (2008) in the context of the applications of Δ_p to image recognition, and for the evolutionary problem in Does (2011). In Le Gruyer (1998, 2007) the asymptotic mean value expansion for the case $p = \infty$ has been developed and the uniform convergence of solutions to: $u(x) = \frac{1}{2}(\inf_{B_\epsilon(x)} + \sup_{B_\epsilon(x)})u$, to the viscosity solution of $\Delta_\infty u = 0$ has been shown.

The connection of the Dirichlet problem for Δ_∞ to the Tug-of-War games has been first analysed in the fundamental paper by Peres et al. (2009), followed by the extension to games with no terminal state and the Neuman boundary condition in Antunovic et al. (2012), and to the biased games in presence of the drift term: $\beta|\nabla u| + \Delta_\infty u = 0$ in Peres et al. (2010). Armstrong and Smart (2012) further used a boundary-biased modification of the original construction and obtained estimates for the rate of convergence of the finite difference solutions, to the infinitely harmonic functions under Dirichlet boundary condition.

The second seminal paper by Peres and Sheffield (2008) treated the case of $p \in (1, \infty)$, albeit via dynamic programming principle and the implicated Tug-of-War game with noise distinct than the expansions (3.13), (3.14) presented in this section. Papers by Manfredi et al. (2012b,a) introduced a version of the Tug-of-War game modelled on (3.13) for $p \geq 2$ and proved the uniform convergence of game values to the unique viscosity solution of the associated Dirichlet problem for smooth domains. Convergence rates of solutions u_ϵ to the limiting u in case of ∇u vanishing in finitely many points, have been developed in Luiro and Parviainen (2017). In Luiro et al. (2014) existence and uniqueness of solutions to (3.18) has been shown for $p \geq 2$, while the well-posedness of a more general class of dynamic programming principles has been proved in Liu and Schikorra (2015).

The asymptotic mean value formula similar to that in (3.13) and applicable in the context of the parabolic problem: $u_t = |\nabla u|^{2-p}\Delta_p u$, has been studied in Manfredi et al. (2010), where equivalence of the viscosity solutions and the viscous asymptotic solutions of the aforementioned version of (3.13) has been proved, via the Tug of War game for $p \geq 2$. The parabolic case with varying exponent $p(x)$ was considered in Parviainen and Ruosteenoja (2016). In a similar context, a paper by Nyström and Parviainen (2017) developed an option-pricing model based on a Tug-of-War game for the financial market.

In Charro et al. (2009), a variation of the Tug-of-War game from Peres et al. (2009) has been adapted to treat the problem $\Delta_\infty u = 0$ subject to the mixed boundary conditions: Dirichlet with Lipschitz data on a portion of the boundary, and Neuman on the remaining C^1 portion of the boundary. Another variant may be found in Gomez and Rossi (2013), leading to the dynamic programming principle: $u_\epsilon(x) = \frac{1}{2}\big(\inf_{y\in\epsilon A(x)} + \sup_{y\in\epsilon A(x)}\big)u_\epsilon(x + \epsilon y)$ whose solutions are shown to converge uniformly to the viscosity solutions to: $\langle\nabla^2 u : \operatorname{argmin}_{z\in A(x)}\langle\nabla v(x), z\rangle = 0$. Towards extending this result to the parabolic problem, some work had been done in Gomez and Rossi (2013).

Numerical schemes based on the averaging principles in Theorems 3.4 and 3.5 were derived in: Oberman (2005) for Δ_∞, and in Casas and Torres (1996); Falcone et al. (2013); Codenotti et al. (2017) for Δ_p.

We also mention that the limiting case $p = 1$ (not studied in these Course Notes) is related to the level-set formulation of motion by mean curvature and a variant of Spencer's "pusher-chooser" game, discussed in Kohn and Serfaty (2006) (see also Buckdahn et al. (2001) for the corresponding representation formula). More generally, an approach of realizing a PDE via the Hamilton–Jacobi equation of a deterministic two-persons game has been implemented for a broad class of fully nonlinear second order parabolic or elliptic problems in Kohn and Serfaty (2010).

In preparing the material in Sects. 3.6* and 3.7*, the author has benefited from the book by Mörters and Peres (2010) and from discussions with Y. Peres.

Chapter 4
Boundary Aware Tug-of-War with Noise: Case $p \in (2, \infty)$

The aim of this chapter is to improve the regularity of the mean value characterized approximate solutions to p-harmonic functions, by implementing the dynamic programming principle (3.14) instead of (3.13) used in Chap. 3 and by further interpolating between the averaging operator and the boundary condition. Our discussion will be presented in the exponent range $p \in (2, \infty)$; however, the linear case $p = 2$ is still formally valid and can be easily covered by approximation or by obvious modifications of the arguments below.

In Sect. 4.1 we show that the *mean value equation* modelled on (3.14) has a unique solution u_ϵ on $\mathcal{D} \subset \mathbb{R}^N$, that is continuous (resp. Lipschitz) for continuous (resp. Lipschitz) boundary data F. In Sect. 4.2 we prove that if F is already a restriction of a smooth p-harmonic function with nonvanishing gradient, then the resulting family $\{u_\epsilon\}_{\epsilon\to 0}$ converges to F uniformly, at a linear rate in ϵ. The same result is reproved in Sect. 4.4, based on the interpretation of u_ϵ as the value of a two-players *boundary aware Tug-of-War with noise* introduced in Sect. 4.3. In this game, the token which is initially placed at $x_0 \in \mathcal{D}$ is then advanced to further consecutive positions according to the following rule: at each step of the game, either of the two players (each acting with probability $\frac{1}{3}$) shifts the token by the distance at most $r\epsilon$ away from its current position, which is followed by a random shift in the ball of radius ϵ. The scaling radius factor r depends on p and N. The game terminates with probability proportional to: 1 minus the distance of the token's previous location from $\mathbb{R}^N \setminus \mathcal{D}$, in the ϵ-neighbourhood of $\partial\mathcal{D}$. Then, $u^\epsilon(x_0)$ is defined as the expected value of F (which is first continuously extended on the said neighbourhood) at the stopping position x_τ, subject to both players playing optimally. As in Chap. 3, the optimality criterion is based on the game rule that Player II pays to Player I the value $F(x_\tau)$.

In the last Sect. 4.5* we show that at the linear exponent $p = 2$, the described above process is a discrete realization along Brownian motion paths, whereas the stopping positions x_τ remain close to those paths exiting positions, with high probability. We conclude that the family $\{u^\epsilon\}_{\epsilon\to 0}$, as in Chap. 3, converges pointwise

M. Lewicka, *A Course on Tug-of-War Games with Random Noise*, Universitext,
https://doi.org/10.1007/978-3-030-46209-3_4

to the *Brownian motion harmonic extension* u of F on $\mathcal{D}$, while for regular domains this convergence is uniform, guaranteeing that $u_{|\partial\mathcal{D}} = F$. This section is more involved as it uses material from Appendix B, and may be skipped at first reading.

4.1 The Second Averaging Principle

In this section we study the averaging principle (3.14) under a particular choice of weights $\alpha = \frac{1}{3}$, $\beta = \frac{2}{3}$. As we shall see, solutions to the related averaging equation (4.1) below inherit the continuity/Lipschitz continuity properties of the boundary data. This is essentially due to the presence of the integral average in every term of (4.1) and its further interpolation to the boundary data.

Theorem 4.1 *Let $\mathcal{D}$, Γ_{out} and $\mathcal{D}^\diamond$ be as in Definition 3.8. Fix $p > 2$ and $r = \sqrt{\frac{3(p-2)}{2(N+2)}}$, and let $\epsilon > 0$ be such that $(1+r)\epsilon < 1$. Then, given a bounded, Borel function $F : \Gamma_{out} \cup \Gamma_\epsilon \to \mathbb{R}$, there exists a unique bounded, Borel function $u_\epsilon : \mathcal{D}^\diamond \to \mathbb{R}$ which is the solution to:*

$$u_\epsilon(x) = d_\epsilon(x)\Big(\frac{1}{3}\mathcal{A}_\epsilon u_\epsilon(x) + \frac{1}{3}\inf_{y\in B_{r\epsilon}(x)} \mathcal{A}_\epsilon u_\epsilon(y) + \frac{1}{3}\sup_{y\in B_{r\epsilon}(x)} \mathcal{A}_\epsilon u_\epsilon(y)\Big) + (1 - d_\epsilon(x))\,F(x) \qquad \textit{for all } x \in \mathcal{D}^\diamond, \tag{4.1}$$

where we denote:

$$d_\epsilon(x) = \frac{1}{\epsilon}\min\{\epsilon, \operatorname{dist}(x, \Gamma_{out})\}. \tag{4.2}$$

Proof

1. We remark that, by continuity of the averaging functions $y \mapsto \mathcal{A}_\epsilon u_\epsilon(y)$ in (4.1), we could be writing: $\min_{y\in\bar{B}_{r\epsilon}(x)} \mathcal{A}_\epsilon u_\epsilon(y)$ and $\max_{y\in\bar{B}_{r\epsilon}(x)} \mathcal{A}_\epsilon u_\epsilon(y)$, instead of: $\inf_{y\in B_{r\epsilon}(x)} \mathcal{A}_\epsilon u_\epsilon(y)$ and $\sup_{y\in B_{r\epsilon}(x)} \mathcal{A}_\epsilon u_\epsilon(y)$. We prefer to keep the latter notation for consistency (Fig. 4.1).

   The proof follows the same steps of the proof of Theorem 3.9. To ease the notation, we drop the subscript ϵ in the solution to (4.1) and write u instead of u_ϵ. Define the operators T and S, which to any bounded, Borel function $v : \mathcal{D}^\diamond \to \mathbb{R}$ associate the continuous function $Sv : \mathcal{D} \to \mathbb{R}$ and the Borel function

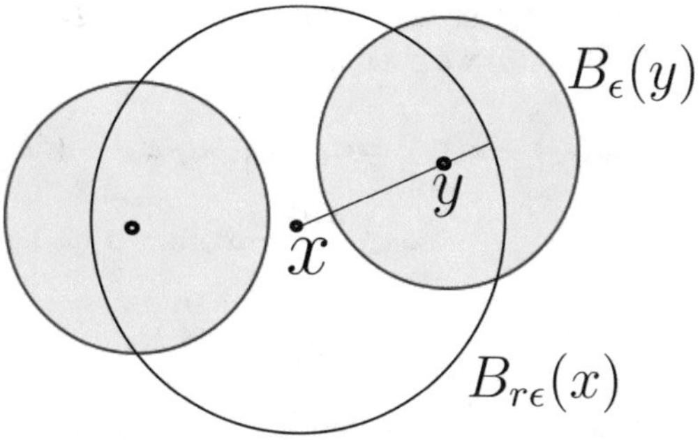

Fig. 4.1 Sampling sets in the averaging term $\inf_{y\in B_{r\epsilon}(x)} \mathcal{A}_\epsilon u_\epsilon(y)$

$Tv : \mathcal{D}^\diamond \to \mathbb{R}$ in:

$$\begin{aligned}(Sv)(x) &= \frac{1}{3}\Big(\mathcal{A}_\epsilon v(x) + \inf_{y\in B_{r\epsilon}(x)} \mathcal{A}_\epsilon v(y) + \sup_{y\in B_{r\epsilon}(x)} \mathcal{A}_\epsilon v(y)\Big)\\ Tv &= d_\epsilon Sv + (1-d_\epsilon)F.\end{aligned} \tag{4.3}$$

Clearly, S and T are monotone, that is: $Sv \le S\bar{v}$ and $Tv \le T\bar{v}$ if $v \le \bar{v}$. Observe that, for any two bounded, Borel functions $v, \bar{v} : \mathcal{D}^\diamond \to \mathbb{R}$ and any $x \in \mathcal{D}$, we have:

$$\begin{aligned}&|(Sv)(x) - (S\bar{v})(x)|\\ &\quad\le \frac{1}{3}\Big(|\mathcal{A}_\epsilon(v-\bar{v})(x)| + \Big|\inf_{y\in B_{r\epsilon}(x)} \mathcal{A}_\epsilon v(y) - \inf_{y\in B_{r\epsilon}(x)} \mathcal{A}_\epsilon \bar{v}(y)\Big|\\ &\qquad\qquad + \Big|\sup_{y\in B_{r\epsilon}(x)} \mathcal{A}_\epsilon v(y) - \sup_{y\in B_{r\epsilon}(x)} \mathcal{A}_\epsilon \bar{v}(y)\Big|\Big)\\ &\quad\le \frac{1}{3}\mathcal{A}_\epsilon|v-\bar{v}|(x) + \frac{2}{3}\sup_{y\in B_{r\epsilon}(x)} \mathcal{A}_\epsilon|v-\bar{v}|(y)\\ &\quad\le \sup_{y\in B_{r\epsilon}(x)} \mathcal{A}_\epsilon|v-\bar{v}|(y).\end{aligned} \tag{4.4}$$

The solution u of (4.1) is obtained as the limit of iterations $u_{n+1} = Tu_n$, where we set $u_0 \equiv \text{const} \le \inf F$ and observe that $u_1 = Tu_0 \ge u_0$ in $\mathcal{D}^\diamond$. By the monotonicity of T, the sequence of Borel functions $\{u_n\}_{n=1}^\infty$ is bounded, nondecreasing and it converges pointwise. Its limit $u : \mathcal{D}^\diamond \to \mathbb{R}$ is a bounded, Borel function that must be a fixed point of T and thus a solution to (4.1).

2. For uniqueness, assume by contradiction that $u \ne \bar{u}$ satisfy (4.1) and set:

$$M = \sup_{x\in\mathcal{D}^\diamond} |u(x) - \bar{u}(x)| = \sup_{x\in\mathcal{D}} |u(x) - \bar{u}(x)| > 0.$$

Let $\{x_n\}_{n=1}^{\infty}$ be a sequence of points in $\mathcal{D}$, such that $\lim_{n\to\infty} |u(x_n) - \bar{u}(x_n)| = M$. By (4.4) we obtain:

$$\begin{aligned}|u(x_n) - \bar{u}(x_n)| &= d_\epsilon(x_n)|(Su)(x_n) - (S\bar{u})(x_n)| \\ &\leq \frac{d_\epsilon(x_n)}{3}\mathcal{A}_\epsilon|u - \bar{u}|(x_n) + \frac{2d_\epsilon(x_n)}{3} \sup_{y\in B_{r\epsilon}(x_n)} \mathcal{A}_\epsilon|u - \bar{u}|(y) \\ &\leq d_\epsilon(x_n)M \leq M.\end{aligned}$$

Now, passing to the limit with $n \to \infty$ where $\lim_{n\to\infty} x_n = x \in \bar{\mathcal{D}}$, we get:

$$\frac{d_\epsilon(x)}{3}\mathcal{A}_\epsilon|u - \bar{u}|(x) + \frac{2d_\epsilon(x)}{3}M = M,$$

which yields:

$$d_\epsilon(x) = 1 \quad \text{and} \quad \mathcal{A}_\epsilon|u - \bar{u}|(x) = M,$$

so that in particular $|u - \bar{u}| = M$ almost everywhere in $B_\epsilon(x) \subset \mathcal{D}^\diamond$. The set $D_M = \{x \in \mathcal{D};\ \mathcal{A}_\epsilon|u - \bar{u}|(x) = M\}$ is thus nonempty. Clearly, D_M is relatively closed in $\mathcal{D}$. But it is also open, because for every $x \in D_M$, each point in $B_\epsilon(x) \cap \mathcal{D}$ is a limit of some sequence $\{x_n\}_{n=1}^{\infty}$ with the property that $|u(x_n) - \bar{u}(x_n)| = M$, and the same argument as before yields: $B_\epsilon(x) \cap \mathcal{D} \subset D_M$. It now follows that $D_M = \mathcal{D}$, which contradicts $u = \bar{u}$ on Γ_{out}, as in the proof of Theorem 3.9.

□

We further have the following easy observations:

Corollary 4.2 *In the setting of Theorem 4.1, let u_ϵ and $\bar{u}_\epsilon$ be the unique solutions to (4.1) with the respective Borel, bounded boundary data F and $\bar{F}$. If $F \leq \bar{F}$, then $u_\epsilon \leq \bar{u}_\epsilon$ in $\mathcal{D}^\diamond$.*

Corollary 4.3 *In the setting of Theorem 4.1, let u_0 be any bounded, Borel function on $\mathcal{D}^\diamond$. Then the sequence $\{u_n\}_{n=1}^{\infty}$, defined recursively by: $u_{n+1} = Tu_n$ where T is as in (4.3), converges uniformly to the unique solution u_ϵ to (4.1).*

Proof The proof is the same as that of Corollary 3.11. To show uniform convergence, we observe that by (4.4):

$$\begin{aligned}|u_{n+1}(x) - u(x)| &= |(Tu_n)(x) - (Tu)(x)| \leq |(Su_n)(x) - (Su)(x)| \\ &\leq \sup_{y\in B_{r\epsilon}(x)} \mathcal{A}_\epsilon|u_n - u_\epsilon|(y) \leq \frac{1}{|B_\epsilon(0)|}\int_{\mathcal{D}^\diamond} |u_n(z) - u(z)|\, dz,\end{aligned}$$

for all $x \in \mathcal{D}$. This yields the result by the dominated convergence theorem and in view of the pointwise convergence of $\{u_n\}_{n=1}^{\infty}$ to u_ϵ. □

We now prove the claimed regularity property:

Theorem 4.4 *In the setting of Theorem 4.1, if F is continuous (resp. Lipschitz continuous) on $\Gamma_{out} \cup \Gamma_\epsilon$, then u_ϵ is also continuous (resp. Lipschitz continuous) on its domain $\mathcal{D}^\diamond$.*

Proof

1. Since the function Sv is continuous in $\mathcal{D}$ for any Borel, bounded $v : \mathcal{D}^\diamond \to \mathbb{R}$, it follows that each u_n is also continuous as a linear combination of Su_{n-1} and a continuous F, by means of a continuous weight function d_ϵ. Since a uniform limit of continuous functions is continuous, the claim follows.
2. We now argue that the function Sv is always Lipschitz continuous in $\mathcal{D}$, with the Lipschitz constant depending on ϵ, N and $\|v\|_{L^\infty}$. Note first that for every $y, \bar{y} \in \mathcal{D} + B_{r\epsilon}(0)$ there holds:

$$
\begin{aligned}
|\mathcal{A}_\epsilon v(y) - \mathcal{A}_\epsilon v(\bar{y})| &\leq \frac{|B_\epsilon(y) \,\triangle\, B_\epsilon(\bar{y})|}{|B_\epsilon(0)|}\|v\|_{L^\infty} \\
&\leq 2\frac{V_{N-1}}{V_N} \cdot \frac{|y-\bar{y}|}{\epsilon}\|v\|_{L^\infty},
\end{aligned}
\tag{4.5}
$$

where we denote $V_N = |B_1(0)|$ and V_{N-1} is, likewise, the volume of the unit ball in $\mathbb{R}^{N-1}$. Indeed, for any $y_0 \in \mathbb{R}^N$ we have:

$$
\begin{aligned}
|B_1(0) \,\triangle\, B_1(y_0)| &= 2|B_1(0) \setminus B_1(y_0)| \\
&\leq 2\Big|\big\{z + ty_0 \in \mathbb{R}^n;\ |z| = 1,\ \langle z, y_0\rangle \leq 0,\ t \in [0,1]\big\}\Big| \\
&= 2|y_0|V_{N-1},
\end{aligned}
$$

and then (4.5) follows through a simple rescaling argument:

$$
\begin{aligned}
|B_\epsilon(y) \,\triangle\, B_\epsilon(\bar{y})| &= \big|B_\epsilon(0) \,\triangle\, B_\epsilon(\bar{y} - y)\big| = \epsilon^N \Big|B_1(0) \,\triangle\, B_1\Big(\frac{\bar{y} - y}{\epsilon}\Big)\Big| \\
&\leq \frac{|B_\epsilon(0)|}{V_N} \cdot \frac{2|y - \bar{y}|}{\epsilon} V_{N-1}.
\end{aligned}
$$

Since (4.5) establishes the Lipschitz continuity of $y \mapsto \mathcal{A}_\epsilon v(y)$, it follows that both the infimum and supremum of this function on $B_{r\epsilon}(x)$ are also Lipschitz (in x), with the same Lipschitz constant. Consequently, for every $x, \bar{x} \in \mathcal{D}$:

$$
\big|(Sv)(x) - (Sv)(\bar{x})\big| \leq 6\frac{V_{N-1}}{V_N} \cdot \frac{\|v\|_{L^\infty}}{\epsilon}|x - \bar{x}|. \tag{4.6}
$$

3. Let now F be Lipschitz continuous on Γ, with Lipschitz constant Lip F. We claim that the functions in the sequence:

$$u_n = T^n u_0, \qquad u_0 \equiv \inf F,$$

must also be Lipschitz continuous, with uniformly controlled Lipschitz constants. More precisely, we will show that:

$$\begin{aligned} \operatorname{Lip} u_n &\leq L \\ &= \max\left\{4\Big(\frac{V_{N-1}}{V_N}+1\Big)\frac{\|F\|_{L^\infty}}{\epsilon}, 2\operatorname{Lip} F\right\} \qquad \text{for all } n \geq 0. \end{aligned} \tag{4.7}$$

Since u_ϵ is a uniform limit $\{u_n\}_{n=1}^\infty$, it will thus follow that u_ϵ itself is Lipschitz, with Lipschitz constant obeying the same bound as in (4.7).

We proceed by induction. For $n = 0$ the claim is trivial, as u_0 is constant. Assume that (4.7) holds for u_n. Observe that: $\|u_n\|_{L^\infty} \leq \|F\|_{L^\infty}$ and $\|Su_n\|_{L^\infty} \leq 3\|F\|_{L^\infty}$ and also that $\operatorname{Lip} d_\epsilon = \frac{1}{\epsilon}$. Then, for any $x, \bar{x} \in \mathcal{D}^\diamond$ we have:

$$\begin{aligned} |u_{n+1}(x) - u_{n+1}(\bar{x})| &\leq d_\epsilon(x)\big|Su_n(x) - Su_n(\bar{x})\big| + |d_\epsilon(x) - d_\epsilon(\bar{x})| \cdot \|Su_n\|_{L^\infty} \\ &\quad + (1 - d_\epsilon(x))|F(x) - F(\bar{x})| + |d_\epsilon(x) - d_\epsilon(\bar{x})| \cdot \|F\|_{L^\infty} \\ &\leq \Big(2\frac{V_{N-1}}{V_N}\frac{\|F\|_{L^\infty}}{\epsilon} + \frac{\|F\|_{L^\infty}}{\epsilon} + \frac{L}{2} + \frac{\|F\|_{L^\infty}}{\epsilon}\Big)|x - \bar{x}| \\ &= \Big(\frac{L}{2} + 2\Big(\frac{V_{N-1}}{V_N} + 1\Big)\frac{\|F\|_{L^\infty}}{\epsilon}\Big)|x - \bar{x}| \leq L|x - \bar{x}|, \end{aligned}$$

where in the first bound we have used (4.6). This completes the induction step in the proof of (4.7). The proof of Theorem 4.4 is done. □

Exercise 4.5

(i) Modify the proof of Theorem 4.4 and show that if F is Lipschitz continuous, then u_ϵ solving (3.21) in Exercise 3.12 must be Lipschitz as well.

(ii) Let $F : \partial\mathcal{D} \to \mathbb{R}$ be a Lipschitz function. Show that the formula:

$$F(x) = \inf_{y \in \partial\mathcal{D}} \big(F(y) + (\operatorname{Lip} F)|y - x|\big) \qquad \text{for } x \in \mathbb{R}^N,$$

defines a Lipschitz continuous extension of F on $\mathbb{R}^N$, with the same Lipschitz constant Lip F. Modify the above definition to obtain another Lipschitz extension $\tilde{F}$ that still preserves the Lipschitz constant, and also obeys the bound: $\inf_{\partial\mathcal{D}} F \leq \tilde{F}(x) \leq \sup_{\partial\mathcal{D}} F$.

We note that this is a particular case of the classical Kirszbraun theorem, that ensures existence of Lipschitz continuous extension $\tilde{F} : H_1 \to H_2$ of any

$F : D \to H_2$ defined on some subset D of a Hilbert space H_1 and taking values in another Hilbert space H_2, such that Lip $\tilde{F}$ = Lip F.

4.2 The Basic Convergence Theorem

Ultimately, our goal is to show that solutions u_ϵ of (4.1) converge uniformly to the unique viscosity solution u to the p-Laplace equation (5.1) with $f = 0$ and the prescribed continuous boundary data F. This will be accomplished in Sects. 5.2, 5.5 and 5.6, for a quite large family of admissible domains $\mathcal{D}$ (so-called game-regular), using the game-theoretical approach. As we shall see below, when F is already the restriction of a given p-harmonic function u on $\mathcal{D}^\diamond$, the same result follows directly from the averaging expansions of u.

Theorem 4.6 *Let $\mathcal{D}$, $\mathcal{D}^\diamond$ be as in Definition 3.8. Given $p > 2$, assume that a (necessarily smooth) bounded function $u : int\, \mathcal{D}^\diamond \to \mathbb{R}$ satisfies:*

$$\Delta_p u = 0 \quad \text{and} \quad \nabla u \neq 0 \qquad \text{in } int\, \mathcal{D}^\diamond. \tag{4.8}$$

Then, solutions u_ϵ of (4.1) with the boundary data $F = u_{|\Gamma_{out} \cup \Gamma_\epsilon}$ and with the parameter $r = \sqrt{\frac{3(p-2)}{2(N+2)}}$, converge to u uniformly as $\epsilon \to 0$. More precisely:

$$\|u_\epsilon - u\|_{C(\mathcal{D})} \leq C\epsilon, \tag{4.9}$$

for all $\epsilon < \frac{1}{1+r}$, with a constant C depending on N, p, u, $\mathcal{D}^\diamond$ but not on ϵ.

Below, we give a proof of (4.9) that has an analytic flavour, modelled on the $p = 2$ case where a quadratic correction is used. An alternative probabilistic proof will be given in Sect. 4.4, after developing the Tug-of-War interpretation of (4.1) as the dynamic programming principle for a stochastic process described in Sect. 4.3.

The analytic proof necessitates the two Lemmas below. The first one allows to push a given p-harmonic nonsingular function u in the region of strict p-subharmonicity, by adding a polynomial correction. Observe that when $p = 2$ then $u + \epsilon|x|^2$ is clearly subharmonic. A similar observation remains valid for $p > 2$:

Lemma 4.7 *Let $\mathcal{D}$ and $\mathcal{D}^\diamond$ be as in Definition 3.8. Assume that $B_1(0) \cap \bar{\mathcal{D}} = \emptyset$ and let $u : int\, \mathcal{D}^\diamond \to \mathbb{R}$ satisfy (4.8). Then, there exists $q \geq 2$ and $\hat{\epsilon} \in (0, 1)$ such that the smooth functions $v_\epsilon : int\, \mathcal{D}^\diamond \to \mathbb{R}$, defined for $\epsilon < \hat{\epsilon}$ by:*

$$v_\epsilon(x) = u(x) + \epsilon|x|^q,$$

satisfy:

$$\Delta_p v_\epsilon(x) \geq \epsilon q |\nabla v_\epsilon(x)|^{p-2} \quad \textit{and} \quad \nabla v_\epsilon(x) \neq 0 \qquad \textit{for all } x \in \bar{\mathcal{D}}. \tag{4.10}$$

Proof Observe first the following easy formulas:

$$\nabla |x|^q = q|x|^{q-2}x, \qquad \nabla^2 |x|^q = q(q-2)|x|^{q-4} x \otimes x + q|x|^{q-2}\mathrm{Id},$$
$$\Delta |x|^q = q(q-2+N)|x|^{q-2}.$$

Fix $x \in \bar{\mathcal{D}}$ and denote $a = \nabla v_\epsilon(x)$ and $b = \nabla u(x)$. Then, by (4.8) we have:

$$\begin{aligned}\Delta_p v_\epsilon(x) &= |\nabla v_\epsilon(x)|^{p-2}\Big(\epsilon \Delta |x|^q + (p-2)\Big(\epsilon\Big\langle \nabla^2 |x|^q : \frac{a}{|a|} \otimes \frac{a}{|a|}\Big\rangle \\ &\qquad + \Big\langle \nabla^2 u(x) : \frac{a}{|a|} \otimes \frac{a}{|a|} - \frac{b}{|b|} \otimes \frac{b}{|b|}\Big\rangle\Big)\Big) \\ &\geq |\nabla v_\epsilon(x)|^{p-2} \epsilon q |x|^{q-2}\Big(q - 2 + N + (p-2)\Big(1 - \frac{4|\nabla^2 u(x)|}{|\nabla u(x)|}|x|\Big)\Big).\end{aligned} \tag{4.11}$$

Above, we have also used the bound:

$$\Big\langle \nabla^2 |x|^q : \frac{a}{|a|} \otimes \frac{a}{|a|}\Big\rangle = q(q-2)|x|^{q-2}\Big\langle \frac{a}{|a|}, \frac{x}{|x|}\Big\rangle^2 + q|x|^{q-2} \geq q|x|^{q-2},$$

together with the following straightforward estimate:

$$\Big|\frac{a}{|a|} \otimes \frac{a}{|a|} - \frac{b}{|b|} \otimes \frac{b}{|b|}\Big| \leq 4\frac{|a-b|}{|b|}. \tag{4.12}$$

The result now follows by fixing an exponent q that satisfies:

$$q \geq 3 - N + (p-2)\Big(\max_{x \in \bar{\mathcal{D}}} \frac{|\nabla^2 u(x)|}{|\nabla u(x)|} - 1\Big),$$

so that the quantity in the last line parentheses in (4.11) is greater than 1, and further taking $\epsilon > 0$ small enough to have: $\min_{x \in \bar{\mathcal{D}}} |\nabla v_\epsilon(x)| > 0$. □

The second preliminary result is the comparison principle for the corrected p-harmonic functions v_ϵ:

Lemma 4.8 *In the setting of Lemma 4.7, let $p > 2$, $r = \sqrt{\frac{3(p-2)}{2(N+2)}}$ and denote:*

$$S_\epsilon v_\epsilon(x) = \frac{1}{3}\mathcal{A}_\epsilon v_\epsilon(x) + \frac{1}{3}\inf_{y \in B_{r\epsilon}(x)} \mathcal{A}_\epsilon v_\epsilon(y) + \frac{1}{3}\sup_{y \in B_{r\epsilon}(x)} \mathcal{A}_\epsilon v_\epsilon(y),$$

Then there exists $q \geq 2$ and $\hat{\epsilon} \in (0, 1)$ such that for all $\epsilon < \hat{\epsilon}$ there holds (4.10), together with:

$$v_\epsilon(x) \leq S_\epsilon v_\epsilon(x) \qquad \textit{for all } x \in \bar{\mathcal{D}}. \tag{4.13}$$

Proof By translation of $\mathcal{D}$, we may without loss of generality, assume that $B_1(0) \cap \mathcal{D}^\diamond = \emptyset$. We now apply Exercise 3.7 (iv) to each v_ϵ, with the parameters: $\alpha = \frac{1}{3}$, $\beta = \frac{2}{3}$. Since the averaging operator in the right-hand side of (4.1) equals S_ϵ, the estimate in (4.10) yields:

$$\begin{aligned} v_\epsilon(x) - S_\epsilon v_\epsilon(x) \leq & -\frac{\epsilon^3}{2(N+2)} q \\ & + C\epsilon^2 \Big(\omega_{\nabla^2 v_\epsilon}(B_{(1+r)\epsilon}(x)) + \epsilon \frac{\|\nabla^2 v_\epsilon\|^2_{C^0(B_\epsilon(x))} + \omega_{\nabla^2 v_\epsilon}(B_\epsilon(x))^2}{|\nabla v_\epsilon(x)|} \Big), \end{aligned}$$

where C depends only on N and p. Clearly, the right-hand side above is negative for q sufficiently large, provided that $\epsilon < \hat{\epsilon}$ is sufficiently small. This proves the claim. □

An Analytical Proof of Theorem 4.6

1. By translation of $\mathcal{D}$, we may without loss of generality, assume that $B_1(0) \cap \mathcal{D}^\diamond = \emptyset$. Consider the differences:

$$\phi_\epsilon(x) = v_\epsilon(x) - u_\epsilon(x) = u(x) - u_\epsilon(x) + \epsilon|x|^q,$$

where the functions v_ϵ are as in Lemma 4.7, defined for q and $\epsilon < \hat{\epsilon}$ as in Lemma 4.8. By (4.1) and (4.13), we obtain:

$$\begin{aligned} \phi_\epsilon(x) &\leq d_\epsilon(x)\big(v_\epsilon(x) - S_\epsilon u_\epsilon(x)\big) + (1 - d_\epsilon(x))\big(v_\epsilon(x) - u(x)\big) \\ &\leq d_\epsilon(x)\big(S_\epsilon v_\epsilon(x) - S_\epsilon u_\epsilon(x)\big) + (1 - d_\epsilon(x))\big(v_\epsilon(x) - u(x)\big) \\ &\leq d_\epsilon(x)\Big(\frac{1}{3}\mathcal{A}_\epsilon \phi_\epsilon(x) + \frac{2}{3} \sup_{y \in B_{(1+r)\epsilon}(x)} \phi_\epsilon(y)\Big) \\ &\quad + (1 - d_\epsilon(x))\big(v_\epsilon(x) - u(x)\big) \qquad \text{for all } x \in \bar{\mathcal{D}}. \end{aligned} \tag{4.14}$$

Note that $\phi_\epsilon \in C(int\, \mathcal{D}^\diamond)$, whereas on the open neighbourhood of $\partial\mathcal{D}^\diamond$ in $\mathcal{D}^\diamond$:

$$(\partial\mathcal{D}^\diamond + B_{1/2}(0)) \cap \mathcal{D}^\diamond \subset \Gamma_{out},$$

we have $\phi_\epsilon(x) = \epsilon|x|^q$. Thus $\phi_\epsilon \in C(\mathcal{D}^\diamond)$ and, consequently, it is possible to define:

$$M_\epsilon = \max_{x \in \mathcal{D}^\diamond} \phi_\epsilon(x).$$

2. We first claim that:

$$\exists x \in \Gamma_{out} \cup \Gamma_\epsilon \qquad \phi_\epsilon(x) = M_\epsilon. \tag{4.15}$$

To this end, we will prove that if $\phi_\epsilon(y) = M_\epsilon$ for some point $y \in \mathcal{D}$ with the property that $\mathrm{dist}(y, \partial\mathcal{D}) \geq \epsilon$, then in fact we also have:

$$\exists x \in \mathcal{D} \qquad \mathrm{dist}(x, \partial\mathcal{D}) < \epsilon \quad \text{and} \quad \phi_\epsilon(x) = M_\epsilon. \tag{4.16}$$

Let D_0 be the connected component of the set $\{x \in \mathcal{D};\ \mathrm{dist}(x, \partial\mathcal{D}) \geq \epsilon\}$ containing y. Consider its subset:

$$D_M = \{z \in D_0;\ \phi_\epsilon(z) = M_\epsilon\}.$$

Clearly, D_M is nonempty and closed in D_0. To prove that D_M is open in D_0, let $z \in D$. Since $d_\epsilon(z) = 1$, we get by (4.14):

$$M_\epsilon = \phi_\epsilon(z) \leq \frac{1}{3}\mathcal{A}_\epsilon\phi_\epsilon(z) + \frac{2}{3}\sup_{w \in B_{(1+r)\epsilon}(z)} \phi_\epsilon(w) \leq \frac{1}{3}\mathcal{A}_\epsilon\phi_\epsilon(z) + \frac{2}{3}M_\epsilon,$$

which results in:

$$M_\epsilon \leq \mathcal{A}_\epsilon\phi_\epsilon(z) \leq M_\epsilon,$$

implying that $\phi_\epsilon \equiv M_\epsilon$ on $B_\epsilon(z) \cap D_0$ and proving the openness property of D_M. In conclusion, there must be $D_M = D_0$ so that $(\phi_\epsilon)_{|D_0} \equiv M_\epsilon$ and in particular $\phi_\epsilon(z) = M_\epsilon$ for all $z \in \partial D_0 \subset \partial\{x \in \mathcal{D};\ \mathrm{dist}(x, \partial\mathcal{D}) \geq \epsilon\}$.

By the same argument as above, we get $\phi_\epsilon \equiv M_\epsilon$ on the ball $B_\epsilon(z)$ that must therefore contain a point x as in (4.16), proving (4.15).

3. Having established (4.15), we now deduce a bound on M_ϵ. We distinguish two cases. In the first case, the maximum $M_\epsilon = \phi_\epsilon(x)$ is attained at some $x \in \mathcal{D}$ with $\mathrm{dist}(x, \partial\mathcal{D}) < \epsilon$, so that $d_\epsilon(x) < 1$. Then (4.14) yields:

$$M_\epsilon = \phi_\epsilon(x) \leq d_\epsilon(x)M_\epsilon + (1 - d_\epsilon(x))\big(v_\epsilon(x) - u(x)\big),$$

immediately implying that:

$$M_\epsilon \leq v_\epsilon(x) - u(x) = \epsilon|x|^q.$$

In the second case, $M_\epsilon = \phi_\epsilon(x)$ for some $x \in \Gamma_{out}$, so by $u_\epsilon = u$ on Γ, likewise:

$$M_\epsilon = \phi_\epsilon(x) = \epsilon|x|^q.$$

In either case, there follows the one-sided inequality:

$$\max_{x\in\bar{\mathcal{D}}} \big(u(x) - u_\epsilon(x)\big) \leq \max_{x\in\bar{\mathcal{D}}} \phi_\epsilon(x) + C\epsilon = M_\epsilon + C\epsilon \leq 2C\epsilon,$$

where $C = \max_{x\in\bar{\mathcal{D}}} |x|^q$ depends on $\mathcal{D}$, p, N and u but not on ϵ. Applying the same argument to the p-harmonic function $(-u)$ and noting $(-u)_\epsilon = -u_\epsilon$, we arrive at:

$$\min_{x\in\bar{\mathcal{D}}} \big(u(x) - u_\epsilon(x)\big) = -\max_{x\in\bar{\mathcal{D}}} \big(u_\epsilon(x) - u(x)\big) \geq -2C\epsilon.$$

This concludes the proof of the bound (4.9). □

Exercise 4.9

(i) Show that inequality (4.12) is valid for any $a, b \in \mathbb{R}^N \setminus \{0\}$.
(ii) Modify the proof of Theorem 4.6 for the case of the averaging operator (3.18). Namely, let $\mathcal{D}$ and Γ_ϵ be as in Definition 3.8. Fix $p \geq 2$ and consider the solutions u_ϵ to (3.18), with coefficients $\alpha = \alpha_{N,p}$, $\beta = \beta_{N,p}$ as in (3.12), and where the boundary data $F = u_{|\Gamma_\epsilon}$ is given by a smooth function $u : \mathcal{D}^\Diamond \to \mathbb{R}$ satisfying (4.8). Prove that: $\|u_\epsilon - u\|_{C(\mathcal{D})} \leq C\epsilon$, for all small $\epsilon > 0$ and a constant C depending on N, p, u and $\mathcal{D}^\Diamond$ but not on ϵ.
(iii) Give an example of an open, bounded, connected set $\mathcal{D} \subset \mathbb{R}^N$, such that for every small $\epsilon > 0$ the set $\{x \in \mathcal{D};\ \mathrm{dist}(x, \partial\mathcal{D}) \geq \epsilon\}$ is disconnected.

Exercise 4.10 Given a p-harmonic function $u : int\, \mathcal{D}^\Diamond \to \mathbb{R}$ with nonvanishing gradient (i.e. (4.8) is satisfied), Lemma 4.7 proves that the variation $x \mapsto |x|^q$ that shifts u into the region of sub-p-harmonicity. Carry out calculations below for an alternative construction.

(i) For every $A > 0$, let $g_A : (0, \infty) \to (0, \infty)$ be a continuous, increasing function given by:

$$g_A(t) = \log\big(A(e^t - 1) + 1\big).$$

Note that $g_1(t) = t$. For $A < 1$, show that: $g_A(t) - t \in (\log A, 0)$ and $g_A'(t) - 1 \in (A - 1, 0)$ for all $t > 0$. For $A > 1$, show that: $g_A(t) - t \in (0, \log A)$ and $g_A'(t) - 1 \in (0, A - 1)$ for all $t > 0$. Moreover, when $A < 1$, we have:

$$g_A''(t) = (1 - A)\frac{Ae^t}{(A(e^t - 1) + 1)^2} > 0.$$

The approximation function g_A has been studied in Juutinen et al. (2001) and also in Lindqvist and Lukkari (2010) as a tool for proving the comparison principle for a variation of the ∞-Laplace equation.
(ii) Prove that $(g_A)^{-1} = g_{1/A}$.

(iii) Let $u : int\,\mathcal{D}^\diamond \to \mathbb{R}$ be a smooth, positive function with nonvanishing gradient. For every $\epsilon \in (0, 1)$, define $v_\epsilon = g_{1-\epsilon} \circ u$. It then follows that:

$$\log(1-\epsilon) < v_\epsilon(x) - u(x) < 0 \quad \text{for all } x \in int\,\mathcal{D}^\diamond$$

and we have:

$$\Delta_p v_\epsilon(x) = \Big|g'_{1-\epsilon}(u(x))\Big|^{p-2}\Big(g'_{1-\epsilon}(u(x))\Delta_p u(x) + (p-1)g''_{1-\epsilon}(u(x))|\nabla u(x)|^p\Big).$$

Thus, if $\Delta_p u = 0$ in $int\,\mathcal{D}^\diamond$, we obtain $\Delta_p v_\epsilon(x) > 0$. More precisely:

$$\begin{aligned}\Delta_p v_\epsilon(x) &= (p-1)\Big|g'_{1-\epsilon}(u(x))\Big|^{p-2} g''_{1-\epsilon}(u(x))|\nabla u(x)|^p\\ &\geq (p-1)\frac{\epsilon}{2^{p-2}e^{u(x)}}|\nabla u(x)|^p,\\ |\nabla v_\epsilon(x)| &\geq \frac{1}{2}|\nabla u(x)|,\end{aligned}$$

valid for all $\epsilon < \frac{1}{2}$ and all $x \in int\,\mathcal{D}^\diamond$.

4.3 Playing Boundary Aware Tug-of-War with Noise

Below we develop the probability setting similar to that of Sect. 3.4, consistent with the averaging principle (4.1) instead of (3.18). We use notation of Definition 3.8 and fix $p > 2$, $r = \sqrt{\frac{3(p-2)}{2(N+2)}}$, $\epsilon > 0$ as in Theorem 4.1.

1. Consider the product probability space $(\Omega_1, \mathcal{F}_1, \mathbb{P}_1)$ on:

$$\Omega_1 = B_1(0) \times \{1, 2, 3\} \times (0, 1),$$

equipped with the σ-algebra $\mathcal{F}_1$ which is the smallest σ-algebra containing all the products $D \times A \times B \subset \Omega_1$ where $D \subset B_1(0) \subset \mathbb{R}^N$ and $B \subset (0, 1)$ are Borel, and $A \subset \{1, 2, 3\}$. The probability measure $\mathbb{P}_1$ is given as the product of: the normalized Lebesgue measure on $B_1(0)$, the uniform counting measure on $\{1, 2, 3\}$, and the Lebesgue measure on $(0, 1)$ (see Example A.2):

$$\mathbb{P}_1(D \times A \times B) = \frac{|D|}{|B_1(0)|} \cdot \frac{|A|}{3} \cdot |B|.$$

For each $n \in \mathbb{N}$, we denote by $\Omega_n = (\Omega_1)^n$ the Cartesian product of n copies of Ω_1, and we let $(\Omega_n, \mathcal{F}_n, \mathbb{P}_n)$ be the product probability space. Further:

$$(\Omega, \mathcal{F}, \mathbb{P}) = (\Omega_1, \mathcal{F}_1, \mathbb{P}_1)^{\mathbb{N}}$$

denotes the probability space on the countable product:

$$\Omega = (\Omega_1)^{\mathbb{N}} = \prod_{i=1}^{\infty} \Omega_1 = \big(B_1(0) \times \{1, 2, 3\} \times (0, 1)\big)^{\mathbb{N}},$$

defined by means of Theorem A.12 (compare Example A.13 (ii)). For each $n \in \mathbb{N}$, the σ-algebra $\mathcal{F}_n$ is identified with the sub-σ-algebra of $\mathcal{F}$, consisting of sets of the form: $F \times \prod_{i=n+1}^{\infty} \Omega_1$ for all $F \in \mathcal{F}_n$. We also define $\mathcal{F}_0 = \{\emptyset, \Omega\}$ and observe that the sequence $\{\mathcal{F}_n\}_{n=1}^{\infty}$ is a filtration of $\mathcal{F}$. The elements of Ω_n are the n-tuples $\{(w_i, a_i, b_i)\}_{i=1}^{n}$, while the elements of Ω are sequences $\{(w_i, a_i, b_i)\}_{i=1}^{\infty}$, where $w_i \in B_1(0)$, $a_i \in \{1, 2, 3\}$ and $b_i \in (0, 1)$ for all $i \in \mathbb{N}$.

2. We now describe the strategies $\sigma_I = \{\sigma_I^n\}_{n=0}^{\infty}$ and $\sigma_{II} = \{\sigma_{II}^n\}_{n=0}^{\infty}$ of Players I and II. For every $n \geq 0$, these are the functions:

$$\sigma_I^n, \sigma_{II}^n : H_n \to B_r(0) \subset \mathbb{R}^N,$$

defined on the spaces of "finite histories" $H_n = \mathbb{R}^N \times (\mathbb{R}^N \times \Omega_1)^n$, assumed to be measurable with respect to the (target) Borel σ-algebra $\mathcal{B}(B_r(0))$ and the (domain) product σ-algebra in H_n.

3. Given an initial point $x_0 \in \mathcal{D}^\diamond$ and the strategies σ_I and σ_{II}, we inductively define a sequence of vector-valued random variables:

$$X_n^{x_0,\sigma_I,\sigma_{II}} : \Omega \to \mathcal{D}^\diamond \qquad \text{for } n = 0, 1, \ldots.$$

For simplicity of notation, we suppress the superscripts $x_0, \sigma_I, \sigma_{II}$ and write X_n instead of $X_n^{x_0,\sigma_I,\sigma_{II}}$. Firstly, we set $X_0 \equiv x_0$ and we write $h_0 \equiv x_0 \in H_0$. The sequence $\{X_n\}_{n=0}^{\infty}$ will be adapted to the filtration $\{\mathcal{F}_n\}_{n=0}^{\infty}$ of $\mathcal{F}$, and thus each X_n for $n \geq 1$ is effectively defined on Ω_n. Recall that the scaled distance from the complement of $\mathcal{D}$ in $\mathcal{D}^\diamond$ is:

$$d_\epsilon(x) = \frac{1}{\epsilon} \min \big\{\epsilon, \operatorname{dist}(x, \Gamma_{out})\big\}.$$

We now set:

$$X_n\big((w_1, a_1, b_1), \ldots, (w_n, a_n, b_n)\big) =$$
$$= \begin{cases} x_{n-1} + \epsilon w_n + \epsilon \sigma_I^{n-1}(h_{n-1}) & \text{for } a_n = 1 \text{ and } d_\epsilon(x_{n-1}) \geq b_n \\ x_{n-1} + \epsilon w_n + \epsilon \sigma_{II}^{n-1}(h_{n-1}) & \text{for } a_n = 2 \text{ and } d_\epsilon(x_{n-1}) \geq b_n \\ x_{n-1} + \epsilon w_n & \text{for } a_n = 3 \text{ and } d_\epsilon(x_{n-1}) \geq b_n \\ x_{n-1} & \text{for } d_\epsilon(x_{n-1}) < b_n. \end{cases} \tag{4.17}$$

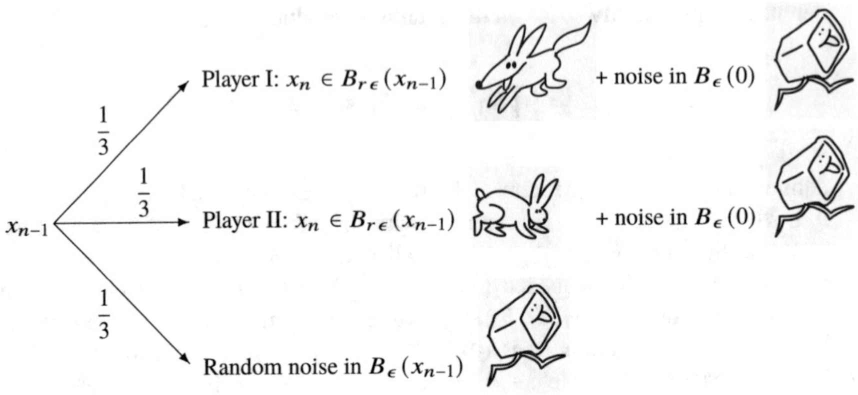

Fig. 4.2 Player I, Player II and random noise in the boundary aware Tug-of-War

The n-th position x_n and the n-th augmented history h_n:

$$x_n = X_n\big((w_1, a_1, b_1), \dots, (w_n, a_n, b_n)\big)$$
$$h_n = \big(x_0, (x_1, w_1, a_1, b_1), \dots, (x_n, w_n, a_n, b_n)\big) \in H_n$$

are then measurable functions of the argument $\big((w_1, a_1, b_1), \dots, (w_n, a_n, b_n)\big) \in \Omega_n$. It is clear that X_n is $\mathcal{F}_n$-measurable and that it takes values in $\mathcal{D}^\diamond$.

In this game, the token position is advanced by random shifts ϵw_n of length at most ϵ, preceded by the scaled deterministic shifts $\epsilon\sigma_I^{n-1}(h_{n-1})$ and $\epsilon\sigma_{II}^{n-1}(h_{n-1})$, which are activated according to the value of the equally probable outcomes $a_n \in \{1, 2, 3\}$. Namely, $a_n = 1$ corresponds to activating σ_I and $a_n = 2$ to activating σ_{II}, whereas $a_n = 3$ results in not activating any of the deterministic strategies (Fig. 4.2). The token at $x_n \in \Gamma_{out} \cup \Gamma_\epsilon$ with $d_\epsilon(x_n) < b_{n+1}$ is not moved.

4. The variables $b_n \in (0, 1)$ serve to set a threshold for reading the value of the game from the prescribed boundary data F. We now define the random variable $\tau^{x_0,\sigma_I,\sigma_{II}} : \Omega \to \mathbb{N} \cup \{+\infty\}$:

$$\tau^{x_0,\sigma_I,\sigma_{II}}\big((w_1, a_1, b_1), (w_2, a_2, b_2), \dots\big) = \min\{n \geq 1;\ b_n > d_\epsilon(x_{n-1})\}, \tag{4.18}$$

where, as before, $x_n = X_n\big((w_1, a_1, b_1), \dots, (w_n, a_n, b_n)\big)$. We drop the superscript $x_0, \sigma_I, \sigma_{II}$ and write τ instead of $\tau^{x_0,\sigma_I,\sigma_{II}}$ if no ambiguity arises. Clearly, τ is $\mathcal{F}$-measurable and, in fact, it is a stopping time relative to the filtration $\{\mathcal{F}_n\}_{n=0}^\infty$, similarly as in Lemma 3.13 (the proof is left as an exercise):

Lemma 4.11 *In the above setting, we have:* $\mathbb{P}(\tau < +\infty) = 1$.

5. Given now a starting position $x_0 \in \mathcal{D}^\diamond$ and two strategies σ_I and σ_{II}, the $\mathcal{F}$-measurable vector-valued random variable $\big(X^{x_0,\sigma_I,\sigma_{II}}\big)_{\tau^{x_0}} : \Omega \to \mathcal{D}^\diamond$ is:

$$\big(X^{x_0,\sigma_I,\sigma_{II}}\big)_{\tau^{x_0,\sigma_I,\sigma_{II}}}(\omega) = X^{x_0,\sigma_I,\sigma_{II}}_{\tau^{x_0,\sigma_I,\sigma_{II}}(\omega)}(\omega) \qquad \mathbb{P} - \text{a.s. in } \Omega.$$

Let $F : \Gamma_{out} \cup \Gamma_\epsilon \to \mathbb{R}$ be a bounded, Borel function. Define:

$$\begin{aligned} u_I^\epsilon(x_0) &= \sup_{\sigma_I} \inf_{\sigma_{II}} \mathbb{E}\Big[F \circ \big(X^{x_0,\sigma_I,\sigma_{II}}\big)_{\tau^{x_0,\sigma_I,\sigma_{II}}}\Big] \\ u_{II}^\epsilon(x_0) &= \inf_{\sigma_{II}} \sup_{\sigma_I} \mathbb{E}\Big[F \circ \big(X^{x_0,\sigma_I,\sigma_{II}}\big)_{\tau^{x_0,\sigma_I,\sigma_{II}}}\Big], \end{aligned} \tag{4.19}$$

where sup and inf are taken over all strategies as above.

The main result in this context is the following version of Theorem 3.14:

Theorem 4.12 *Given a bounded, Borel function* $F : \Gamma_{out} \cup \Gamma_\epsilon \to \mathbb{R}$ *and letting* $p, r, \epsilon, u_\epsilon$ *be as in Theorem 4.1, we have:*

$$u_I^\epsilon = u_\epsilon = u_{II}^\epsilon \qquad \textit{in } \mathcal{D}^\diamond.$$

Proof

1. We drop the sub/superscript ϵ to ease the notation in the proof. To show that $u_{II} \leq u$ in $\mathcal{D}^\diamond$, fix $x_0 \in \mathcal{D}^\diamond$ and let $\eta > 0$. By Lemma 3.15, there exists a strategy $\sigma_{0,II}$ for Player II, such that $\sigma^n_{0,II}(h_n) = \sigma^n_{0,II}(x_n)$ and that, for every $h_n \in H_n$ we have:

$$\begin{aligned} &\mathcal{A}_\epsilon u(x_n + \epsilon\sigma^n_{0,II}(x_n)) \leq \inf_{y \in B_{r\epsilon}(x_n)} \mathcal{A}_\epsilon u(y) + \frac{\eta}{2^{n+1}} \qquad \text{if } x_n \in \mathcal{D} \\ &\sigma^n_{0,II}(x_n) = x_n \qquad \text{if } x_n \notin \mathcal{D}. \end{aligned} \tag{4.20}$$

Fix a strategy σ_I of Player I and consider the following sequence of random variables $M_n : \Omega \to \mathbb{R}$:

$$M_n = (u \circ X_n)\mathbb{1}_{\tau > n} + (F \circ X_\tau)\mathbb{1}_{\tau \leq n} + \frac{\eta}{2^n}.$$

As usual, we have dropped the superscripts in $X_n^{x_0,\sigma_I,\sigma_{0,II}}$ and $\tau^{x_0,\sigma_I,\sigma_{II}}$.[1] Note that each M_n is well defined; the range of X_n is contained in $\mathcal{D}^\diamond$ that is the domain of u, and on the other hand when $\tau \leq n$ then for some $1 \leq k \leq n$ we have: $b_k > d_\epsilon(x_{k-1})$. In this case, $x_k = x_{k-1} \in \Gamma$ and so $F(x_k)$ is well defined. Clearly, each M_n is also $\mathcal{F}_n$-measurable.

We now show that $\{M_n\}_{n=0}^\infty$ is a supermartingale with respect to the filtration $\{\mathcal{F}_n\}_{n=0}^\infty$. We write:

$$\begin{aligned}\mathbb{E}(M_n \mid \mathcal{F}_{n-1}) = {} & \mathbb{E}\big((u \circ X_n)\mathbb{1}_{\tau>n} \mid \mathcal{F}_{n-1}\big) + \mathbb{E}\big((F \circ X_n)\mathbb{1}_{\tau=n} \mid \mathcal{F}_{n-1}\big) \\ & + \mathbb{E}\big((F \circ X_\tau)\mathbb{1}_{\tau<n} \mid \mathcal{F}_{n-1}\big) + \frac{\eta}{2^n} \quad \text{a.s.}\end{aligned} \tag{4.21}$$

and compute each of the terms in the above sum.

2. Since $(F \circ X_\tau)\mathbb{1}_{\tau<n} = (F \circ X_{\tau-1})\mathbb{1}_{\tau<n}$ is $\mathcal{F}_{n-1}$-measurable, it follows that:

$$\mathbb{E}\big((F \circ X_\tau)\mathbb{1}_{\tau<n} \mid \mathcal{F}_{n-1}\big) = (F \circ X_\tau)\mathbb{1}_{\tau<n} \quad \text{a.s.} \tag{4.22}$$

Further, writing: $\mathbb{1}_{\tau=n} = \mathbb{1}_{\tau\geq n}\mathbb{1}_{b_n>d_\epsilon(x_{n-1})}$, we obtain in view of Exercise A.16 (v) and Lemma A.17 (ii), that:

$$\begin{aligned}\mathbb{E}\big((F \circ X_n)\mathbb{1}_{\tau=n} \mid \mathcal{F}_{n-1}\big) &= \mathbb{E}\big((F \circ X_{n-1})\mathbb{1}_{\tau\geq n}\mathbb{1}_{b_n>d_\epsilon(x_{n-1})} \mid \mathcal{F}_{n-1}\big) \\ &= \mathbb{E}\big(\mathbb{1}_{b_n>d_\epsilon(x_{n-1})} \mid \mathcal{F}_{n-1}\big)\big(F \circ X_{n-1}\big)\mathbb{1}_{\tau\geq n} \\ &= \mathbb{P}_1\big(\{b_n > d_\epsilon(x_{n-1})\}\big)\big(F \circ X_{n-1}\big)\mathbb{1}_{\tau\geq n} \\ &= (1 - d_\epsilon(x_{n-1}))\big(F \circ X_{n-1}\big)\mathbb{1}_{\tau\geq n} \qquad \text{a.s.}\end{aligned} \tag{4.23}$$

We now compute the first term in the right-hand side of (4.21). Similarly as above, write: $\mathbb{1}_{\tau>n} = \mathbb{1}_{\tau\geq n}\mathbb{1}_{b_n\leq d_\epsilon(x_{n-1})}$, so that:

$$\mathbb{E}\big((u\circ X_n)\mathbb{1}_{\tau>n} \mid \mathcal{F}_{n-1}\big) = \mathbb{E}\big((u\circ X_n)\mathbb{1}_{b_n\leq d_\epsilon(x_{n-1})} \mid \mathcal{F}_{n-1}\big)\mathbb{1}_{\tau\geq n} \quad \text{a.s.} \tag{4.24}$$

[1] More precisely, we write:

$$M_n = \big(u \circ X_n^{x_0,\sigma_I,\sigma_{0,II}}\big)\mathbb{1}_{\tau^{x_0,\sigma_I,\sigma_{II}}>n} + F \circ \big(X^{x_0,\sigma_I,\sigma_{0,II}}\big)_{\tau^{x_0,\sigma_I,\sigma_{II}}}\mathbb{1}_{\tau^{x_0,\sigma_I,\sigma_{II}}\leq n} + \frac{\eta}{2^n}.$$

Observe that $(u \circ X_n)\mathbb{1}_{b_n \leq d_\epsilon(x_{n-1})}$ is $\mathcal{F}_n$-measurable, so by Lemma A.17 (ii) and the defining property (4.20) of $\sigma_{0,II}$, we get:

$$\mathbb{E}\big((u \circ X_n)\mathbb{1}_{b_n \leq d_\epsilon(x_{n-1})} \mid \mathcal{F}_{n-1}\big) = \int_{\Omega_1} (u \circ X_n)\mathbb{1}_{b_n \leq d_\epsilon(x_{n-1})} \, \mathrm{d}\mathbb{P}_1$$
$$= d_\epsilon(x_{n-1})\frac{1}{3}\Big(\mathcal{A}_\epsilon u(x_{n-1}) + \mathcal{A}_\epsilon u(x_{n-1} + \epsilon\sigma_I^{n-1}) + \mathcal{A}_\epsilon u(x_{n-1} + \epsilon\sigma_{0,II}^{n-1})\Big)$$
$$\leq d_\epsilon(x_{n-1})\big(S \circ X_{n-1} + \frac{\eta}{2^n}\big) \quad \text{a.s.} \tag{4.25}$$

Consequently, (4.24) and (4.25) result in:

$$\mathbb{E}\big((u \circ X_n)\mathbb{1}_{\tau > n} \mid \mathcal{F}_{n-1}\big) \leq d_\epsilon(x_{n-1})\big(S \circ X_{n-1}\big)\mathbb{1}_{\tau \geq n} + \frac{\eta}{2^n} \qquad \text{a.s.}$$

Together with (4.22) and (4.23), this implies:

$$\begin{aligned}\mathbb{E}(M_n \mid \mathcal{F}_{n-1}) \leq {} & \Big(d_\epsilon(x_{n-1})(S \circ X_{n-1}) + (1 - d_\epsilon(x_{n-1}))(F \circ X_{n-1})\mathbb{1}_{\tau \geq n}\Big)\mathbb{1}_{\tau \geq n} \\ & + (F \circ X_{\tau-1})\mathbb{1}_{\tau < n} + \frac{\eta}{2^{n-1}} \\ = {} & (u \circ X_{n-1})\mathbb{1}_{\tau > n-1} + (F \circ X_{\tau-1})\mathbb{1}_{\tau \leq n-1} + \frac{\eta}{2^{n-1}} = M_{n-1} \quad \text{a.s.}\end{aligned}$$

where we have used (4.1) to $u = u_\epsilon$.

3. The supermartingale property being established, we now conclude that:

$$\mathbb{E}[M_\tau] \leq \mathbb{E}[M_0] = u(x_0) + \eta,$$

in view of Doob's theorem (Theorem A.34 (ii)) and the boundedness of F and u, implying the uniform boundedness of $\{M_n\}_{n=0}^\infty$. Finally, since:

$$F \circ X_\tau = M_\tau - \frac{\eta}{2^\tau},$$

we get:

$$u_{II}(x_0) \leq \sup_{\sigma_I} \mathbb{E}\big[F \circ X_\tau\big] \leq \sup_{\sigma_I} \mathbb{E}\big[M_\tau\big] \leq u(x_0) + \eta.$$

As $\eta > 0$ was arbitrary, we obtain the claimed comparison:

$$u_{II}(x_0) \leq u(x_0).$$

To show the remaining inequality $u(x_0) \leq u_I(x_0)$, we argue exactly as above. Take σ_{II} to be an arbitrary strategy, while choose $\sigma_{0,I}$ so that $\sigma_{0,I}^n(h_n) = \sigma_{0,I}^n(x_n)$ and that, for every $h_n \in H_n$ we have:

$$\mathcal{A}_\epsilon u(x_n + \sigma_{0,I}^n(x_n)) \geq \sup_{y \in B_{r\epsilon}(x_n)} \mathcal{A}_\epsilon u(y) - \frac{\eta}{2^{n+1}} \qquad \text{if } x_n \in \mathcal{D}$$

$$\sigma_{0,I}(x_n) = x_n \qquad \text{if } x_n \notin \mathcal{D}.$$

Then, the following sequence of random variables $\{\bar{M}_n\}_{n=0}^\infty$ is a submartingale with respect to the filtration $\{\mathcal{F}_n\}_{n=0}^\infty$:

$$\bar{M}_n = (u \circ X_n)\mathbb{1}_{\tau>n} + (F \circ X_\tau)\mathbb{1}_{\tau \leq n} - \frac{\eta}{2^n} \tag{4.26}$$

and the Doob theorem implies $u(x_0) \leq u_I(x_0)$. Since $u_I(x_0) \leq u_{II}(x_0)$ (see Exercise 3.16), we conclude that $u_I = u = u_{II}$ in $\mathcal{D}^\diamond$. We leave these points as an exercise below. The proof of Theorem 4.12 is done. □

Exercise 4.13

(i) Prove Lemma 4.11.
(ii) Show that the sequence defined in (4.26) is a submartingale with respect to the filtration $\{\mathcal{F}_n\}_{n=0}^\infty$. Deduce that $u_\epsilon(x_0) \leq u_I^\epsilon(x_0)$.

4.4 A Probabilistic Proof of the Basic Convergence Theorem

In this section, we make use of the game-theoretical interpretation of solutions to (4.1), that has been developed in Sect. 4.12, to give an alternative proof of Theorem 4.6. The previous proof in Sect. 4.2 used solely analytic methods.

A Game-Theoretical Proof of Theorem 4.6

1. Consider the following "negative gradient strategy" $\sigma_{0,II}$ for Player II that depends only on the last position x_n of the token in $\mathcal{D}$:

$$\sigma_{0,II}^n(h_n) = \sigma_{0,II}^n(x_n) = \begin{cases} -r(1-\epsilon^2)\dfrac{\nabla u(x_n)}{|\nabla u(x_n)|} & \text{if } x_n \in \mathcal{D} \\ 0 & \text{if } x_n \notin \mathcal{D}. \end{cases} \tag{4.27}$$

By the analysis in Theorem 3.4 we obtain:

$$\begin{aligned} \inf_{y \in B_{r\epsilon}(x)} u(y) &\geq u\Big(x - r\epsilon \frac{\nabla u(x_n)}{|\nabla u(x_n)|}\Big) - C\epsilon^3 \\ &\geq u\big(x + \epsilon\sigma_{0,II}^n(x)\big) - C\epsilon^3 \qquad \text{for all } x \in \bar{\mathcal{D}}, \end{aligned}$$

where C above is a universal constant, depending on u, p and $\mathcal{D}^\diamond$ but independent of $x \in \bar{\mathcal{D}}$ and (sufficiently small) $\epsilon > 0$. In view of Exercise 3.7, we also deduce:

$$\inf_{y\in B_{r\epsilon}(x)} \mathcal{A}_\epsilon u(y) \geq \Big(\inf_{y\in B_{r\epsilon}(x)} u(y)\Big) + \frac{\epsilon^2}{2(N+2)}\Big(\inf_{y\in B_{r\epsilon}(x)} \Delta u(y)\Big) + C\epsilon^3,$$

together with the following estimate:

$$u\big(x+\epsilon\sigma_{0,II}^n(x)\big) \geq \mathcal{A}_\epsilon u\big(x+\epsilon\sigma_{0,II}^n(x)\big) - \frac{\epsilon^2}{2(N+2)}\Delta u(x+\epsilon\sigma_{0,II}^n(x)) - C\epsilon^3,$$

which result in:

$$\begin{aligned}\inf_{y\in B_{r\epsilon}(x)} \mathcal{A}_\epsilon u(y) &\geq \mathcal{A}_\epsilon u\big(x+\epsilon\sigma_{0,II}^n(x)\big) + \frac{\epsilon^2}{2(N+2)}\cdot \sup_{y\in B_{r\epsilon}(x)} \Delta u(y)\\ &\quad - \Delta u(x+\epsilon\sigma_{0,II}^n(x)) - C\epsilon^3\\ &\geq \mathcal{A}_\epsilon u\big(x+\epsilon\sigma_{0,II}^n(x)\big) - C\epsilon^3 \qquad \text{for all } x \in \bar{\mathcal{D}}.\end{aligned} \tag{4.28}$$

2. Fix now an initial position $x_0 \in \mathcal{D}^\diamond$ and fix a strategy σ_I for Player I. Using the definition of the process $\big\{X_n = X_n^{x_0,\sigma_I,\sigma_{0,II}}\big\}_{n=0}^\infty$ in (4.17) and of the stopping time $\tau = \tau^{x_0,\sigma_I,\sigma_{0,II}}$ in (4.18), similarly as in the proof of Theorem 4.12 we compute:

$$\begin{aligned}&\mathbb{E}\big(u\circ X_{\tau\wedge n} \mid \mathcal{F}_{n-1}\big)\\ &\quad= \mathbb{E}\Big((u\circ X_n)\mathbb{1}_{b_n\leq d_\epsilon(x_{n-1})}\mathbb{1}_{\tau\geq n} \mid \mathcal{F}_{n-1}\Big) + \mathbb{E}\Big((u\circ X_n)\mathbb{1}_{b_n> d_\epsilon(x_{n-1})}\mathbb{1}_{\tau\geq n} \mid \mathcal{F}_{n-1}\Big)\\ &\quad\quad+\mathbb{E}\Big((u\circ X_\tau)\mathbb{1}_{\tau<n} \mid \mathcal{F}_{n-1}\Big)\\ &\quad= d_\epsilon(x_{n-1})\frac{1}{3}\Big(\mathcal{A}_\epsilon u(x_{n-1}) + \mathcal{A}_\epsilon u(x_{n-1}+\epsilon\sigma_I^{n-1}) + \mathcal{A}_\epsilon u(x_{n-1}+\epsilon\sigma_{0,II}^{n-1})\Big)\mathbb{1}_{\tau\geq n}\\ &\quad\quad+\big(1-d_\epsilon(x_{n-1})\big)u(x_{n-1})\mathbb{1}_{\tau\geq n} + u(x_\tau)\mathbb{1}_{\tau<n} \qquad \text{a.s.}\end{aligned}$$

Since $d_\epsilon(x_{n-1}) > 0$ implies $x_{n-1} \in \mathcal{D}$, we may apply the bound (4.28) in the first term of the right-hand side above, to get:

$$\begin{aligned}&\mathbb{E}\big(u\circ X_{\tau\wedge n} \mid \mathcal{F}_{n-1}\big)\\ &\quad\leq d_\epsilon(x_{n-1})\frac{1}{3}\Big(\mathcal{A}_\epsilon u(x_{n-1}) + \sup_{y\in B_{r\epsilon}(x_{n-1})} \mathcal{A}_\epsilon u(y) + \inf_{y\in B_{r\epsilon}(x)} \mathcal{A}_\epsilon u(y)\Big)\mathbb{1}_{\tau\geq n}\\ &\quad\quad- d_\epsilon(x_{n-1})u(x_{n-1})\mathbb{1}_{\tau\geq n} + u(x_{n-1})\mathbb{1}_{\tau\geq n} + u(x_\tau)\mathbb{1}_{\tau<n} + C\epsilon^3 d_\epsilon(x_{n-1})\mathbb{1}_{\tau\geq n} \ \text{a.s.}\end{aligned}$$

Finally, using Exercise 3.7 we arrive at:

$$\begin{aligned}\mathbb{E}\big(u \circ X_{\tau\wedge n} \mid \mathcal{F}_{n-1}\big) &\leq u(x_{n-1})\mathbb{1}_{\tau\geq n} + u(x_\tau)\mathbb{1}_{\tau<n} + C\epsilon^3 d_\epsilon(x_{n-1})\mathbb{1}_{\tau\geq n}\\ &= u \circ X_{\tau\wedge(n-1)} + C\epsilon^3 d_\epsilon(x_{n-1})\mathbb{1}_{\tau\geq n} \qquad \text{a.s.}\end{aligned} \tag{4.29}$$

where, as usual, $C > 0$ denotes a constant that may vary from line to line of the calculation above, but it is independent of n, x_0, σ_I and ϵ. Consequently, the following sequence $\{M_n\}_{n=0}^\infty$ of random variables:

$$M_n = u \circ X_{\tau\wedge n} - C\epsilon^3\big((\tau - 1) \wedge n\big) \tag{4.30}$$

is a supermartingale on $(\Omega, \mathcal{F}, \mathbb{P})$ with respect to the filtration $\{\mathcal{F}_n\}_{n=0}^\infty$. Indeed:

$$\begin{aligned}&\mathbb{E}\big((\tau - 1) \wedge n \mid \mathcal{F}_{n-1}\big)\\ &\quad= \mathbb{E}\big(n\mathbb{1}_{b_n\leq d_\epsilon(x_{n-1})}\mathbb{1}_{\tau\geq n} \mid \mathcal{F}_{n-1}\big)\\ &\qquad+ \mathbb{E}\big((n-1)\mathbb{1}_{b_n>d_\epsilon(x_{n-1})}\mathbb{1}_{\tau\geq n} \mid \mathcal{F}_{n-1}\big) + \mathbb{E}\big((\tau-1)\mathbb{1}_{\tau<n} \mid \mathcal{F}_{n-1}\big)\\ &\quad= nd_\epsilon(x_{n-1})\mathbb{1}_{\tau\geq n} + (n-1)\big(1 - d_\epsilon(x_{n-1})\big)\mathbb{1}_{\tau\geq n} + (\tau-1)\mathbb{1}_{\tau<n}\\ &\quad= (n-1)\mathbb{1}_{\tau\geq n} + (\tau-1)\mathbb{1}_{\tau<n} + d_\epsilon(x_{n-1})\mathbb{1}_{\tau\geq n}\\ &\quad= (\tau-1) \wedge (n-1) + d_\epsilon(x_{n-1})\mathbb{1}_{\tau\geq n} \qquad \text{a.s.}\end{aligned} \tag{4.31}$$

and so, by (4.29):

$$\begin{aligned}\mathbb{E}\big(M_n \mid \mathcal{F}_{n-1}\big) &\leq u \circ X_{\tau\wedge(n-1)} + C\epsilon^3 d_\epsilon(x_{n-1})\mathbb{1}_{\tau\geq n}\\ &\quad - C\epsilon^3(\tau-1) \wedge (n-1) - C\epsilon^3 d_\epsilon(x_{n-1})\mathbb{1}_{\tau\geq n} = M_{n-1} \qquad \text{a.s.}\end{aligned}$$

proving the supermartingale property.

3. Below, we will show the following bound, valid uniformly in $n \in \mathbb{N}$:

$$0 \leq \mathbb{E}\big[(\tau-1) \wedge n\big] \leq \frac{C}{\epsilon^2}. \tag{4.32}$$

Since the expectation decreases along a supermartingale, (4.32) results in:

$$u(x_0) = \mathbb{E}[M_0] \geq \mathbb{E}[M_n] \geq \mathbb{E}\big[u \circ X_{\tau\wedge n}\big] - C\epsilon \qquad \text{for all } n \geq 0.$$

Observe that $\lim_{n\to 0}\mathbb{E}[u \circ X_{\tau\wedge n}] = \mathbb{E}[u \circ X_\tau]$ by the Lebesgue dominated convergence theorem. Concluding, we have that for all strategies σ_I:

$$u(x_0) \geq \mathbb{E}\Big[u \circ \big(X^{x_0,\sigma_I,\sigma_0,II}\big)_{\tau^{x_0,\sigma_I,\sigma_0,II}}\Big] - C\epsilon.$$

Taking now the supremum over all σ_I-s and recalling the definition (4.19) and Theorem 4.12, we arrive at:

$$u(x_0) \geq u^\epsilon_{II}(x_0) - C\epsilon = u_\epsilon(x_0) - C\epsilon. \tag{4.33}$$

4. We now aim at proving the estimate (4.32), under the special assumption:

$$u \leq 0 \quad \text{in } \ int\, \mathcal{D}^\diamond. \tag{4.34}$$

We first find a bound on the conditional expectation of the nonnegative random variable $\big(u \circ X_{\tau\wedge n} - u \circ X_{\tau\wedge(n-1)}\big)^2$. It is nonzero only in the event $\{\tau > n\} = \{b_n \leq d_\epsilon(x_{n-1})\} \cap \{\tau \geq n\}$. By definition (4.17) and using Jenssen's inequality, we obtain:

$$\begin{aligned}
&\mathbb{E}\Big(\big(u \circ X_{\tau\wedge n} - u \circ X_{\tau\wedge(n-1)}\big)^2 \mid \mathcal{F}_{n-1}\Big) \\
&\quad = \mathbb{E}\Big(\big(u \circ X_n - u \circ X_{n-1}\big)^2 \mathbb{1}_{b_n \leq d_\epsilon(x_{n-1})} \mathbb{1}_{\tau\geq n} \mid \mathcal{F}_{n-1}\Big) \\
&\quad \geq d_\epsilon(x_{n-1})\frac{1}{3} \fint_{B_\epsilon(0)} \Big(u(x_{n-1} + \epsilon w_n + \epsilon\sigma_{0,II}^{n-1}) - u(x_{n-1})\Big)^2 \, \mathrm{d}w_n \, \mathbb{1}_{\tau\geq n} \\
&\quad \geq d_\epsilon(x_{n-1})\frac{1}{3} \Big(\fint_{B_\epsilon(0)} u(x_{n-1} + \epsilon w_n + \epsilon\sigma_{0,II}^{n-1}) - u(x_{n-1}) \, \mathrm{d}w_n\Big)^2 \mathbb{1}_{\tau\geq n} \\
&\quad = d_\epsilon(x_{n-1})\frac{1}{3} \Big(\mathcal{A}_\epsilon u(x_{n-1} + \epsilon\sigma_{0,II}^{n-1}) - u(x_{n-1})\Big)^2 \mathbb{1}_{\tau\geq n} \qquad \text{a.s.}
\end{aligned}$$

We now observe that for $x_{n-1} \in \mathcal{D}$, the estimates in Exercise 3.7 and the choice of the gradient strategy in (4.27) yield:

$$\begin{aligned}
\Big|\mathcal{A}_\epsilon u(x_{n-1} + \epsilon\sigma_{0,II}^{n-1}) - u(x_{n-1})\Big| &\geq \big|u(x_{n-1} + \epsilon\sigma_{0,II}^{n-1}) - u(x_{n-1}\big| - C\epsilon^2 \\
&\geq r\epsilon|\nabla u(x_{n-1})| - C\epsilon^2 \geq C\epsilon.
\end{aligned}$$

Hence, it follows that:

$$\mathbb{E}\Big(\big(u \circ X_{\tau\wedge n} - u \circ X_{\tau\wedge(n-1)}\big)^2 \mid \mathcal{F}_{n-1}\Big) \geq C\epsilon^2 d_\epsilon(x_{n-1})\mathbb{1}_{\tau\geq n} \qquad \text{a.s.} \tag{4.35}$$

We further deduce that the sequence $\{\bar{M}_n\}_{n=0}^\infty$, defined by:

$$\bar{M}_n = \big(u \circ X_{\tau\wedge n}\big)^2 - C\epsilon^2\big((\tau - 1) \wedge n\big),$$

is a submartingale with respect to the filtration $\{\mathcal{F}_n\}_{n=0}^{\infty}$. Indeed, by (4.35), (4.29) and the assumed nonpositivity of u on $\mathcal{D}$, we get:

$$\begin{aligned}
&\mathbb{E}\Big((u\circ X_{\tau\wedge n})^2 - (u\circ X_{\tau\wedge(n-1)})^2 \mid \mathcal{F}_{n-1}\Big)\\
&\quad = \mathbb{E}\Big((u\circ X_{\tau\wedge n} - u\circ X_{\tau\wedge(n-1)})^2 \mid \mathcal{F}_{n-1}\Big)\\
&\qquad\quad + 2\mathbb{E}\Big((u\circ X_{\tau\wedge n} - u\circ X_{\tau\wedge(n-1)})(u\circ X_{\tau\wedge(n-1)}) \mid \mathcal{F}_{n-1}\Big)\\
&\quad \geq C\epsilon^2 d_\epsilon(x_{n-1})\mathbb{1}_{\tau\geq n} + 2(u\circ X_{\tau\wedge(n-1)})\mathbb{E}\Big(u\circ X_{\tau\wedge n} - u\circ X_{\tau\wedge(n-1)} \mid \mathcal{F}_{n-1}\Big)\\
&\quad \geq C\epsilon^2 d_\epsilon(x_{n-1})\mathbb{1}_{\tau\geq n} \qquad \text{a.s.}
\end{aligned}$$

The submartingale property of $\{\bar{M}_n\}_{n=0}^{\infty}$ follows now in view of (4.31):

$$\begin{aligned}
&\mathbb{E}\Big(\bar{M}_n - \bar{M}_{n-1} \mid \mathcal{F}_{n-1}\Big)\\
&\quad \geq C\epsilon^2 d_\epsilon(x_{n-1})\mathbb{1}_{\tau\geq n} - C\epsilon^2\mathbb{E}\Big((\tau-1)\wedge n - (\tau-1)\wedge(n-1) \mid \mathcal{F}_{n-1}\Big)\\
&\quad = 0 \qquad \text{a.s.}
\end{aligned}$$

Consequently, for every $n \in \mathbb{N}$ we have:

$$0 \leq u(x_0)^2 = \mathbb{E}[\bar{M}_0] \leq \mathbb{E}[\bar{M}_n] = \mathbb{E}\big[(u\circ X_{\tau\wedge n})^2\big] - C\epsilon^2\mathbb{E}\big[(\tau-1)\wedge n\big],$$

so that, in view of boundedness of u:

$$\mathbb{E}\big[(\tau-1)\wedge n\big] \leq \frac{1}{C\epsilon^2}\mathbb{E}\big[(u\circ X_{\tau\wedge n})^2\big] \leq \frac{C}{\epsilon^2},$$

proving (4.32).

5. The above steps establish (4.33) in the nonpositive regime (4.34). Note, however, that every bounded function u can be modified, by subtracting a constant, to satisfy (4.34), without violating the properties as in the statement of the Theorem. This modification results in the same additive modification of the solution u_ϵ to (4.1) with the boundary data $F = u_{|\Gamma_{out}\cup\Gamma_\epsilon}$. It thus follows that (4.33) holds for any u, without assuming (4.34).

   In order to show the reverse inequality:

$$u(x_0) \leq u_\epsilon(x_0) + C\epsilon, \tag{4.36}$$

one may use a symmetric argument, adopting the "positive gradient strategy" $\sigma_{0,I}$ for Player I and showing that $\big\{u\circ X_{\tau\wedge n} + C\epsilon^3\big((\tau-1)\wedge n\big)\big\}_{n=0}^{\infty}$ is a submartingale, whereas $\big\{(u\circ X_{\tau\wedge n})^2 - C\epsilon^2\big((\tau-1)\wedge n\big)\big\}_{n=0}^{\infty}$ is again a submartingale again, under the extra requirement that $u \geq 0$ in $int\,\mathcal{D}^{\diamond}$. Consequently, (4.36) follows for such nonnegative p-harmonic functions u with nonzero gradient, and by translation, for any u as well (see Exercise 4.14).

Alternatively, one can deduce (4.36) directly from (4.33) by replacing u with $-u$ and switching the roles of the players. □

Exercise 4.14 In the setting of proofs above, use the outline given in Step 5 and prove that: $u(x_0) \leq u_I^\epsilon(x_0) + C\epsilon$, which implies the bound (4.36).

4.5* The Boundary Aware Process at $p = 2$ and Brownian Trajectories

Similarly to Sects. 2.7* and 3.6*, in this section we compare the Tug-of-War game process (4.17) corresponding to $p = 2$ (in the limiting sense) with a discrete realization of the Brownian motion, specified below.

We will use notation of Sect. 2.7*, for $(\bar{\Omega}, \bar{\mathcal{F}}, \bar{\mathbb{P}}) = (\Omega_{\mathcal{B}}, \mathcal{F}_{\mathcal{B}}, \mathbb{P}_{\mathcal{B}}) \times (\Omega, \mathcal{F}, \mathbb{P})$ being the product probability space where $(\Omega, \mathcal{F}, \mathbb{P})$ denotes the space in Sect. 4.3, and for the filtration $\{\bar{\mathcal{F}}_t\}_{t\geq 0}$. We also refer to the standard Brownian motion $\{\mathcal{B}_t^N\}_{t\geq 0}$ discussed in Appendix B. For $\mathcal{D} \subset \mathbb{R}^N$ open, bounded, connected, and given a starting position $x_0 \in \mathcal{D}$, we have the exit time:

$$\bar{\tau}(\omega_{\mathcal{B}}) = \min\big\{t \geq 0;\ x_0 + \mathcal{B}_t^N(\omega_{\mathcal{B}}) \in \partial\mathcal{D}\big\},$$

and for each fixed $\epsilon \in (0, 1)$ we set, as in (3.35):

$$\begin{aligned} &\bar{\tau}_0 = 0,\\ &\bar{\tau}_{n+1}\big(\omega_{\mathcal{B}}, \{w_i\}_{i=1}^{\infty}\big) = \min\Big\{t \geq \bar{\tau}_n;\ \big|\mathcal{B}_t^N(\omega_{\mathcal{B}}) - \mathcal{B}_{\bar{\tau}_n(\omega_{\mathcal{B}},\omega)}^N(\omega_{\mathcal{B}})\big| = \epsilon|w_{n+1}|\Big\}. \end{aligned} \tag{4.37}$$

Recall now the setting of the boundary aware Tug-of-War in Sect. 4.3, where the relevant stopping time is:

$$\tau_2^{\epsilon,x_0} = \min\big\{n \geq 1;\ b_n > d_\epsilon(X_{n-1}^{\epsilon,x_0})\big\},$$

and where (3.36) is replaced by:

$$u^\epsilon(x_0) = \int_\Omega F \circ X_{\tau_2-1}^{\epsilon,x_0}\,\mathrm{d}\mathbb{P}. \tag{4.38}$$

The following is the main result of this section:

Theorem 4.15 *In the above context, $\{u^\epsilon\}_{\epsilon\to 0}$ in (4.38) converge pointwise on $\mathcal{D}$, to the harmonic extension of $F_{|\partial\mathcal{D}}$ in (3.37). When $\mathcal{D}$ is regular according to (3.44), namely when every $y_0 \in \partial\mathcal{D}$ satisfies:*

$$\mathbb{P}_{\mathcal{B}}\Big(\inf\big\{t > 0;\ y_0 + \mathcal{B}_t^N \notin \mathcal{D}\big\} = 0\Big) = 1, \tag{4.39}$$

then convergence of $\{u^\epsilon\}_{\epsilon\to 0}$ to u is uniform.

The below given proof is more involved than that in Sect. 3.6*, as the stopping positions X_{τ_2-1} may well occur outside of $\mathcal{D}$ or, more generally, after the Brownian path has exited $\mathcal{D}$. In particular, it uses the classical result that the boundary points y_0 not satisfying (4.39) are polar for the Brownian motion (see Theorem 8.13 in Mörters and Peres 2010):

$$\forall x \in \mathbb{R}^N \quad \mathbb{P}_{\mathcal{B}}\Big(\exists t > 0;\ x + \mathcal{B}_t^N \in \partial\mathcal{D} \text{ where (4.39) does not hold}\Big) = 0. \tag{4.40}$$

We will also necessitate an elementary bound on τ_2 when close to the boundary:

Lemma 4.16 *There exists a constant $c_N > 0$ depending only on the dimension N, such that the following bound is valid for the stopping of the discrete process $\{X_n^{\epsilon,x_0}\}_{n=0}^\infty$:*

$$\mathbb{P}\big(\tau_2^{\epsilon,x_0} - 1 \leq 2\big) \geq c_N \quad \textit{for all } x_0 \in \mathcal{D} \textit{ with } \operatorname{dist}(x_0, \partial\mathcal{D}) \leq 2\epsilon.$$

Proof Let $x_0 \in \partial\mathcal{D} + B_{2\epsilon}(0)$ and let $y_0 \in \partial\mathcal{D}$ satisfy $|x_0 - y_0| < 2\epsilon$. Then, for all $a_1, a_2 \in A := B_{1/7}\big(\frac{y_0 - x_0}{3\epsilon}\big) \subset B_1(0)$, there holds:

$$\Big|a_1 + a_2 - \frac{y_0 - x_0}{\epsilon}\Big| \leq \frac{y_0 - x_0}{3\epsilon} + \frac{2}{7} < \frac{20}{21}.$$

Hence, for every $x_1 \in \epsilon A + x_0$ and $x_2 \in \epsilon A + x_1$ we have:

$$1 - d_\epsilon(x_2) \geq 1 - \frac{1}{\epsilon}\operatorname{dist}(x_2, \mathbb{R}^N \setminus \mathcal{D}) \geq 1 - \frac{1}{\epsilon}|x_2 - y_0| \geq \frac{1}{21}.$$

Consequently:

$$\mathbb{P}(\tau_2 \leq 3) \geq \mathbb{P}\big(b_3 > d_\epsilon(X_2^{\epsilon,x_0})\big) \geq \frac{1}{21}\mathbb{P}\Big(\{X_1^{\epsilon,x_0} \in \epsilon A + x_0\} \cap \{X_2^{\epsilon,x_0} \in \epsilon A + X_1\}\Big)$$
$$= \frac{1}{21}\Big(\frac{|A|}{|B_1(0)|}\Big)^2 = c_N,$$

as claimed. □

Proof of Theorem 4.15

1. This first step serves as a preparation for the main argument of the proof. For each $x_0 \in \mathcal{D}$ and $\epsilon \in (0,1)$, define the following sequence of stopping times on $(\Omega_{\mathcal{B}}, \mathcal{F}_{\mathcal{B}}, \mathbb{P}_{\mathcal{B}})$ for $\{\mathcal{B}_t^N\}_{t\geq 0}$:

$$\begin{aligned} &\lambda_0^{\epsilon,x_0} = \bar{\tau}, \qquad \lambda_{n+1}^{\epsilon,x_0} = \min\big\{t \geq \bar{\lambda}_n^{\epsilon,x_0};\ x_0 + \mathcal{B}_t^N \not\in \mathcal{D}\big\},\\ &\text{where } \bar{\lambda}_n^{\epsilon,x_0} = \min\big\{t > \lambda_n^{\epsilon,x_0};\ |\mathcal{B}_t^N - \mathcal{B}^N_{\lambda_n^{\epsilon,x_0}}| \geq 2\epsilon\big\}. \end{aligned} \tag{4.41}$$

It is straightforward that $\{\lambda_n\}_{n=0}^\infty$ and $\{\bar{\lambda}_n\}_{n=0}^\infty$ above are indeed $\mathbb{P}_{\mathcal{B}}$-a.s. finite stopping times. For all $\delta > 2\epsilon$ consider now the events (that are decreasing as n increases):

$$A_n^\epsilon(\delta) = \big\{\lambda_n^{\epsilon,x_0} < \tau_\delta\big\}, \quad \text{where } \tau_\delta = \min\big\{t > \bar{\tau};\ |\mathcal{B}_t^N - \mathcal{B}_{\bar{\tau}}^N| \geq \delta\big\}.$$

We claim that:

$$\forall \eta > 0,\ n \geq 1,\ \delta > 0 \quad \exists \hat{\epsilon} > 0 \quad \forall \epsilon \in (0, \hat{\epsilon}) \qquad \mathbb{P}_{\mathcal{B}}\big(A_n^\epsilon(\delta)\big) \geq 1 - \eta. \tag{4.42}$$

To show the statement in (4.42), observe that:

$$\begin{aligned} A_n^\epsilon(\delta) \supset C_n^\epsilon(\delta) := \Big\{&\exists \bar{\tau} = t_0 < t_1 < \ldots < t_n < \tau_\delta;\\ &|\mathcal{B}^N_{t_{i+1}} - \mathcal{B}^N_{t_i}| > 4\epsilon \text{ and}\\ &x_0 + \mathcal{B}^N_{t_{i+1}} \not\in \bar{\mathcal{D}} + \bar{B}_\epsilon(0)\ \ \forall i = 0\ldots n-1\Big\}. \end{aligned} \tag{4.43}$$

The inclusion above follows by checking inductively that for every $\omega_B \in C_n^\epsilon(\delta)$ there holds: $\lambda_j^{\epsilon,x_0}(\omega_{\mathcal{B}}) \leq t_j$ for $j = 0\ldots n$ (see Exercise 4.17). Also, $\{C_k^\epsilon(\delta) \in \mathcal{F}\}_{\epsilon\to 0}$ is increasing and there holds:

$$\bigcup_{\epsilon>0} C_n^\epsilon(\delta) \supset \Big\{x_0 + \mathcal{B}_{\bar{\tau}}^N \quad \text{satisfies (4.39)}\Big\}.$$

In view of (4.40), the event in the right-hand side has probability 1. Thus $\mathbb{P}_{\mathcal{B}}(C_n^\epsilon(\delta)) \geq 1 - \eta$ if only $\epsilon \in (0, \hat{\epsilon})$ is small enough (relative to the fixed parameters η, δ and n).

2. Fix $x_0 \in \mathcal{D}$, $\epsilon \in (0, 1)$ and define the stopping time on $(\bar{\Omega}, \bar{\mathcal{F}}, \bar{\mathbb{P}})$ by referring to the discrete stopping times in (4.37):

$$T_2^{\epsilon,x_0} = \min\big\{\bar{\tau}_n \geq 0;\ b_{n+1} > d_\epsilon(x_0 + \mathcal{B}^N_{\bar{\tau}_n})\big\}.$$

As in Step 1 of proof of Theorem 3.19, it follows that for every $f \in C_c(\mathbb{R}^N)$:

$$\int_{\bar{\Omega}} f \circ (x_0 + \mathcal{B}^N_{T_2})\, \mathrm{d}\bar{\mathbb{P}} = \int_{\Omega} f \circ X^{\epsilon,x_0}_{\tau_2 - 1}\, \mathrm{d}\mathbb{P}. \tag{4.44}$$

The proof of Theorem 4.15 will be based on the observation that:

$$\forall \eta, \delta > 0 \quad \exists \hat{\epsilon} > 0 \quad \forall \epsilon \in (0, \hat{\epsilon}) \qquad \bar{\mathbb{P}}\big(T_2^{\epsilon,x_0} < \tau_\delta\big) \geq 1 - 2\eta, \tag{4.45}$$

where we recall that $\tau_\delta = \min\big\{t > \bar{\tau};\ |\mathcal{B}^N_t - \mathcal{B}^N_{\bar{\tau}}| \geq \delta\big\}$. Consider the following sequence of random variables, via the discrete stopping times $\{\bar{\tau}_n\}_{n=0}^\infty$ in (4.37) and the stopping times $\{\lambda_n\}_{n=0}^\infty$ in (4.41):

$$\mu_n^{\epsilon,x_0}(\omega_{\mathcal{B}}, \omega) = \min\Big\{\bar{\tau}_j;\ \bar{\tau}_j(\omega_{\mathcal{B}}, \omega) \geq \lambda_n^{\epsilon,x_0}(\omega_{\mathcal{B}})\Big\}. \tag{4.46}$$

As in the proof of Theorem 3.19, it is elementary to check that $\{\mu_n^{\epsilon,x_0}\}_{n=0}^\infty$ are a.s. well defined stopping times on $(\bar{\Omega}, \bar{\mathcal{F}}, \bar{\mathbb{P}})$ for $\{\mathcal{B}^N_t\}_{t\geq 0}$. We also note that:

$$\big|\mathcal{B}^N_{\mu_n} - \mathcal{B}^N_{\lambda_{i_n}}\big| < 2\epsilon \qquad \text{for all } n \geq 0,\ \bar{\mathbb{P}} - \text{a.s.}, \tag{4.47}$$

where by $\{i_n\}_{n=0}^\infty$ we denoted the increasing sequence of indices satisfying $\mu_n^{\epsilon,x_0} = \bar{\tau}_{i_n}$. There clearly holds: $A_n^\epsilon(\delta) \subset \{\mu_{n-1}^{\epsilon,x_0} < \tau_\delta\}$ for all $k \geq 1$.

We are ready to conclude (4.45). Since $i_{3n} + 2 < i_{3(n+1)}$ by construction in (4.46), it follows that choosing $n > 1$ large enough for $(1 - c_N)^n < \eta$ and further choosing $\hat{\epsilon} > 0$ so that (4.42) holds with n replaced by $3n + 1$, we get for all $\epsilon \in (0, \hat{\epsilon})$:

$$\begin{aligned}\bar{\mathbb{P}}\big(T_2^{\epsilon,x_0} \geq \tau_\delta\big) &\leq \mathbb{P}_{\mathcal{B}}\big(\Omega_{\mathcal{B}} \setminus A^\epsilon_{3n+1}(\delta)\big) \\ &\quad + \bar{\mathbb{P}}\Big(\bigcup_{k=0}^{n} \{\bar{\tau}_{i_{3k}} < T_2^{\epsilon,x_0}\} \cup \{\bar{\tau}_{i_{3k}+1} < T_2^{\epsilon,x_0}\} \cup \{\bar{\tau}_{i_{3k}+2} < T_2^{\epsilon,x_0}\}\Big) \\ &\leq \eta + (1 - c_N)^n \leq 2\eta,\end{aligned}$$

where we have used the strong Markov property and Lemma 4.16 to bound the second probability term in the above sum.

3. To deduce the pointwise convergence of $\{u^\epsilon(x_0)\}_{\epsilon\to 0}$, fix $\delta_0 > 0$ and let $\delta > 0$ be given by the uniform continuity of F, so that:

$$\forall x, y \in \mathbb{R}^N \qquad |x - y| < \delta \Rightarrow |F(x) - F(y)| \leq \frac{\delta_0}{3}. \tag{4.48}$$

We analyse three events in the equality: $\{T_2^{\epsilon,x_0} \geq \tau_\delta\} \cup \{\bar{\tau} < T_2^{\epsilon,x_0} < \tau_\delta\} \cup \{T_2^{\epsilon,x_0} \leq \bar{\tau}\} = \bar{\Omega}$, that is valid up to sets of $\bar{\mathbb{P}}$- measure 0. By (4.44) we obtain:

$$\begin{aligned} |u^\epsilon(x_0) - u(x_0)| &= \Big| \int_\Omega F \circ X^{\epsilon,x_0}_{\tau_2 - 1}\, \mathrm{d}\mathbb{P} - \int_{\Omega_B} F \circ \big(x_0 + \mathcal{B}^N_{\bar{\tau}}\big)\, \mathrm{d}\mathbb{P}_{\mathcal{B}} \Big| \\ &\leq \int_{\bar{\Omega}} \big|F \circ (x_0 + \mathcal{B}^N_{T_2}) - F \circ \big(x_0 + \mathcal{B}^N_{\bar{\tau}}\big)\big|\, \mathrm{d}\bar{\mathbb{P}} \\ &\leq 2\|F\|_\infty \cdot \bar{\mathbb{P}}\big(T_2^{\epsilon,x_0} \geq \tau_\delta\big) \\ &\quad + \int_{\{\bar{\tau} < T_2^{\epsilon,x_0} < \tau_\delta\}} \big|F \circ (x_0 + \mathcal{B}^N_{T_2}) - F \circ \big(x_0 + \mathcal{B}^N_{\bar{\tau}}\big)\big|\, \mathrm{d}\bar{\mathbb{P}} \\ &\quad + \int_{\{T_2^{\epsilon,x_0} \leq \bar{\tau}\}} \big|F \circ (x_0 + \mathcal{B}^N_{T_2}) - F \circ \big(x_0 + \mathcal{B}^N_{\bar{\tau}}\big)\big|\, \mathrm{d}\bar{\mathbb{P}}. \end{aligned} \tag{4.49}$$

Applying (4.45) with $\eta = \frac{\delta_0}{12\|F\|_\infty + 1}$, we estimate the first term in the right-hand side above by $\frac{\delta_0}{3}$, for all $\epsilon \in (0, \hat{\epsilon})$. On the other hand, the second term is bounded by $\frac{\delta_0}{3} \cdot \bar{\mathbb{P}}\big(T_2^{\epsilon,x_0} < \tau_\delta\big) \leq \frac{\delta_0}{3}$, in view of (4.48). Finally, we write the third term as:

$$\begin{aligned} &\int_{\{T_2^{\epsilon,x_0} \leq \bar{\tau}\}} \big|F \circ (x_0 + \mathcal{B}^N_{T_2}) - F \circ \big(x_0 + \mathcal{B}^N_{\bar{\tau}}\big)\big|\, \mathrm{d}\bar{\mathbb{P}} \\ &\qquad = \int_{\bar{\Omega}} \big|F \circ (x_0 + \mathcal{B}^N_{T_2^{\epsilon,x_0} \wedge \bar{\tau}}) - F \circ \big(x_0 + \mathcal{B}^N_{\bar{\tau}}\big)\big|\, \mathrm{d}\bar{\mathbb{P}}. \end{aligned}$$

Similarly to Step 2 of the proof of Theorem 3.19, we note that the family of stopping times $\{T_2^{\epsilon,x_0} \wedge \bar{\tau}\}_{\epsilon\to 0}$ converges to $\bar{\tau}$, implying that the term above is bounded by $\frac{\delta_0}{3}$ for all $\epsilon \in (0, \hat{\hat{\epsilon}})$ is small enough. In conclusion, (4.49) becomes:

$$|u^\epsilon(x_0) - u(x_0)| \leq \delta_0 \qquad \text{for all } \epsilon \in (0, \hat{\epsilon} \wedge \hat{\hat{\epsilon}}),$$

proving the claim.

4. To show the uniform convergence in presence of (4.39), we first recall that by Theorem 3.21 in Chap. 3, a stronger condition (3.45) actually holds for every $y_0 \in \partial\mathcal{D}$. Due to compactness of $\partial\mathcal{D}$, this condition in automatically uniform in

y_0, to the effect that:

$$\begin{aligned} &\forall \eta, \delta > 0 \quad \exists \hat\delta \in (0, \delta) \quad \forall y_0 \in \partial\mathcal{D}, \quad x_0 \in B_{\hat\delta}(y_0) \cap \mathcal{D} \\ &\qquad \mathbb{P}_{\mathcal{B}}\big(\bar\tau < \min\{t \geq 0;\ x_0 + \mathcal{B}_t^N \not\in B_\delta(y_0)\}\big) \geq 1 - \eta. \end{aligned} \tag{4.50}$$

Fix $\delta_0 > 0$, let δ be as in (4.48) and let $\hat\delta \in (0, \delta)$ be as in (4.50) applied to δ and $\eta = \frac{\delta_0}{18\|F\|_\infty + 1}$. Denote $\xi^{\epsilon, x_0} = \min\big\{t \geq 0;\ \mathrm{dist}(x_0 + \mathcal{B}_t^N, \partial\mathcal{D}) \leq \frac{\hat\delta}{2}\big\}$. Then the strong Markov property yields that for every $x_0 \in \mathcal{D}$ we have:

$$\begin{aligned} &\int_{\bar\Omega} \big|F \circ (x_0 + \mathcal{B}^N_{T_2^{\epsilon, x_0}}) - F \circ \big(x_0 + \mathcal{B}^N_{\bar\tau}\big)\big| \,\mathrm{d}\bar{\mathbb{P}} \\ &\qquad = \int_\Omega \int_{\bar\Omega} \big|F \circ \big(x_0 + \mathcal{B}^N_{T_2^{\epsilon, x_0}}(\omega) + \mathcal{B}^N_{T_2^{\epsilon, x_0 + T^{\epsilon, x_0}(\omega)}}(\bar\omega)\big) \\ &\qquad\qquad - F \circ \big(x_0 + \mathcal{B}^N_{T_2^{\epsilon, x_0}}(\omega) + \mathcal{B}^N_{\bar\tau^{x_0 + T^{\epsilon, x_0}(\omega)}}(\bar\omega)\big)\big| \,\mathrm{d}\bar{\mathbb{P}}\,\mathrm{d}\mathbb{P}_{\mathcal{B}}. \end{aligned}$$

To estimate the internal integral above, for $x \in \mathcal{D}$ satisfying: $\mathrm{dist}(x, \partial\mathcal{D}) \leq \frac{\hat\delta}{2}$, let $y_0 \in \partial\mathcal{D}$ be such that $|x - y_0| \leq \frac{\hat\delta}{2}$. As in (4.49) and invoking (4.50), we get:

$$\begin{aligned} &\int_{\bar\Omega} \big|F \circ (x + \mathcal{B}^N_{T_2^{\epsilon, x}}) - F \circ \big(x + \mathcal{B}^N_{\bar\tau^x}\big)\big| \,\mathrm{d}\bar{\mathbb{P}} \\ &\quad \leq 2\|F\|_\infty \bar{\mathbb{P}}\big(T_2^{\epsilon, x} \geq \tau_\delta\big) \\ &\qquad + 2\|F\|_\infty \cdot \bar{\mathbb{P}}\big(\{T_2^{\epsilon, x} \leq \bar\tau^x\} \cap \big\{\bar\tau^x \geq \min\{t \geq 0;\ x + \mathcal{B}_t^N \not\in B_\delta(y_0)\big\}\big) \\ &\qquad + \int_{\{< T_2^{\epsilon, x_0} \in B_{2\epsilon}(x + \mathcal{B}^N_{\bar\tau^x})\}} \big|F \circ (x + \mathcal{B}^N_{T_2^{\epsilon, x}}) - F \circ \big(x + \mathcal{B}^N_{\bar\tau^x}\big)\big| \,\mathrm{d}\bar{\mathbb{P}} \\ &\quad \leq 2\|F\|_\infty \cdot 3\eta + \frac{2\delta_0}{3} \leq \delta_0, \end{aligned}$$

where we have used (4.45) and Exercise 4.17. This ends the proof. □

Exercise 4.17 Prove the inclusion in (4.43). Show that each event $C_k^\epsilon(\delta) \in \mathcal{F}$ has its defining property in (4.43) valid for all starting positions in a small neighbourhood of a given x_0. Deduce that for all $\eta, \delta > 0$ and $n \geq 1$ there exists $\hat\epsilon > 0$ such that for all $\epsilon \in (0, \hat\epsilon)$ and all $x_0 \in \bar{\mathcal{D}}$, there holds: $\mathbb{P}_{\mathcal{B}}\big(A_n^\epsilon(\delta)\big) \geq 1 - \eta$.

Exercise 4.18 Modify the arguments in Sect. 3.7 to show that (4.39) is equivalent to the following game-regularity condition, at $y_0 \in \partial\mathcal{D}$:

$$\begin{aligned} &\forall \eta, \delta > 0 \quad \exists \hat\delta \in (0, \delta) \quad \forall x_0 \in B_{\hat\delta}(y_0) \cap \mathcal{D} \\ &\qquad \mathbb{P}\big(X^{\epsilon, x_0}_{\tau_2 - 1} \in B_\delta(y_0)\big) \geq 1 - \eta. \end{aligned} \tag{4.51}$$

4.6 Bibliographical Notes

Paper by Manfredi et al. (2012b) analysed the Tug-of-War game modelled on (3.13) for $p \geq 2$ and proved the uniform convergence of the game values to the unique viscosity solution of the associated Dirichlet problem in domains satisfying the external cone condition. The inhomogeneous case and the game with running pay-off has been treated in Ruosteenoja (2014).

In Arroyo et al. (2017) an alternative dynamic programming principle, valid for all continuous distributions of exponents $p(x) \in (1, \infty)$ that are bounded away from 1 and ∞, has been introduced. It involved averaging on codimension 1 sets, and convergence of its solutions has been shown in domains $\mathcal{D}$ satisfying the exterior corkscrew condition. For p that is Hölder, Arroyo et al. (2018) showed the asymptotic local Lipschiz continuity of the game values $\{u_\epsilon\}_{\epsilon\to 0}$. Paper by Luiro and Parviainen (2018) concerned more general averaging principles and the asymptotic Hölder regularity of their solutions.

The averaging principle involving randomized radius of sampling was analysed in Attouchi et al. (2019), where the associated game values were shown to possess a gradient that weakly converges to the gradient of the p-harmonic limiting solution.

Convergence of game values related to the parabolic problem $u_t = |\nabla u|^{2-p}\Delta_p u$ has been shown in Manfredi et al. (2010) and in Parviainen and Ruosteenoja (2016) for the varying exponent $p(x)$.

A version of the averaging principle and the Tug-of-War game converging to the viscosity solutions of the obstacle problem for infinity-Laplacian was analysed in Manfredi et al. (2015), for p-Laplacian and the Tug-of-War with noise in Lewicka and Manfredi (2014), and for the double obstacle problem with $p \geq 2$ in Codenotti et al. (2017). Further, in Lewicka et al. (2019) the averaging principles and games for p-Laplacian in the Heisenberg group were developed.

The paper by Blanc et al. (2018) studied a related optimal control problem (a "one-player game") in the description of the viscosity solutions to the Dirichlet problem for the maximal Pucci operator: $\Lambda \sum \lambda^+ + \lambda \sum \lambda^- = f$, where λ-s stand for the eigenvalues of $\nabla^2 u$. In Blanc and Rossi (2018), a game-theoretical interpretation of viscosity solutions to the Dirichlet problem associated with the following j-th eigenvalue problem: $\lambda_j(\nabla^2 u) = 0$ has been developed, based on the Rayleigh quotient formula. A nonlocal Tug-of-War related to the infinity fractional Laplacian has been indicated in Bjorland et al. (2012).

We also mention that a paper by del Teso et al. (2018) shows convergence of the dynamic programming principles for the p-Laplacian, without using probability arguments. In Luiro et al. (2013) Harnack inequality has been shown for the game based on (3.13) and $p > 2$, and in Parviainen and Ruosteenoja (2016) for the time-dependent game informing the parabolic problem with variable exponents $p(x)$.

Chapter 5
Game-Regularity and Convergence: Case $p \in (2, \infty)$

In this chapter we study the question of convergence of solutions $\{u_\epsilon\}_{\epsilon\to 0}$ to the *mean value equation* (4.1), discussed in Chap. 4. Based on the analysis in Sects. 2.7*, 3.6* and 4.5* valid in the linear case $p = 2$, it is similarly expected that the family $\{u_\epsilon\}_{\epsilon\to 0}$ or, equivalently, the family of the boundary aware Tug-of-War game values $\{u^\epsilon\}_{\epsilon\to 0}$ converges pointwise in $\mathcal{D}$ to the Perron solution of the Dirichlet problem: $\Delta_p u = 0$ in $\mathcal{D}$, $u = F$ on $\partial\mathcal{D}$ (see Definition C.51 in Appendix C), for any continuous boundary data F. At the same time, this convergence should be uniform for regular boundary, implying then consistency with boundary data, in the sense that: $u_{|\partial\mathcal{D}} = F$. Since the former result is not yet available at the time when these Course Notes are written, we will directly concentrate on the latter one and on the regular case.

In Sect. 5.1, we show that any uniform limit of u_ϵ-s as $\epsilon \to 0$ must be the *viscosity solution* of the aforementioned Dirichlet problem. There, we also briefly recall the definition of viscosity solutions, which is the notion fitted to studying a larger class of fully nonlinear PDEs, including equations in non-divergence form (see Appendix C). Since the family $\{u_\epsilon\}_{\epsilon\in(0,1)}$ is automatically equibounded (by $\|F\|_{\mathcal{C}^0}$), proving the uniform convergence of its subsequence is equivalent to proving its equicontinuity, in virtue of the Ascoli–Arzelá theorem. We further observe that this last condition is implied by a seemingly weaker requirement of equicontinuity close to the boundary points. In Sect. 5.2 we introduce the notion of *game-regularity*, which is essentially equivalent to such local equicontinuity at the boundary. This notion extends the walk-regularity studied in Chap. 2, and both notions may be seen as natural extensions of Doob's regularity for Brownian motion (see Sect. 3.7*), in the discrete setting. Sections 5.3 and 5.4 serve as a technical preparation for the subsequent justification of two sufficient conditions for game-regularity, which are: the *exterior cone condition* in Sect. 5.5 (a result directly extending the statement in case $p = 2$ of Chap. 2), and $p > N$ in Sect. 5.6.

The arguments towards both these results are essentially taken from the seminal paper by Peres and Sheffield (2008). Here, we significantly expanded the discussion

M. Lewicka, *A Course on Tug-of-War Games with Random Noise*, Universitext,
https://doi.org/10.1007/978-3-030-46209-3_5

and carefully provided all the previously omitted details, for the benefit of the reader with limited probability-oriented background.

5.1 Convergence to p-Harmonic Functions

Recall first the definition of viscosity solution to the p-Laplace equation, studied in more detail in Sect. C.8 of Appendix C.

Definition 5.1 Let $\mathcal{D} \subset \mathbb{R}^N$ be an open, bounded set and let $F : \partial\mathcal{D} \to \mathbb{R}$ be a continuous function. We say that a continuous $u : \bar{\mathcal{D}} \to \mathbb{R}$ is a *viscosity solution* to the following problem in $\bar{\mathcal{D}}$:

$$\Delta_p u = 0 \text{ in } \mathcal{D} \qquad \text{and} \qquad u = F \text{ on } \partial\mathcal{D}, \tag{5.1}$$

if $u = F$ on $\partial\mathcal{D}$ and:

(i) for every $x_0 \in \mathcal{D}$ and every $\phi \in C^2(\bar{\mathcal{D}})$ such that:

$$\phi(x_0) = u(x_0), \quad \phi < u \text{ in } \bar{\mathcal{D}} \setminus \{x_0\} \quad \text{and} \quad \nabla\phi(x_0) \neq 0, \tag{5.2}$$

there holds: $\Delta_p \phi(x_0) \leq 0$,

(ii) for every $x_0 \in \mathcal{D}$ and every $\phi \in C^2(\bar{\mathcal{D}})$ such that:

$$\phi(x_0) = u(x_0), \quad \phi > u \text{ in } \bar{\mathcal{D}} \setminus \{x_0\} \quad \text{and} \quad \nabla\phi(x_0) \neq 0, \tag{5.3}$$

there holds: $\Delta_p \phi(x_0) \geq 0$.

We show that any uniform limit of solutions to the dynamic programming principle studied in the previous chapter is automatically a viscosity solution as above:

Theorem 5.2 *Let $\mathcal{D}$, Γ_{out}, $\mathcal{D}^\diamond$ be as in Definition 3.8, and let $F : \Gamma_\epsilon \cup \Gamma_{out} \to \mathbb{R}$ be a given continuous function. Fix $p > 2$ and recall that:*

$$r = \sqrt{\frac{3(p-2)}{2(N+2)}}.$$

Assume that $J \subset (0, 1)$ is a sequence decreasing to 0, such that the sequence $\{u_\epsilon\}_{\epsilon \in J}$ of solutions to (4.1) converges uniformly as $\epsilon \to 0, \epsilon \in J$ to some limit $u : \mathcal{D}^\diamond \to \mathbb{R}$. Then u must be the viscosity solution to (5.1) in $\bar{\mathcal{D}}$.

Proof

1. Clearly, $u = F$ on $\partial\mathcal{D}$ because $u_\epsilon = F$ on Γ_{out} for each ϵ. Recall also that the functions u_ϵ are continuous, in view of Theorem 4.1.
   Fix $x_0 \in \mathcal{D}$ and let ϕ be a test function as in (5.2). We first claim that there exists a sequence $\{x_\epsilon\}_{\epsilon\in J} \in \mathcal{D}$, such that:
$$\lim_{\epsilon\to 0, \epsilon\in J} x_\epsilon = x_0 \quad \text{and} \quad u_\epsilon(x_\epsilon) - \phi(x_\epsilon) = \min_{\bar{\mathcal{D}}} (u_\epsilon - \phi). \tag{5.4}$$
   To prove the above, for every $j \in \mathbb{N}$ define $a_j > 0$ and $\epsilon_j > 0$ such that:
$$a_j = \min_{\bar{\mathcal{D}}\setminus B_{1/j}(x_0)} (u - \phi) \qquad \text{and} \qquad \|u_\epsilon - u\|_{C^0(\bar{\mathcal{D}})} \leq \frac{1}{2}a_j \quad \text{for all } \epsilon \leq \epsilon_j.$$
   Without loss of generality, the sequence $\{\epsilon_j\}_{j=1}^\infty$ is decreasing to 0 as $j \to \infty$. Now, for $\epsilon \in (\epsilon_{j+1}, \epsilon_j] \cap J$, let $x_\epsilon \in \bar{B}_{1/j}(x_0)$ satisfy:
$$u_\epsilon(x_\epsilon) - \phi(x_\epsilon) = \min_{\bar{B}_{1/j}(x_0)} (u_\epsilon - \phi).$$
   Observing that the following bound is valid for every $x \in \bar{\mathcal{D}} \setminus B_{1/j}(x_0)$:
$$\begin{aligned} u_\epsilon(x) - \phi(x) &\geq u(x) - \phi(x) - \|u_\epsilon - u\|_{C^0(\bar{\mathcal{D}})} \geq a_j - \frac{1}{2}a_j \geq \|u_\epsilon - u\|_{C^0(\bar{\mathcal{D}})} \\ &\geq u_\epsilon(x_0) - \phi(x_0) \geq \min_{\bar{B}_{1/j}(x_0)} (u_\epsilon - \phi), \end{aligned}$$
   proves (5.4).
2. Since by (5.4) we have: $\phi(x) \leq u_\epsilon(x) + \big(\phi(x_\epsilon) - u_\epsilon(x_\epsilon)\big)$ for all $x \in \bar{\mathcal{D}}$, it follows that:
$$\begin{aligned} &\frac{1}{3}\Big(\mathcal{A}_\epsilon\phi(x_\epsilon) + \inf_{y\in B_{r\epsilon}(x_\epsilon)} \mathcal{A}_\epsilon\phi(y) + \sup_{y\in B_{r\epsilon}(x_\epsilon)} \mathcal{A}_\epsilon\phi(y)\Big) - \phi(x_\epsilon) \\ &\quad \leq \frac{1}{3}\Big(\mathcal{A}_\epsilon u_\epsilon(x_\epsilon) + \inf_{y\in B_{r\epsilon}(x_\epsilon)} \mathcal{A}_\epsilon u_\epsilon(y) + \sup_{y\in B_{r\epsilon}(x_\epsilon)} \mathcal{A}_\epsilon u_\epsilon(y)\Big) \\ &\qquad\qquad + \big(\phi(x_\epsilon) - u_\epsilon(x_\epsilon)\big) - \phi(x_\epsilon) = 0, \end{aligned} \tag{5.5}$$
   for all ϵ small enough to guarantee that $d_\epsilon(x_\epsilon) = 1$ in (4.1). On the other hand, (3.16) in Theorem 3.5 yields:
$$\begin{aligned} &\frac{1}{3}\Big(\mathcal{A}_\epsilon\phi(x_\epsilon) + \inf_{y\in B_{r\epsilon}(x_\epsilon)} \mathcal{A}_\epsilon\phi(y) + \sup_{y\in B_{r\epsilon}(x_\epsilon)} \mathcal{A}_\epsilon\phi(y)\Big) - \phi(x_\epsilon) \\ &\qquad = \frac{\epsilon^2}{2(N+2)} \frac{1}{|\nabla\phi(x_\epsilon)|^{p-2}} \Delta_p\phi(x_\epsilon) + o(\epsilon^2), \end{aligned} \tag{5.6}$$

for ϵ small enough to get $\nabla\phi(x_\epsilon) \neq 0$. Combining (5.5) and (5.6) gives:

$$\Delta_p\phi(x_\epsilon) \leq o(1).$$

Passing to the limit with $\epsilon \to 0, \epsilon \in J$ establishes the desired inequality $\Delta_p\phi(x_0) \leq 0$ and proves part (i) of Definition 5.1. The verification of part (ii) is done similarly and the result follows by uniqueness in Corollary C.63. □

Recall that the sequence $\{u_\epsilon\}_{\epsilon\in J}$ is equicontinuous, if for every $\eta > 0$ there exists $\delta > 0$, such that for all $\epsilon \in J$:

$$|u_\epsilon(x_0) - u_\epsilon(y_0)| \leq \eta \quad \text{for all } x_0, y_0 \in \mathcal{D}^\diamond \text{ with } |x_0 - y_0| \leq \delta.$$

We now prove that, in the present context, equicontinuity is equivalent to the seemingly weaker property, i.e. equicontinuity close to the boundary points:

Theorem 5.3 *Let $\mathcal{D}$, Γ_{out} and $\mathcal{D}^\diamond$ be as in Definition 3.8, and let F, p and r, J be as in Theorem 5.2. Assume that $\{u_\epsilon\}_{\epsilon\in J}$ is a sequence of solutions to (4.1), namely:*

$$\begin{aligned} u_\epsilon(x) = d_\epsilon(x)\Big(\frac{1}{3}\mathcal{A}_\epsilon u_\epsilon(x) + \frac{1}{3}\inf_{y\in B_{r\epsilon}(x)} \mathcal{A}_\epsilon u_\epsilon(y) + \frac{1}{3}\sup_{y\in B_{r\epsilon}(x)} \mathcal{A}_\epsilon u_\epsilon(y)\Big) \\ + (1 - d_\epsilon(x))\,F(x) \qquad \text{for all } x \in \mathcal{D}^\diamond, \end{aligned}$$

satisfying: for every $\eta > 0$ there exists $\delta > 0$ such that for all $\epsilon \in J$:

$$\begin{aligned} |u_\epsilon(x_0) - u_\epsilon(y_0)| \leq \eta \quad &\text{for all } x_0 \in \bar{\mathcal{D}},\ y_0 \in \partial\mathcal{D} \\ &\text{with } |x_0 - y_0| \leq \delta. \end{aligned} \tag{5.7}$$

Then the sequence $\{u_\epsilon\}_{\epsilon\in J}$ is equicontinuous.

We will give two independent proofs of the above result: an analytical and a probabilistic one, based on the game-theoretical interpretation of solutions u_ϵ as specified in Theorem 4.12.

An Analytical Proof of Theorem 5.3

1. Fix $\eta > 0$ and let $\delta \ll 1$ be such that (5.7) holds with $\frac{\eta}{3}$ instead of η and that a similar implication is valid on Γ_{out}, in view of the uniform continuity of F. Define the new boundary region $\Gamma_{\delta/3} = \partial\mathcal{D} + \bar{B}_{\delta/3}(0) \subset \Gamma$. We conclude that for all $\epsilon \in J$:

$$|u_\epsilon(x_0) - u_\epsilon(y_0)| \leq \eta \quad \text{for all } x_0, y_0 \in \Gamma_{\delta/3} \text{ with } |x_0 - y_0| \leq \frac{\delta}{3}. \tag{5.8}$$

2. Define now a new domain $\tilde{\mathcal{D}}$ together with its boundary $\tilde{\Gamma}$:

$$\tilde{\mathcal{D}} = \Big\{z \in \mathcal{D};\ \mathrm{dist}(z, \partial\mathcal{D}) > \frac{\delta}{6}\Big\}, \qquad \tilde{\Gamma} = \partial\tilde{\mathcal{D}} + \bar{B}_{\delta/8}(0).$$

Note that for any two points x_0, y_0 and $\epsilon > 0$ which satisfy:

$$x_0, y_0 \in \tilde{\mathcal{D}} \text{ with } |x_0 - y_0| \leq \frac{\delta}{48} \quad \text{and} \quad \epsilon < \frac{\delta}{48(1+r)}, \tag{5.9}$$

it results that: $\mathrm{dist}\big(z - (x_0 - y_0), \Gamma_{out}\big) \geq \frac{\delta}{48} > \epsilon$ for every $z \in \tilde{\mathcal{D}}^{\diamondsuit} \doteq \tilde{\mathcal{D}} \cup \tilde{\Gamma}$. Consequently $d_\epsilon(z - (x_0 - y_0)) = 1$ and thus (4.1) yields:

$$\tilde{u}_\epsilon(z) = \frac{1}{3}\Big(\mathcal{A}_\epsilon\tilde{u}_\epsilon(z) + \inf_{y\in B_{r\epsilon}(z)} \mathcal{A}_\epsilon\tilde{u}_\epsilon(y) + \sup_{y\in B_{r\epsilon}(z)} \mathcal{A}_\epsilon\tilde{u}_\epsilon(y)\Big),$$

where we define a translated copy of each u_ϵ by:

$$\tilde{u}_\epsilon(z) = u_\epsilon\big(z - (x_0 - y_0)\big) + \eta.$$

Calling $\tilde{d}_\epsilon(z) = \frac{1}{\epsilon}\min\{\epsilon, \mathrm{dist}(z, \tilde{\Gamma} \setminus \tilde{\mathcal{D}})\}$, we thus observe:

$$\begin{aligned}\tilde{u}_\epsilon(z) = \tilde{d}_\epsilon(z)\Big(\frac{1}{3}\mathcal{A}_\epsilon\tilde{u}_\epsilon(z) + \frac{1}{3}\inf_{y\in B_{r\epsilon}(z)} \mathcal{A}_\epsilon\tilde{u}_\epsilon(y) + \frac{1}{3}\sup_{y\in B_{r\epsilon}(z)} \mathcal{A}_\epsilon\tilde{u}_\epsilon(y)\Big) \\ + \big(1 - \tilde{d}_\epsilon(z)\big)\tilde{u}_\epsilon(z) \qquad \text{for all } z \in \tilde{\mathcal{D}}^{\diamondsuit}.\end{aligned} \tag{5.10}$$

On the other hand, u_ϵ similarly solves the same problem (5.10) in $\tilde{\mathcal{D}}^{\diamondsuit}$, subject to its own boundary data u_ϵ on $\tilde{\Gamma}$. Note now that $\tilde{u}_\epsilon \geq u_\epsilon$ in $\tilde{\Gamma}$, because:

$$\tilde{u}_\epsilon(z) - u_\epsilon(z) = u_\epsilon(z - (x_0 - y_0)) - u_\epsilon(z) + \eta \geq -\eta + \eta = 0,$$

where we used (5.8) in view of: $z, z-(x_0-y_0) \in \Gamma_{\delta/3}$ and $|(z-(x_0-y_0))-z| \leq \frac{\delta}{3}$.

Consequently, by the comparison result in Corollary 4.2, it follows that $\tilde{u}_\epsilon \geq u_\epsilon$ in $\tilde{\mathcal{D}}^{\diamondsuit}$. In particular, we get:

$$u_\epsilon(x_0) - u_\epsilon(y_0) = u_\epsilon(x_0) - \tilde{u}_\epsilon(x_0) + \eta \leq \eta.$$

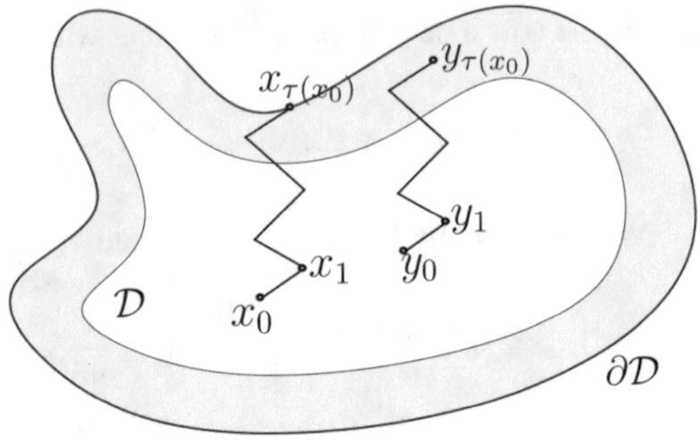

Fig. 5.1 The "mirror strategies" shift the token in the x_0 and y_0 games in parallel

3. Exchanging x_0 with y_0, the same argument as above yields: $|u_\epsilon(x_0) - u_\epsilon(y_0)| \leq \eta$ for all x_0, y_0 and $\epsilon \in J$ satisfying (5.9). Recalling (5.8), we conclude that for all $\epsilon \in J$ with $\epsilon < \frac{\delta}{48(1+r)}$, there holds:

$$|u_\epsilon(x_0) - u_\epsilon(y_0)| \leq \eta \quad \text{for all } x_0, y_0 \in \mathcal{D}^\Diamond \text{ with } |x_0 - y_0| \leq \frac{\delta}{48}.$$

This establishes equicontinuity of $\{u_\epsilon\}_{\epsilon \in J}$ and proves the theorem. □

We now give an alternative proof of the fact that equicontinuity at the boundary extends to equicontinuity in the interior of the domain $\mathcal{D}$, by a translation-type argument. Given two nearby interior initial positions of the token x_0 and y_0, and given two strategies σ_I, σ_{II} of the two players for the game starting at x_0, one considers a new "translated" game, starting at y_0 and utilizing the outcomes $\{(w_i, a_i, b_i)\}_{i=1}^\infty$ of the previous game, with "mirror strategies" $\bar{\sigma}_I, \bar{\sigma}_{II}$ (Fig. 5.1). These mirror strategies are designed so that they shift the token from y_{n-1} to y_n by vector $x_n - x_{n-1}$. We proceed playing the two games in parallel, until one of the positions x_n or y_n reaches the neighbourhood of $\partial\mathcal{D}$, where the assumed boundary equicontinuity may be applied. Since $|X_n^{x_0,\sigma_I,\sigma_{II}} - X_n^{y_0,\bar{\sigma}_I,\bar{\sigma}_{II}}| = |x_0 - y_0| < \delta$ and thus, consequently, $|u_\epsilon(x_n) - u_\epsilon(y_n)| \leq \eta$, the same bound on $|u_\epsilon(x_0) - u_\epsilon(y_0)|$ follows by choosing the strategies σ_I, σ_{II} optimally.

A Game-Theoretical Proof of Theorem 5.3

1. Fixing $\eta > 0$, we may, similarly as in (5.8) in the first step of the analytical proof, deduce that there exists $\delta > 0$ such that for all $\epsilon \in J$:

$$\begin{aligned} |u_\epsilon(x_0) - u_\epsilon(y_0)| \leq \frac{\eta}{3} \quad & \text{for all } x_0, y_0 \in \big(\partial\mathcal{D} + B_\delta(0)\big) \cup \Gamma_{out} \\ & \text{with } |x_0 - y_0| < \frac{\delta}{3}. \end{aligned} \tag{5.11}$$

Assume that:

$$(1+r)\epsilon < \frac{\delta}{3} \tag{5.12}$$

Using Lemma 3.15, we find a strategy $\sigma_{0,I}$ of Player I, satisfying $\sigma_{0,I}^n(h_n) = \sigma_{0,I}^n(x_n)$ for every $h_n \in H_n$ and $n \geq 0$, together with:

$$\mathcal{A}_\epsilon u_\epsilon(x_n + \epsilon\sigma_{0,I}^n(x_n)) \geq \sup_{y \in B_{r\epsilon}(x_n)} \mathcal{A}_\epsilon u_\epsilon(y) - \frac{\eta}{3 \cdot 2^{n+1}} \qquad \text{if } x_n \in \mathcal{D}$$

$$\sigma_{0,I}^n(x_n) = x_n \qquad \text{if } x_n \notin \mathcal{D}.$$

Likewise, let $\sigma_{0,II}$ be a strategy of Player II, depending only on the current position x_n of the token in $\mathcal{D}$, and satisfying:

$$\mathcal{A}_\epsilon u_\epsilon(x_n + \epsilon\sigma_{0,II}^n(x_n)) \leq \inf_{y \in B_{r\epsilon}(x_n)} \mathcal{A}_\epsilon u_\epsilon(y) + \frac{\eta}{3 \cdot 2^{n+1}} \qquad \text{if } x_n \in \mathcal{D}$$

$$\sigma_{0,II}^n(x_n) = x_n \qquad \text{if } x_n \notin \mathcal{D}.$$

2. Let now $x_0, y_0 \in \mathcal{D}$ be such that either $\operatorname{dist}(x_0, \partial\mathcal{D}) \geq \delta$ or $\operatorname{dist}(y_0, \partial\mathcal{D}) \geq \delta$, and that $|x_0 - y_0| < \frac{\delta}{3}$. We define the "mirror strategies":

$$\bar{\sigma}_{0,I}^n(x_n) = \sigma_{0,I}^n(x_n + y_0 - x_0), \qquad \bar{\sigma}_{0,II}^n(x_n) = \sigma_{0,II}^n(x_n + x_0 - y_0),$$

and consider two sequences of random variables:

$$\{X_n = X_n^{x_0, \bar{\sigma}_{0,I}, \sigma_{0,II}}\}_{n=0}^\infty, \qquad \{Y_n = X_n^{y_0, \sigma_{0,I}, \bar{\sigma}_{0,II}}\}_{n=0}^\infty,$$

given by the rule (4.17). We also set the stopping times:

$$\tau^{x_0} = \tau^{x_0, \bar{\sigma}_{0,I}, \sigma_{0,II}}, \qquad \tau^{y_0} = \tau^{y_0, \sigma_{0,I}, \bar{\sigma}_{0,II}}$$

and recall that (see the proof of Theorem 4.12) the following sequence of random variables $\{M_n^{x_0}\}_{n=0}^\infty$ is a supermartingale with respect to the filtration $\{\mathcal{F}_n\}_{n=0}^\infty$:

$$M_n^{x_0} = (u_\epsilon \circ X_n)\mathbb{1}_{\tau^{x_0} > n} + (F \circ X_{\tau^{x_0}})\mathbb{1}_{\tau^{x_0} \leq n} + \frac{\eta}{3 \cdot 2^n},$$

whereas $\{M_n^{y_0}\}_{n=0}^\infty$ below is a submartingale:

$$M_n^{y_0} = (u_\epsilon \circ Y_n)\mathbb{1}_{\tau^{y_0} > n} + (F \circ Y_{\tau^{y_0}})\mathbb{1}_{\tau^{y_0} \leq n} - \frac{\eta}{3 \cdot 2^n}.$$

3. Let $\tau_\delta : \Omega \to \mathbb{N} \cup \{+\infty\}$ be a new stopping time, namely:

$$\begin{aligned} &\tau_\delta\big((w_1, a_1, b_1), (w_2, a_2, b_2), \ldots\big) \\ &\quad = \min\Big\{n \geq 1;\ \operatorname{dist}(x_n, \partial\mathcal{D}) < \frac{2\delta}{3} \text{ or } \operatorname{dist}(y_n, \partial\mathcal{D}) < \frac{2\delta}{3}\Big\}, \end{aligned}$$

where x_n and y_n denote the consecutive token positions in the games starting, respectively, at x_0 and y_0, that is: $x_n = X_n\big((w_1, a_1, b_1), \ldots, (w_n, a_n, b_n)\big)$ and $y_n = Y_n\big((w_1, a_1, b_1), \ldots, (w_n, a_n, b_n)\big)$. In view of (5.12), it follows that:

$$\tau_\delta < \min\{\tau^{x_0}, \tau^{y_0}\},$$

so in particular $\mathbb{P}(\tau_\delta < +\infty) = 1$. Also, directly from the definition of τ_δ and the choice of the mirror strategies, we observe:

$$\mathrm{dist}(X_{\tau_\delta}, \partial\mathcal{D}) < \delta, \quad \mathrm{dist}(Y_{\tau_\delta}, \partial\mathcal{D}) < \delta \quad \text{and} \quad X_{\tau_\delta} - Y_{\tau_\delta} = x_0 - y_0 \qquad \text{a.s.}$$

Consequently, by (5.11) we obtain:

$$|u_\epsilon \circ X_{\tau_\delta} - u_\epsilon \circ Y_{\tau_\delta}| \leq \frac{\eta}{3} \qquad \text{a.s.} \tag{5.13}$$

and further, in view of Theorem A.34 and using (5.12), we get:

$$\begin{aligned} u_\epsilon(y_0) &= \mathbb{E}[M_0^{y_0}] + \frac{\eta}{3} \leq \mathbb{E}[M_{\tau_\delta}^{y_0}] + \frac{\eta}{3} = \mathbb{E}\Big[u_\epsilon \circ Y_{\tau_\delta} - \frac{\eta}{2^{\tau_\delta}}\Big] + \frac{\eta}{3} \\ &\leq \mathbb{E}[u_\epsilon \circ Y_{\tau_\delta}] + \frac{\eta}{3} \leq \mathbb{E}\Big[u_\epsilon \circ X_{\tau_\delta} - \frac{\eta}{2^{\tau_\delta}}\Big] + \frac{2\eta}{3} = \mathbb{E}\Big[M_{\tau_\delta}^{x_0} - \frac{\eta}{2^{\tau_\delta}}\Big] + \frac{2\eta}{3} \\ &\leq \mathbb{E}[M_{\tau_\delta}^{x_0}] + \frac{2\eta}{3} \leq \mathbb{E}[M_0^{x_0}] + \frac{2\eta}{3} = u_\epsilon(x_0) + \eta. \end{aligned}$$

By a symmetric argument, it follows that $u_\epsilon(x_0) \leq u_\epsilon(y_0) + \eta$. Finally, for all $\epsilon \in J$ satisfying (5.12) there holds:

$$|u_\epsilon(x_0) - u_\epsilon(y_0)| \leq \eta \quad \text{for all } \ x_0, y_0 \in \mathcal{D}^\diamond \ \text{ with } \ |x_0 - y_0| < \frac{\delta}{3}.$$

Since the remaining $\big\{u_\epsilon;\ \epsilon \in J \text{ and } 3(1+r)\epsilon > \delta\big\}$ consists of finitely many continuous functions, it follows that $\{u_\epsilon\}_{\epsilon \in J}$ is equicontinuous, as claimed. □

Exercise 5.4

(i) In the setting of Theorem 5.2, prove that for any $x_0 \in \mathcal{D}$ and any test function $\phi \in C^2(\bar{\mathcal{D}})$ satisfying (5.3), there holds $\Delta_p \phi(x_0) \geq 0$.

(ii) Let $\mathcal{D}$, ϵ_0 and Γ_{out} be as in Definition 3.8 and let $F : \Gamma_{out} \to \mathbb{R}$ be a given continuous function. Fix $\alpha \in (0, 1]$ and $\beta = 1 - \alpha$. In the setting of Theorem 3.9, assume that a sequence $\{u_\epsilon\}_{\epsilon \in J}$ of solutions to (3.18) converges uniformly in $\mathcal{D}^\diamond$ to a continuous limit function $u : \mathcal{D}^\diamond \to \mathbb{R}$. Prove that u is a viscosity solution to the homogeneous problem (5.1) in $\bar{\mathcal{D}}$, with $p = \frac{N+2}{\alpha} - N$ (and $f = 0$).

(iii) In the same setting of Theorem 3.9, prove that the asymptotic equicontinuity near the boundary implies asymptotic equicontinuity throughout $\mathcal{D}^\diamond$ for solutions of (3.18). Namely, let $F : \Gamma_{out} \to \mathbb{R}$ be a continuous function and assume that the following holds for some sequence $\{u_\epsilon\}_{\epsilon\in J}$ that satisfies (3.18). For every $\eta > 0$ there exists $\hat{\epsilon} \in (0, \epsilon_0)$ and $\delta > 0$ such that for all $\epsilon \in (0, \hat{\epsilon}) \cap J$:

$$|u_\epsilon(x_0) - u_\epsilon(y_0)| \leq \eta \quad \text{for all } x_0 \in \bar{\mathcal{D}}, \ y_0 \in \partial\mathcal{D} \text{ with } |x_0 - y_0| \leq \delta.$$

Prove that a stronger property of the sequence $\{u_\epsilon\}_{\epsilon\in J}$ is automatically valid:

$$\left[\begin{array}{l} \text{For every } \eta > 0, \text{ there exists } \hat{\epsilon} \in (0, \epsilon_0), \ \delta > 0 \text{ such that:} \\ |u_\epsilon(x_0) - u_\epsilon(y_0)| \leq \eta \text{ for all } \epsilon \in (0, \hat{\epsilon}) \cap J \\ \qquad\qquad\qquad \text{and all } x_0, y_0 \in \mathcal{D}^\diamond \text{ with } |x_0 - y_0| \leq \delta. \end{array}\right. \tag{5.14}$$

(iv) Prove the following version of the Ascoli–Arzelà theorem for discontinuous functions. Let $\{u_\epsilon\}_{\epsilon\in J}$ be an equibounded sequence of functions $u_\epsilon : \mathcal{D}^\diamond \to \mathbb{R}$ on a compact set $\mathcal{D}^\diamond \subset \mathbb{R}^N$, and assume that the *asymptotic equicontinuity* condition (5.14) holds. Then there exists a subsequence of $\{u_\epsilon\}_{\epsilon\in J}$, converging uniformly, as $\epsilon \to 0$, to a continuous function $u : \mathcal{D}^\diamond \to \mathbb{R}$.

5.2 Game-Regularity and Convergence

In this section, we prove that the coinciding game values $u_I^\epsilon = u_{II}^\epsilon$ converge as $\epsilon \to 0$, to the unique p-harmonic function that solves the Dirichlet problem (5.1) with a given continuous boundary data F.

We now introduce the regularity condition on $\partial\mathcal{D}$, assuring such result in the framework of the Tug-of-War game corresponding to the averaging principle (4.1). A point $y_0 \in \partial\mathcal{D}$ will be called *game-regular* if, whenever the game starts near y_0, Player I has a strategy for making the game terminate near the same y_0, with high probability. More precisely:

Definition 5.5 Let $\mathcal{D}$, ϵ_0 and $\mathcal{D}^\diamond$ be as in Definition 3.8. Fix $p > 2$, $r = \sqrt{\frac{3(p-2)}{2(N+2)}}$ and for each $\epsilon \in (0, \frac{1}{1+r})$ consider the boundary-aware Tug-of-War game with noise according to (4.1), as defined in Sect. 4.3.

(i) We say that a point $y_0 \in \partial\mathcal{D}$ is *game-regular* if for every $\eta, \delta > 0$ there exist $\hat{\delta} \in (0, \delta)$ and $\hat{\epsilon} \in (0, \frac{1}{1+r})$ such that the following holds. Fix $\epsilon \in (0, \hat{\epsilon})$ and choose an initial token position $x_0 \in B_{\hat{\delta}}(y_0)$; there exists then a strategy $\sigma_{0,I}$ of Player I with the property that for every strategy σ_{II} of Player II we have:

$$\mathbb{P}\big(X_\tau \in B_\delta(y_0)\big) \geq 1 - \eta, \tag{5.15}$$

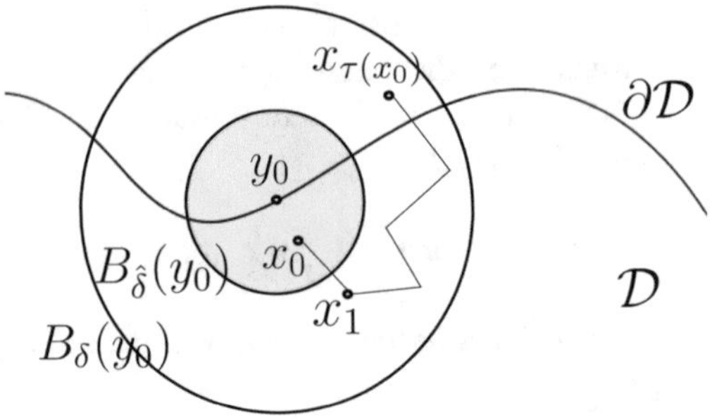

Fig. 5.2 Game-regularity of a boundary point $y_0 \in \partial\mathcal{D}$

where X_τ is the shorthand for the random variable $(X^{x_0,\sigma_{0,I},\sigma_{II}})_{\tau^{x_0,\sigma_{0,I},\sigma_{II}}}$ and where the sequence of random variables $\{X_n^{x_0,\sigma_{0,I},\sigma_{II}}\}_{n=0}^\infty$ records the positions of the token in the ϵ-game in (4.17), while $\tau = \tau^{x_0,\sigma_{0,I},\sigma_{II}}$ is the stopping time in (4.18) (Fig. 5.2).

(ii) We say that the *domain $\mathcal{D}$ is game-regular* if every boundary point $y_0 \in \partial\mathcal{D}$ is game-regular.

Observe that game-regularity is *symmetric* with respect to strategies σ_I and σ_{II}. Namely, game-regularity of $y_0 \in \partial\mathcal{D}$ is equivalent to existence of $\hat\delta \in (0, \delta)$ and $\hat\epsilon > 0$, depending on the prescribed $\eta, \delta > 0$, with the following property. For every $\epsilon \in (0, \hat\epsilon)$ and every $x_0 \in B_{\hat\delta}(y_0)$ there exists a strategy $\sigma_{0,II}$ of Player II such that for every strategy σ_I of Player I there holds:

$$\mathbb{P}\Big((X^{x_0,\sigma_I,\sigma_{0,II}})_{\tau^{x_0,\sigma_I,\sigma_{0,II}}} \in B_\delta(y_0)\Big) \geq 1 - \eta. \tag{5.16}$$

Lemma 5.6 *If $\mathcal{D}$ is game-regular, then $\hat\delta$ and $\hat\epsilon$ in Definition 5.5 (a) can be chosen independently of y_0 (i.e., they depend only on the prescribed thresholds η, δ).*

Proof For a fixed $\eta, \delta > 0$ and each $y_0 \in \partial\mathcal{D}$, choose $\hat\delta(y_0) < \frac{\delta}{2}$ and $\hat\epsilon(y_0) > 0$ sufficiently small and such that Definition 5.5 (a) holds for η and $\frac{\delta}{2}$. By compactness, the boundary $\partial\mathcal{D}$ is covered by finitely many balls $B_{\hat\delta(y_0)}(y_0)$, namely:

$$\partial\mathcal{D} \subset \bigcup_{i=1}^n B_{\hat\delta(y_{0,i})}(y_{0,i}),$$

with $y_{0,i} \in \partial\mathcal{D}$, $i = 1 \ldots n$. Let $\hat\epsilon = \min_{i=1\ldots n} \hat\epsilon(y_{0,i})$ and let $\hat\delta$ be such that for every $y_0 \in \partial\mathcal{D}$, the ball $B_{\hat\delta}(y_0) \subset B_{\hat\delta(y_{0,i})}(y_{0,i})$ for some $i = 1 \ldots n$. Let $\epsilon \in (0, \hat\epsilon)$ and choose an initial token position $x_0 \in B_{\hat\delta}(y_0)$. Then, according to (5.15), there exists $\sigma_{0,I}$ such that for every σ_{II} there holds:

$$\mathbb{P}\Big((X^{x_0,\sigma_{0,I},\sigma_{II}})_{\tau^{x_0,\sigma_{0,I},\sigma_{II}}} \in B_\delta(y_0)\Big)$$
$$\geq \mathbb{P}\Big((X^{x_0,\sigma_{0,I},\sigma_{II}})_{\tau^{x_0,\sigma_{0,I},\sigma_{II}}} \in B_{\delta/2}(y_{0,i})\Big) \geq 1 - \eta,$$

proving the claim in view of $B_{\delta/2}(y_{0,i}) \subset B_\delta(y_0)$. □

Theorem 5.7 *Assume that $\mathcal{D}$ is game-regular. Then, for every continuous function $F : \Gamma_\epsilon \cup \Gamma_{out} \to \mathbb{R}$, the solutions $u_\epsilon : \mathcal{D}^\diamond \to \mathbb{R}$ of (4.1) converge uniformly, as $\epsilon \to 0$, to a continuous $u : \mathcal{D}^\diamond \to \mathbb{R}$ that is the unique viscosity solution of (5.1) in $\bar{\mathcal{D}}$.*

Proof

1. We will show that every sequence $\{u_\epsilon\}_{\epsilon\in J}$, of solutions to (4.1) with the prescribed continuous boundary data $F : \Gamma_\epsilon \cup \Gamma_{out} \to \mathbb{R}$, where $J \subset (0, 1)$ is a sequence decreasing to 0, has a subsequence that converges uniformly. By Theorem 5.2 and Theorem C.63, it follows that the limit of such subsequence is the unique viscosity solution u of (5.1), with the boundary condition $F_{|\partial\mathcal{D}}$. Thus the entire family $\{u_\epsilon\}_{\epsilon\to 0}$ must converge uniformly to u.

   Since, according to Theorem 4.4 and Corollary 4.2, solutions to (4.1) are continuous and equibounded, it suffices to check their equicontinuity in $\mathcal{D}^\diamond$. Equivalently, in virtue of Theorem 5.3, we will prove the equicontinuity of $\{u_\epsilon\}_{\epsilon\in J}$ at the boundary. To this end, fix $\eta > 0$ and let $\delta > 0$ be such that:

$$|F(x) - F(y)| \le \frac{\eta}{3} \quad \text{for all } x, y \in \Gamma \text{ with } |x - y| < \delta. \tag{5.17}$$

   By Lemma 5.6 and the observation after Definition 5.5 it follows that we may choose $\hat{\delta} < \delta$ and $\hat{\epsilon} > 0$ such that for every $\epsilon \in (0, \hat{\epsilon})$, $y_0 \in \partial\mathcal{D}$ and $x_0 \in B_{\hat{\delta}}(y_0)$, there exists a strategy $\sigma_{0,II}$ with the property that for all σ_I we have:

$$\mathbb{P}\big(X_\tau \in B_\delta(y_0)\big) \ge 1 - \frac{\eta}{6\|F\|_{\mathcal{C}^0(\Gamma)} + 1}, \tag{5.18}$$

   where above we denoted: $X_\tau = (X^{x_0,\sigma_I,\sigma_{0,II}})_{\tau^{x_0,\sigma_I,\sigma_{0,II}}}$. Let $y_0 \in \partial\mathcal{D}$ and $x_0 \in \bar{\mathcal{D}}$ satisfy: $|x_0 - y_0| \le \hat{\delta}$. Then, in virtue of Theorem 4.12, we observe that:

$$\begin{aligned} u_\epsilon(x_0) - u_\epsilon(y_0) &= u_{II}^\epsilon(x_0) - F(y_0) \le \sup_{\sigma_I} \mathbb{E}\big[F \circ X_\tau\big] - F(y_0) \\ &\le \mathbb{E}\big[F \circ X_\tau - F(y_0)\big] + \frac{\eta}{3}, \end{aligned}$$

   for some fixed strategy σ_I and with the notation: $X_\tau = \big(X^{x_0,\sigma_I,\sigma_{0,II}}\big)_{\tau^{x_0,\sigma_I,\sigma_{0,II}}}$. Thus, by (5.18) and (5.17), we further get:

$$\begin{aligned} u_\epsilon(x_0) - u_\epsilon(y_0) &\le \int_{\{X_\tau\in B_\delta(y_0)\}} |F(x_\tau) - F(y_0)| \,\mathrm{d}\mathbb{P} \\ &\quad + \int_{\{X_\tau\notin B_\delta(y_0)\}} |F(x_\tau) - F(y_0)| \,\mathrm{d}\mathbb{P} + \frac{\eta}{3} \\ &\le \frac{\eta}{3} + 2\|F\|_{\mathcal{C}^0(\Gamma)}\mathbb{P}\big(\{X_\tau \notin B_\delta(y_0)\}\big) + \frac{\eta}{3} \le \eta. \end{aligned}$$

2. The reverse inequality $u_\epsilon(x_0) - u_\epsilon(y_0) \geq -\eta$ is obtained by taking the strategy $\sigma_{0,I}$ as in Definition 5.5 (i) for the same thresholds $\frac{\eta}{6\|F\|_{C^0(\Gamma)}+1}$ and δ as in (5.17), and using the fact that:

$$u_\epsilon(x_0) = u_I^\epsilon(x_0) \geq \inf_{\sigma_{II}} \mathbb{E}\big[F \circ X_\tau\big],$$

where $X_\tau = \big(X^{x_0,\sigma_{0,I},\sigma_{II}}\big)_{\tau^{x_0,\sigma_{0,I},\sigma_{II}}}$. The proof is done. □

We now show that if some boundary point $y_0 \in \partial\mathcal{D}$ is not game-regular, then there exists a continuous $F : \mathbb{R}^N \to \mathbb{R}$, such that the solutions $u_\epsilon : \mathcal{D}^\diamond \to \mathbb{R}$ of (4.1) do not converge uniformly. Namely, we have the following:

Theorem 5.8 *Assume that for every continuous $F : \mathbb{R}^N \to \mathbb{R}$, the solutions $u_\epsilon : \mathcal{D}^\diamond \to \mathbb{R}$ of (4.1) converge uniformly on $\mathcal{D}^\diamond$ as $\epsilon \to 0$. Then $\mathcal{D}$ is game-regular.*

Proof Choose $y_0 \in \partial\mathcal{D}$ and let $\eta, \delta > 0$. Consider the data function $F : \mathbb{R}^N \to \mathbb{R}$:

$$F(y) = -|y - y_0|.$$

By assumption, the uniform limit of $\{u_\epsilon\}_{\epsilon<\epsilon_0}$ is a continuous function $u : \mathcal{D}^\diamond \to \mathbb{R}$ coinciding with F on Γ_{out}. Define now $\hat{\delta} \in (0, \epsilon_0)$ and $\hat{\epsilon} \in (0, \frac{\delta}{2(1+r)})$ such that:

$$|u(x_0)| < \frac{\eta\delta}{2} \quad \text{for all } x_0 \in B_{\hat{\delta}}(y_0), \qquad \text{and} \qquad \|u_\epsilon - u\|_{C^0(\mathcal{D}^\diamond)} < \frac{\eta\delta}{2} \quad \text{for all } \epsilon < \hat{\epsilon}.$$

Fix $\epsilon < \hat{\epsilon}$ and let $x_0 \in B_{\hat{\delta}}(y_0)$. It follows that $|u_\epsilon(x_0)| < \eta\delta$, so in particular:

$$\sup_{\sigma_I} \inf_{\sigma_{II}} \mathbb{E}\big[F \circ X_\tau\big] = u_I^\epsilon(x_0) = u_\epsilon(x_0) > -\eta\delta,$$

where $X_\tau = \big(X^{x_0,\sigma_I,\sigma_{II}}\big)_{\tau^{x_0,\sigma_I,\sigma_{II}}}$. Consequently, there exists a strategy $\sigma_{0,I}$ with the property that for every σ_{II} there holds: $\mathbb{E}\big[F \circ \big(X^{x_0,\sigma_{0,I},\sigma_{II}}\big)_{\tau^{x_0,\sigma_{0,I},\sigma_{II}}}\big] > -\eta\delta$. Then, in view of the nonpositivity of F we get:

$$\mathbb{P}\big(X_\tau \not\in B_\delta(y_0)\big) \leq -\frac{1}{\delta}\int_\Omega F(x_\tau)\, \mathrm{d}\mathbb{P} < \eta,$$

and we obtain:

$$\mathbb{P}\big(X_\tau \in B_\delta(y_0)\big) = 1 - \mathbb{P}\big(X_\tau \not\in B_\delta(y_0)\big) \geq 1 - \eta,$$

as requested in (5.16). This ends the proof of y_0 being game-regular. □

5.3 Concatenating Strategies

In this section we prove that game-regularity of a boundary point $y_0 \in \partial\mathcal{D}$ is implied by (5.19) below. This condition requires the validity of (5.15) and (5.16) for one fixed $\eta_0 \in (0, 1)$, rather than for all small $\eta > 0$.

Theorem 5.9 *Let $\mathcal{D}$, ϵ_0 and $\mathcal{D}^\diamond$ be as in Definition 3.8. Fix $p > 2$, $r = \sqrt{\frac{3(p-2)}{2(N+2)}}$ and for each $\epsilon \in (0, \frac{\epsilon_0}{1+r})$ consider the boundary-aware Tug-of-War game with noise according to (4.1), as defined in Sect. 4.3.*

For a given boundary point $y_0 \in \partial\mathcal{D}$, assume that there exists $\theta_0 < 1$ such that for every $\delta > 0$ there exists $\hat{\delta} \in (0, \delta)$ and $\hat{\epsilon} \in (0, \frac{1}{1+r})$ with the following property. Fix $\epsilon \in (0, \hat{\epsilon})$ and choose an initial token position $x_0 \in B_{\hat{\delta}}(y_0)$; there exists then a strategy $\sigma_{0,II}$ of Player II in the ϵ-game on $\mathcal{D}^\diamond$ corresponding to (4.1) such that for every strategy σ_I of Player I we have:

$$\mathbb{P}\Big(\exists n \leq \tau \quad X_n \notin B_\delta(y_0)\Big) \leq \theta_0. \tag{5.19}$$

Then y_0 is game-regular.

Under condition (5.19), construction of an optimal strategy realizing the (arbitrarily small) threshold η in (5.16) is carried out by *concatenating* the m optimal strategies corresponding to the achievable threshold η_0, on m concentric balls centred at y_0, where $(1-\eta_0)^m \geq 1-\eta$.

Proof of Theorem 5.9

1. Fix $\eta, \delta > 0$. We want to find $\hat{\epsilon}$ and $\hat{\delta}$ such that (5.16) holds. We first observe that for $\eta \leq 1-\theta_0$ the claim follows directly from (5.19). In the general case, let $m \in \{2, 3, \dots\}$ be such that:

$$\theta_0^m \leq \eta. \tag{5.20}$$

We now define the radii $\{\delta_k\}_{k=1}^m$ and the maximal token shifts $\{\epsilon_k\}_{k=1}^m$, and assign the corresponding $\{\hat{\delta}(\delta_k)\}_{k=1}^m$, $\{\hat{\epsilon}(\delta_k)\}_{k=1}^m$ from the assumed condition (5.19). Namely, for every initial token position in $B_{\hat{\delta}(\delta_k)}(y_0)$ in the Tug-of-War game on $\mathcal{D}^\diamond$ with step less than $\hat{\epsilon}(\delta_k)$, there exists a strategy $\sigma_{0,II,k}$ guaranteeing that the token exits $B_{\delta_k}(y_0)$ (before the game is stopped) with probability at most θ_0. This construction will be achieved through the repeated application of (5.19).

We set $\delta_m = \delta$ and find the quantities $\hat{\delta}(\delta_m)$ and $\hat{\epsilon}(\delta_m)$, with the indicated choice of the strategy $\sigma_{0,II,m}$. Decreasing the value of $\hat{\epsilon}(\delta_m)$ if necessary, we then set:

$$\delta_{m-1} = \hat{\delta}(\delta_m) - (1+r)\hat{\epsilon}(\delta_m) > 0, \qquad \epsilon_{m-1} = \min\left\{\hat{\epsilon}(\delta_m), \frac{\hat{\delta}(\delta_m)}{2(1+r)}\right\} > 0.$$

In the same manner, having constructed $\delta_k > 0$ and $\epsilon_k > 0$, we find $\hat{\delta}(\delta_k)$ and $\hat{\epsilon}(\delta_k)$ and define:

$$\delta_{k-1} = \hat{\delta}(\delta_k) - (1+r)\hat{\epsilon}(\delta_k) > 0, \qquad \epsilon_{k-1} = \min\Big\{\epsilon_k, \hat{\epsilon}(\delta_k), \frac{\hat{\delta}(\delta_k)}{2(1+r)}\Big\} > 0.$$

Eventually, we call:

$$\hat{\delta} = \hat{\delta}(\delta_1), \qquad \hat{\epsilon} = \min\big\{\epsilon_1, \hat{\epsilon}(\delta_1)\big\}.$$

To show that the condition of game-regularity at y_0 is satisfied, we will concatenate strategies $\{\sigma_{0,II,k}\}_{k=1}^m$ by switching to $\sigma_{0,II,k+1}$ immediately after the token exits $B_{\delta_k}(y_0) \subset B_{\hat{\delta}(\delta_{k+1})}(y_0)$ (Fig. 5.3). This is carried out in the next step.

2. Fix $x_0 \in B_{\hat{\delta}}(y_0)$ and let $\epsilon \in (0, \hat{\epsilon})$. Define the strategy $\sigma_{0,II}$ of Player II:

$$\sigma_{0,II}^n = \sigma_{0,II}^n\big(x_0, (x_1, w_1, a_1, b_1), \dots, (x_n, w_n, a_n, b_n)\big) \quad \text{for all } n \geq 0,$$

separately in the following two cases.

Case 1. If $x_k \in B_{\delta_1}(y_0)$ for all $k \leq n$, then we set:

$$\sigma_{0,II}^n = \sigma_{0,II,1}^n\big(x_0, (x_1, w_1, a_1, b_1), \dots, (x_n, w_n, a_n, b_n)\big).$$

Case 2. Otherwise, define:

$$k \doteq k(x_0, x_1, \dots, x_n) = \max\Big\{1 \leq k \leq m-1;\ \exists\, 0 \leq i \leq n \ \ x_i \notin B_{\delta_k}(y_0)\Big\}$$

$$i \doteq \min\Big\{0 \leq i \leq n;\ x_i \notin B_{\delta_k}(y_0)\Big\}.$$

and set:

$$\sigma_{0,II}^n = \sigma_{0,II,k+1}^{n-i}\big(x_i, (x_{i+1}, w_{i+1}, a_{i+1}, b_{i+1}), \dots, (x_n, w_n, a_n, b_n)\big).$$

It is not hard to check that each $\sigma_{0,II}^n : H_n \to B_{r\epsilon}(0) \subset \mathbb{R}^N$ is (Borel) measurable.

Let σ_I be now any strategy of Player I. We will show that:

$$\begin{aligned}&\mathbb{P}\big(\exists n \leq \tau \ \ X_n \notin B_{\delta_k}(y_0)\big)\\ &\quad \leq \theta_0 \cdot \mathbb{P}\big(\exists n \leq \tau \ \ X_n \notin B_{\delta_{k-1}}(y_0)\big) \quad \text{for all } k = 2 \dots m,\end{aligned} \tag{5.21}$$

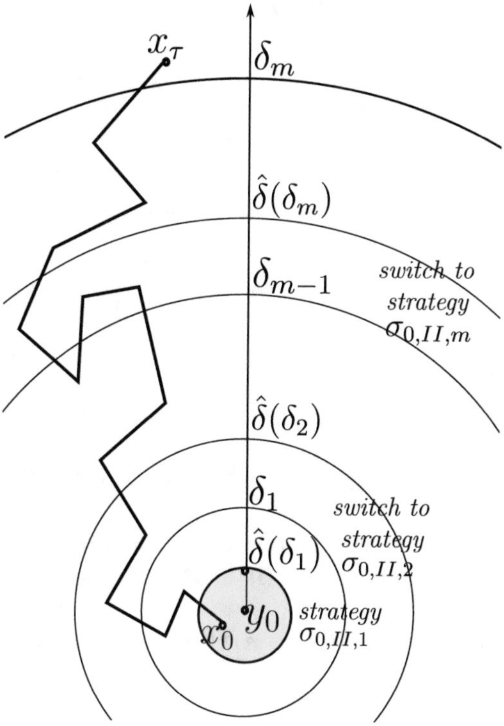

Fig. 5.3 The concatenated strategy $\sigma_{0,II}$ in the proof of Theorem 5.9

where the token position random variables $X_n = X_n^{x_0,\sigma_I,\sigma_{0,II}}$ are defined in (4.17), whereas $\tau = \tau^{x_0,\sigma_I,\sigma_{0,II}}$ is the stopping time in (4.18). Observe that (5.21) implies:

$$\begin{aligned}\mathbb{P}\big(X_\tau \not\in B_\delta(y_0)\big) &\leq \mathbb{P}\big(\exists n \leq \tau \;\; X_n \not\in B_{\delta_m}(y_0)\big)\\ &\leq \theta_0^{m-1}\cdot \mathbb{P}\big(\exists n \leq \tau \;\; X_n \not\in B_{\delta_1}(y_0)\big) \leq \theta_0^m,\end{aligned}$$

for $\hat\delta = \hat\delta(\delta_1)$. This will end the proof of the result by (5.20).

3. In order to show (5.21), we denote:

$$\tilde\Omega = \big\{\exists n \leq \tau \;\; X_n \not\in B_{\delta_{k-1}}(y_0)\big\} \subset \Omega.$$

Without loss of generality, one may assume that $\mathbb{P}(\tilde\Omega) > 0$. Indeed, since: $\mathbb{P}\big(\exists n \leq \tau \;\; X_n \not\in B_{\delta_k}(y_0)\big) \leq \mathbb{P}\big(\exists n \leq \tau \;\; X_n \not\in B_{\delta_{k-1}}(y_0)\big)$, it follows that if $\mathbb{P}(\tilde\Omega) = 0$ then both sides of (5.21) equal 0, and thus the inequality holds.

For $\mathbb{P}(\tilde\Omega) > 0$, we define the probability space $(\tilde\Omega, \tilde{\mathcal{F}}, \tilde{\mathbb{P}})$ by:

$$\tilde{\mathcal{F}} = \big\{A\cap\tilde\Omega;\;\; A\in\mathcal{F}\big\} \quad \text{and} \quad \tilde{\mathbb{P}}(A) = \frac{\mathbb{P}(A)}{\mathbb{P}(\tilde\Omega)} \quad \text{for all } A\in\tilde{\mathcal{F}}.$$

Define also the measurable space $(\Omega_{fin}, \mathcal{F}_{fin})$, by setting $\Omega_{fin} = \bigcup_{n=1}^{\infty} \Omega_n$ and by taking $\mathcal{F}_{fin}$ to be the smallest σ-algebra containing $\bigcup_{n=1}^{\infty} \mathcal{F}_n$ (see Sect. 4.3 for the definition of probability spaces $(\Omega_n, \mathcal{F}_n, \mathbb{P}_n)$).

We now consider the random variables:

$$Y_1 : \tilde{\Omega} \to \Omega_{fin} \qquad Y_1\big(\{(w_n, a_n, b_n)\}_{n=1}^{\infty}\big) \doteq \{(w_n, a_n, b_n)\}_{n=1}^{\tau_k}$$
$$Y_2 : \tilde{\Omega} \to \Omega \qquad Y_2\big(\{(w_n, a_n, b_n)\}_{n=1}^{\infty}\big) \doteq \{(w_n, a_n, b_n)\}_{n=\tau_k+1}^{\infty},$$

where τ_k is the following stopping time on $\tilde{\Omega}$:

$$\tau_k = \min\big\{n = 1, 2, \ldots;\ \ X_n \notin B_{\delta_{k-1}}(y_0)\big\}.$$

We claim that Y_1 and Y_2 are independent, that is:

$$\tilde{\mathbb{P}}\big(\{Y_1 \in A_1\} \cap \{Y_2 \in A_2\}\big) = \tilde{\mathbb{P}}\big(Y_1 \in A_1\big) \cdot \tilde{\mathbb{P}}\big(Y_2 \in A_2\big) \quad \text{for all } A_1 \in \mathcal{F}_{fin},\ A_2 \in \mathcal{F}.$$

To this end, by the definition of σ-algebras $\mathcal{F}_{fin}$ and $\mathcal{F}$, it suffices to check (see Exercise 5.10 (i)) that for every $s, t \in \mathbb{N}$ we have:

$$\mathbb{P}(\tilde{\Omega}) \cdot \mathbb{P}\big(\{Y_1 \in A_1\} \cap \{Y_2 \in A_2\}\big) = \mathbb{P}\big(Y_1 \in A_1\big) \cdot \mathbb{P}\big(Y_2 \in A_2\big) \qquad \text{for all } A_1 \in \mathcal{F}_s,\ A_2 \in \mathcal{F}_t. \tag{5.22}$$

Indeed, computing the probabilities:

$$\mathbb{P}\big(Y_1 \in A_1\big) = \mathbb{P}_s\Big(A_1 \cap \{\tau_k = s\} \cap \bigcap_{i<s} \{b_i \leq d_\epsilon(x_{i-1})\}\Big),$$
$$\mathbb{P}\big(Y_2 \in A_2\big) = \mathbb{P}(\tilde{\Omega}) \cdot \mathbb{P}_t(A_2)$$
$$\mathbb{P}\big(\{Y_1 \in A_1\} \cap \{Y_2 \in A_2\}\big) = \mathbb{P}_s\Big(A_1 \cap \{\tau_k = s\} \cap \bigcap_{i<s} \{b_i \leq d_\epsilon(x_{i-1})\}\Big) \cdot \mathbb{P}_t(A_2),$$

proves (5.22) directly.

4. We now want to apply Lemma A.21 to the independent random variables Y_1, Y_2 as above, and to the indicator function:

$$Z\big(\{(w_n, a_n, b_n)\}_{n=1}^{s}, \{(w_n, a_n, b_n)\}_{n=s+1}^{\infty}\big) = \mathbb{1}_{\big\{\exists n \leq \tau \ \ X_n(\{(w_n, a_n, b_n)\}_{n=1}^{\infty}) \notin B_{\delta_k}(y_0)\big\}}, \tag{5.23}$$

that is a random variable (see Exercise 5.10 (ii)) on the measurable space $\Omega_{fin} \times \Omega$, equipped as usual with the product σ-algebra of $\mathcal{F}_{fin}$ and $\mathcal{F}$. We thus obtain:

$$\mathbb{P}\big(\exists n \leq \tau \quad X_n \notin B_{\delta_k}(y_0)\big) = \int_{\tilde{\Omega}} f(\omega_1)\, \mathrm{d}\tilde{\mathbb{P}}(\omega_1), \tag{5.24}$$

where for a given $\omega_1 = \{(w_n, a_n, b_n)\}_{n=1}^{\infty} \in \tilde{\Omega}$ the integrand function returns:

$$
\begin{aligned}
f(\omega_1) &= \mathbb{P}\Big(\{(\bar{w}_n, \bar{a}_n, \bar{b}_n)\}_{n=1}^{\infty} \in \tilde{\Omega}; \\
&\qquad \exists n \leq \tau \quad X_n\big(\{(w_s, a_s, b_s)\}_{s=1}^{\tau_k}, \{(\bar{w}_s, \bar{a}_s, \bar{b}_s)\}_{s=\tau_k+1}^{\infty}\big) \notin B_{\delta_k}(y_0)\Big) \\
&= \mathbb{P}\Big(\{(\bar{w}_n, \bar{a}_n, \bar{b}_n)\}_{n=1}^{\infty} \in \tilde{\Omega}; \\
&\qquad \exists n \leq \tau \quad X_n^{x_{\tau_k}, \sigma_I, \sigma_{0,II,k}}\big(\{(\bar{w}_s, \bar{a}_s, \bar{b}_s)\}_{s=\tau_k+1}^{\infty}\big) \notin B_{\delta_k}(y_0)\Big).
\end{aligned}
$$

Since $x_{\tau_k} \in B_{\hat{\delta}(\delta_k)}$, by (5.19) it follows that:

$$
f(\omega_1) = \mathbb{P}\Big(\exists n \leq \tau \quad X_n^{x_{\tau_k}, \sigma_I, \sigma_{0,II,k}} \notin B_{\delta_k}(y_0)\Big) \cdot \mathbb{P}(\tilde{\Omega}) \leq \theta_0 \cdot \mathbb{P}(\tilde{\Omega})
$$

for $\tilde{\mathbb{P}}$-a.e. $\omega_1 \in \tilde{\Omega}$. In conclusion, (5.24) implies (5.21) and completes the proof. □

Exercise 5.10

(i) Show that (5.22) implies independence of the random variables Y_1, Y_2.
(ii) Prove that the indicator function F in (5.23) is measurable with respect to the product σ-algebra on $\Omega_{fin} \times \Omega$.

5.4 The Annulus Walk Estimate

The two proofs of game-regularity in the following sections will be based on the concatenating strategies technique in Theorem 5.9 and the analysis of the annulus walk below. Namely, we will derive an estimate on the probability of exiting a given annular domain $\tilde{\mathcal{D}}$ through the external portion of its boundary, while playing the boundary aware game associated with (4.1). We show that when the ratio of the annulus thickness and the distance of the initial token position x_0 from the internal boundary is large enough, then this probability is bounded by a universal constant $\theta_0 < 1$. When $p \geq N$, then θ_0 goes to 0 as the indicated ratio goes to $+\infty$.

Theorem 5.11 *For given radii $0 < R_1 < R_2 < R_3$, consider the annulus $\tilde{\mathcal{D}} = B_{R_3}(0) \setminus \bar{B}_{R_1}(0) \subset \mathbb{R}^N$, together with its thickened boundary:*

$$
\tilde{\Gamma} = \{x \in \mathbb{R}^N;\ \operatorname{dist}(x, \partial\tilde{\mathcal{D}}) < \tilde{\epsilon}_0\},
$$

defined for some $\tilde{\epsilon}_0 \in (0, \frac{R_1}{2})$ (Fig. 5.4).

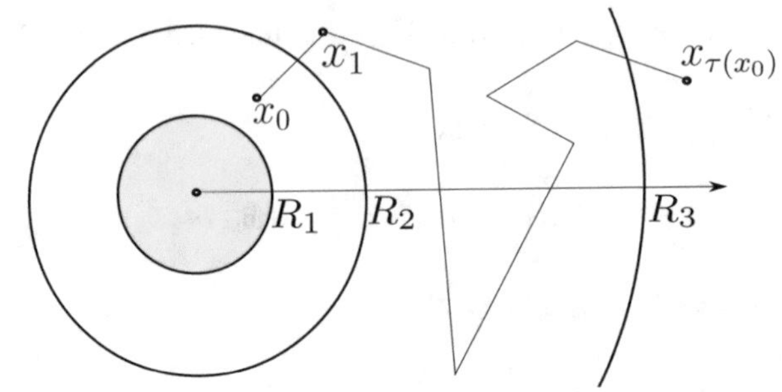

Fig. 5.4 The annulus walk in Theorem 5.11

For every $\xi > 0$, there exists $\hat\epsilon < \frac{\tilde\epsilon_0}{1+r}$, depending on R_1, R_2, R_3 and ξ, p, N, such that for every $x_0 \in \tilde{\mathcal{D}} \cap B_{R_2}(0)$ and every $\epsilon < \hat\epsilon$, Player II has a strategy $\tilde\sigma_{0,II}$ with the property that for every strategy $\tilde\sigma_I$ of Player I there holds:

$$\mathbb{P}\Big(\tilde X_{\tilde\tau} \not\in \bar B_{R_3-\epsilon}(0)\Big) \leq \frac{v(R_2)-v(R_1)}{v(R_3)-v(R_1)} + \xi. \tag{5.25}$$

Here, $v : (0,\infty) \to \mathbb{R}$ is given by:

$$v(t) = \begin{cases} sgn(p-N)\, t^{\frac{p-N}{p-1}} & for\ p \neq N \\ \log t & for\ p = N, \end{cases} \tag{5.26}$$

and $\{\tilde X_n = \tilde X_n^{x_0,\tilde\sigma_I,\tilde\sigma_{0,II}}\}_{n=0}^\infty$ and $\tilde\tau = \tilde\tau^{x_0,\tilde\sigma_I,\tilde\sigma_{0,II}}$ denote, as usual, the random variables corresponding to the token positions and the stopping time in the boundary aware Tug-of-War game played on $\tilde{\mathcal{D}}^\diamond = \tilde{\mathcal{D}} \cup \tilde\Gamma$, as in Sect. 4.3.

Proof Consider the radial function $u : \mathbb{R}^N \setminus \{0\} \to \mathbb{R}$ given by $u(x) = v(|x|)$, where v is as in (5.26). Recall (see Exercise C.27) that:

$$\Delta_p u = 0 \quad \text{and} \quad \nabla u \neq 0 \qquad \text{in } \mathbb{R}^N \setminus \{0\}. \tag{5.27}$$

Let $\tilde u_\epsilon$ be the family of solutions to (4.1) on $\tilde{\mathcal{D}}^\diamond$ with the data $F = u_{|\tilde\Gamma}$. By Theorem 4.6, there exists a constant $C > 0$, depending only on N, p, u and $\tilde{\mathcal{D}}$, such that:

$$\|\tilde u_\epsilon - u\|_{C^0(\tilde{\mathcal{D}})} \leq C\epsilon,$$

for all small $\epsilon > 0$.

Fix $x_0 \in \tilde{\mathcal{D}} \cap B_{R_2}(0)$ and a sufficiently small $\epsilon < \tilde{\epsilon}_0$. Using Theorem 4.12 to the game value $\tilde{u}^{\epsilon}_{II} = \tilde{u}_{\epsilon}$, we see that there exists a strategy $\tilde{\sigma}_{0,II}$ so that for every $\tilde{\sigma}_I$:

$$\mathbb{E}\big[u \circ \tilde{X}_{\tilde{\tau}}\big] - u(x_0) \leq 2C\epsilon. \tag{5.28}$$

We now estimate:

$$\begin{aligned}
&\mathbb{E}\big[u\circ\tilde{X}_{\tilde{\tau}}\big] - u(x_0) \\
&\quad = \int_{\tilde{X}_{\tilde{\tau}} \notin \bar{B}_{R_3-\epsilon}(0)} u(\tilde{X}_{\tilde{\tau}})\, \mathrm{d}\mathbb{P} + \int_{\tilde{X}_{\tilde{\tau}} \in B_{R_1+\epsilon}(0)} u(\tilde{X}_{\tilde{\tau}})\, \mathrm{d}\mathbb{P} - u(x_0) \\
&\quad \geq \mathbb{P}\big(\tilde{X}_{\tilde{\tau}} \notin \bar{B}_{R_3-\epsilon}(0)\big) v\big(\rho R_3 - \epsilon\big) \\
&\qquad + \Big(1 - \mathbb{P}\big(\tilde{X}_{\tilde{\tau}} \notin \bar{B}_{R_3-\epsilon}(0)\big)\Big) v\big(R_1 - (1+r)\epsilon\big) - v(R_2),
\end{aligned}$$

where we have used the fact that v in (5.26) is always an increasing function. Recalling (5.28), this implies:

$$\mathbb{P}\big(\tilde{X}_{\tilde{\tau}} \notin \bar{B}_{R_3-\epsilon}(0)\big) \leq \frac{v(R_2) - v(R_1 - (1+r)\epsilon) + 2C\epsilon}{v(R_3 - \epsilon) - v(R_1 - (1+r)\epsilon)}. \tag{5.29}$$

The proof of (5.25) is now complete, in view of continuity of the right-hand side function above with respect to ϵ. □

By inspecting the quotient in the right-hand side of (5.25) in Exercise 5.14, we immediately obtain:

Corollary 5.12 *The function v in (5.26) has the following properties, for any fixed $0 < R_1 < R_2$:*

$$(i) \qquad \lim_{R_3 \to \infty} \frac{v(R_2) - v(R_1)}{v(R_3) - v(R_1)} = \begin{cases} 1 - \left(\frac{R_2}{R_1}\right)^{\frac{p-N}{p-1}} & \text{for } 2 < p < N \\ 0 & \text{for } p \geq N, \end{cases}$$

$$(ii) \qquad \lim_{M \to \infty} \frac{v(MR_1) - v(R_1)}{v(M^2R_1) - v(R_1)} = \begin{cases} \frac{1}{2} \text{ for } p = N \\ 0 \text{ for } p > N. \end{cases}$$

Consequently, the estimate (5.25) can be replaced by:

$$\mathbb{P}\Big(\tilde{X}_{\tilde{\tau}} \notin \bar{B}_{R_3-\epsilon}(0)\Big) \leq \theta_0 \tag{5.30}$$

valid for any $\theta_0 > 1 - \left(\frac{R_2}{R_1}\right)^{\frac{p-N}{p-1}}$ *if* $p \in (2, N)$*, and any* $\theta_0 > 0$ *if* $p \geq N$*, upon choosing* R_3 *sufficiently large with respect to* R_1 *and* R_2*. When* $p > N$*, the same bound is valid when setting* $R_2 = MR_1$*,* $R_3 = M^2 R_1$ *with the ratio* M *large enough.*

The results of Theorem 5.11 and Corollary 5.12 are invariant under scaling. More precisely, we have:

Corollary 5.13 *The bounds (5.25) and (5.30) remain true if we replace* R_1, R_2, R_3 *by* $\rho R_1, \rho R_2, \rho R_3$ *and* $\hat{\epsilon}$ *by* $\rho\hat{\epsilon}$*, for any* $\rho > 0$*.*

Proof For $\rho > 0$, consider the annulus $\rho\tilde{\mathcal{D}} = B_{\rho R_3}(0) \setminus \bar{B}_{\rho R_1}(0)$ and the boundary aware Tug-of-War game starting from the position $\rho x_0 \in \rho\tilde{\mathcal{D}} \cap B_{\rho R_2}(0)$. Let $\rho\epsilon < \rho\hat{\epsilon}$ and let $\tilde{\sigma}_{0,II}$ be the strategy of Player II in the statement of Theorem 5.11. Then the strategy $\tilde{\sigma}_{0,II,\rho}$ defined by:

$$\begin{aligned}\tilde{\sigma}^n_{0,II,\rho}\big(\rho x_0, (\rho x_1, \rho w_1, a_1, b_1), \ldots, (\rho x_n, \rho w_n, a_n, b_n)\big)\\ = \tilde{\sigma}^n_{0,II}\big(x_0, (x_1, w_1, a_1, b_1), \ldots, (x_n, w_n, a_n, b_n)\big)\end{aligned}$$

gives:

$$\mathbb{P}\Big(\tilde{X}^{\rho}_{\tilde{\tau}^{\rho}} \not\in \bar{B}_{\rho R_3 - \rho\epsilon}(0)\Big) = \mathbb{P}\Big(\tilde{X}_{\tilde{\tau}} \not\in \bar{B}_{R_3-\epsilon}(0)\Big)$$

for any $\tilde{\sigma}_I$ that naturally induces $\tilde{\sigma}_{I,\rho}$ of Player I in the $\rho\epsilon$-step game on $\rho\tilde{\mathcal{D}}$. The estimate (5.25) is then equivalent to the claimed bound on $\rho\tilde{\mathcal{D}}$. □

Exercise 5.14 Verify the statements (i) and (ii) in Corollary 5.12.

5.5 Sufficient Conditions for Game-Regularity: Exterior Cone Property

The purpose of this section is to show the following regularity result.

Theorem 5.15 *Assume that* $\mathcal{D}$*,* $\mathcal{D}^\diamond$ *and* p, r *are as in Definition 5.5. Let* $y_0 \in \partial\mathcal{D}$ *have the exterior cone property, that is: there exists a finite cone* $C \subset \Gamma_{out}$ *with the tip at* y_0*. Then* y_0 *is game-regular.*

Proof With the help of Theorem 5.11, we will show that the assumption of Theorem 5.9 is satisfied, with probability $\theta_0 < 1$ depending only on p, N and the angle of the external cone at y_0.

The exterior cone condition implies that there exists a constant $R_1 > 0$ such that for all $\rho > 0$ there is $z_0 \in C$ satisfying:

$$|z_0 - y_0| = \rho(1 + R_1) \quad \text{and} \quad B_{2\rho R_1}(z_0) \subset C \subset \mathbb{R}^N \setminus \mathcal{D}. \tag{5.31}$$

Define $R_2 = 2+R_1$ and let $R_3 > R_2$ be such that $\frac{v(R_2)-v(R_1)}{v(R_3)-v(R_1)} \leq \theta_0(R_1, R_2, p, N) < 1$, as in Corollary 5.12 (i). We set:

$$\hat{\delta} = \frac{\delta}{1 + R_1 + R_3}, \qquad \rho = \hat{\delta}.$$

Consider the annulus $\tilde{\mathcal{D}} = B_{\rho R_3}(z_0) \setminus \bar{B}_{\rho R_1}(z_0)$ and observe that:

$$B_{\hat{\delta}}(y_0) \subset B_{\rho R_2}(z_0) \setminus \bar{B}_{\rho R_1}(z_0) \subset \tilde{\mathcal{D}} \subset B_{\rho R_3}(z_0) \subset B_\delta(y_0).$$

Let $\hat{\epsilon} > 0$ be as in Theorem 5.11, applied to the annuli with radii $\rho R_1, \rho R_2, \rho R_3$, centred at z_0 (Fig. 5.5) and, for a given $x_0 \in B_{\hat{\delta}}(y_0)$ and $\epsilon < \hat{\epsilon}$, let $\sigma_{0,II} = \tilde{\sigma}_{0,II}$ be the strategy of Player II ensuring validity of the bound (5.30). For a given strategy σ_I of Player I, we claim that:

$$\begin{aligned} &\Big\{\omega \in \Omega;\ \exists n \leq \tau^{x_0,\sigma_I,\sigma_{0,II}}(\omega) \quad X_n^{x_0,\sigma_I,\sigma_{0,II}}(\omega) \not\in B_\delta(y_0)\Big\} \\ &\qquad\qquad \subset \Big\{\omega \in \Omega;\ \tilde{X}_{\tilde{\tau}}^{x_0,\sigma_I,\sigma_{0,II}}(\omega) \not\in B_{\rho R_3 - \epsilon}(z_0)\Big\}, \end{aligned} \tag{5.32}$$

where $\{X_n\}_{n=0}^\infty$ and τ are the random variables in (4.17), (4.18), and $\{\tilde{X}_n\}_{n=0}^\infty$ and $\tilde{\tau}$ are as in the proof of Theorem 5.11, corresponding to the token positions in the annulus $\tilde{\mathcal{D}}$.

Indeed, if ω belongs to the event set in the left-hand side of (5.32), then there must be $\tilde{\tau}(\omega) \leq n$ and $X_i(\omega) = \tilde{X}_i(\omega) \not\in \bar{B}_{\rho R_1+\epsilon}(z_0)$ for all $i \leq \tilde{\tau}(\omega)$. Thus ω must belong to the right hand side set, proving thus (5.32). The final claim follows by (5.30) and by applying Theorem 5.9. □

Note that the constant θ_0 depends only on the cone C at y_0, on p and N. In particular, when $p \geq N$, the assertions in the proof of Theorem 5.15 are true with any $\theta_0 > 0$, whereas for $p \in (2, N)$ they hold with any $\theta_0 > \theta_0(C, p, N) \in (0, 1)$.

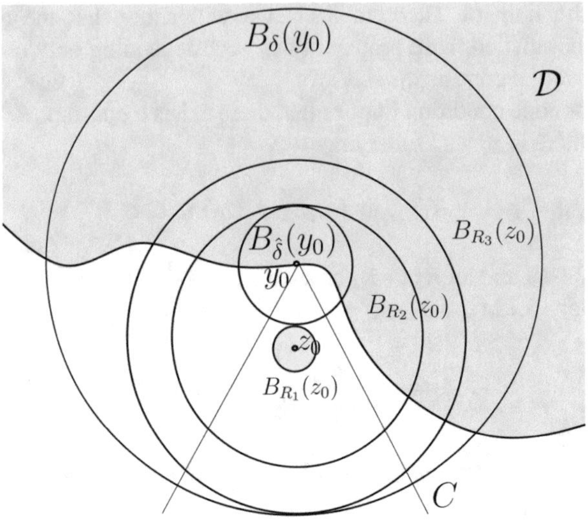

Fig. 5.5 The concentric balls and the annuli in the proof of Theorem 5.15, unscaled to $\hat{\delta} = 1$

5.6 Sufficient Conditions for Game-Regularity: $p > N$

In this section, we show that the geometrical notions of regularity of $\partial\mathcal{D}$ become irrelevant for large p. Namely, every boundary point $y_0 \in \partial\mathcal{D}$ is always game-regular, according to Definition 5.5, when $p > N$. This finding is consistent with the well-known result that points in $\mathbb{R}^N$ have positive p-capacity when $p > N$ (see Sect. C.7 in Appendix C for details).

Theorem 5.16 *Let $\mathcal{D}$, $\mathcal{D}^\diamond$ and p, r be as in Definition 5.5. Assume that the harmonicity exponent p is greater than the dimension N of $\mathcal{D}$:*

$$p > N. \tag{5.33}$$

Then every boundary point $y_0 \in \partial\mathcal{D}$ is game-regular.

We split the proof of Theorem 5.16 in a sequence of lemmas regarding the following random variables, where the condition of the boundary awareness has been suppressed. Namely, for $n = 0, 1, \dots,$ the random variables $Z_n : \Omega \to \mathbb{R}^N$ on the probability space $(\Omega, \mathcal{F}, \mathbb{P})$ are defined recursively, for a given step size ϵ and given strategies σ_I, σ_{II}:

$$Z_n\big((w_1, a_1, b_1), \dots, (w_n, a_n, b_n)\big) = \begin{cases} z_{n-1} + \epsilon w_n + \epsilon\sigma_I^{n-1}(h_{n-1}) & \text{for } a_n = 1 \\ z_{n-1} + \epsilon w_n + \epsilon\sigma_{II}^{n-1}(h_{n-1}) & \text{for } a_n = 2 \\ z_{n-1} + \epsilon w_n & \text{for } a_n = 3. \end{cases}$$

We set a constant $Z_0 \equiv z_0$ and let the n-th history $h_n \in H_n = \mathbb{R}^N \times (\mathbb{R}^N \times \Omega_1)^n$ be given by:

$$h_n = \big(z_0, (z_1, w_1, a_1, b_1), \ldots, (z_n, w_n, a_n, b_n)\big),$$

denoting by z_k the k-th position $Z_k((w_1, a_1, b_1), \ldots, (w_k, a_k, b_k))$.

Lemma 5.17 *Let $\mathcal{D}$, $\mathcal{D}^\diamond$ and p, r be as in Definition 5.5. Assume that, for a boundary point $y_0 \in \partial\mathcal{D}$, the following holds. There exists $\eta_1 > 0$ such that for every $\delta \ll 1$ there is $\hat\delta \in (0,1)$ and $\hat\epsilon \in (0, \frac{\delta}{2+r})$ with the property that for any $\epsilon \in (0, \hat\epsilon)$ and $z_0 \in B_{\hat\delta}(y_0)$, Player II can choose a strategy $\sigma_{0,II}$ satisfying:*

$$\mathbb{P}\Big(\min\big\{n \geq 0;\ Z_n \in B_{\frac{\epsilon}{2}}(y_0)\big\} < \min\big\{n \geq 0;\ Z_n \not\in B_\delta(y_0)\big\}\Big) \geq \eta_1, \tag{5.34}$$

for any strategy σ_I chosen by Player I. Then $y_0 \in \partial\mathcal{D}$ is game-regular.

Proof In virtue of Theorem 5.9, it is enough to show existence of a positive lower bound $\eta_0 > 0$ of the probability that the token remains within a prescribed distance from the boundary point y_0 until the game is stopped. More precisely, we will show that for every $\delta > 0$ there exists $\hat\delta \in (0,1)$ and $\hat\epsilon \in (0, \frac{1}{1+r})$ satisfying the following condition. Fix $\epsilon \in (0, \hat\epsilon)$ and choose an initial token position $x_0 \in B_{\hat\delta}(y_0)$; there exists a strategy $\sigma_{0,II}$ of Player II in the boundary aware Tug-of-War game with step size ϵ, such that for every strategy σ_I there holds:

$$\mathbb{P}\Big(\forall n \leq \tau^{x_0,\sigma_I,\sigma_{0,II}} \quad X_n^{x_0,\sigma_I,\sigma_{0,II}} \in B_\delta(y_0)\Big) \geq \eta_0. \tag{5.35}$$

Without loss of generality, we may assume that $\delta \ll 1$. Consider:

$$\hat\tau = \min\Big\{n \geq 1;\ \Big(Z_{n-1} \in B_{\frac{\epsilon}{2}}(y_0) \text{ and } b_n \geq \frac{1}{2}\Big) \text{ or } \Big(Z_{n-1} \not\in B_\delta(y_0)\Big)\Big\},$$

which is a $\mathcal{F}$-measurable stopping time relative to the filtration $\{\mathcal{F}_n\}_{n=0}^\infty$. The fact that $\mathbb{P}(\hat\tau < +\infty) = 1$ can be proved exactly as in Lemma 3.13. Observe that if we take $z_0 = x_0$, then:

$$\Big\{\forall n \leq \tau \quad X_n \in B_\delta(y_0)\Big\} \supset \Big\{Z_{\hat\tau-1} \in B_{\frac{\epsilon}{2}}(y_0) \text{ and } b_{\hat\tau} \geq \frac{1}{2}\Big\}.$$

Indeed, if $\omega \in \Omega$ belongs to the set in the right-hand side, then $Z_n(\omega) \in B_\delta(y_0)$ for all $n \leq \hat\tau(\omega)$ and $d_\epsilon(Z_{\hat\tau-1}) < \frac{1}{2} \leq b_{\hat\tau}$. Then, there must be $\tau(\omega) \leq \hat\tau(\omega)$, so ω belongs to the set in the left-hand side as well.

Further, we have:

$$\Big\{Z_{\hat{\tau}-1} \in B_{\frac{\epsilon}{2}}(y_0) \ \text{and}\ b_{\hat{\tau}} \geq \frac{1}{2}\Big\} \supset \Big\{\hat{\tau} = 1 + \min\big\{n \geq 0;\ Z_n \in B_{\frac{\epsilon}{2}}(y_0)\big\}\Big\}.$$

and then, by (5.34):

$$\begin{aligned} &\mathbb{P}\Big(\hat{\tau} = 1 + \min\big\{n \geq 0;\ Z_n \in B_{\frac{\epsilon}{2}}(y_0)\big\}\Big) \\ &= \frac{1}{2} \cdot \mathbb{P}\Big(\min\big\{n \geq 0;\ Z_n \in B_{\frac{\epsilon}{2}}(y_0)\big\} < \min\big\{n \geq 0;\ Z_n \not\in B_\delta(y_0)\big\}\Big) \\ &\geq \frac{1}{2}\eta_1, \end{aligned}$$

which completes the proof of (5.35). □

The building block of the proof of (5.34) will be the annulus walk estimate in Theorem 5.11.

Lemma 5.18 *Let p, N, r be as in Definition 5.5. There exists a constant $\bar{p} < \frac{1}{2}$ and a sequence $\{k_i\}_{i=1}^\infty$, depending only on p and N, satisfying:*

$$\begin{aligned} &k_1 > 2(1+r), \\ &k_2 > k_1 + 1 + r, \\ &k_2 + 1 + r < k_3 \leq \frac{k_2(k_2+1)}{k_1} - 1 \end{aligned} \tag{5.36}$$

and:

$$\begin{aligned} &k_{i+1} > k_i + 1 + r, \qquad \frac{k_i}{k_1} \geq \frac{k_{i+1}+1}{k_2+1}, \\ &\text{and} \quad \frac{k_i}{k_1} = \frac{k_{i+2}-r}{k_3 - r} \qquad \text{for all } i \geq 1. \end{aligned} \tag{5.37}$$

Moreover, for any $i \geq 1$, any $\epsilon > 0$ and any z_0 satisfying $k_i\epsilon < |z_0 - y_0| < \frac{k_i}{k_1}(k_2+1)\epsilon$, there exists a strategy $\sigma_{0,II,i+1}$ of Player II such that for every σ_I there holds:

$$\begin{aligned} \mathbb{P}\Big(&\min\{n \geq 0;\ Z_n \not\in \bar{B}_{(k_{i+2}-r)\epsilon}(y_0)\} \\ &< \min\{n \geq 0;\ Z_n \in B_{(k_i+1)\epsilon}(y_0)\}\Big) \leq \bar{p}, \end{aligned} \tag{5.38}$$

in the ϵ-walk $\{Z_n\}_{n=0}^\infty$.

Proof

1. Define first the three positive constants $\kappa_1, \kappa_2, \kappa_3$, depending only on p and N, with the following properties:

$$\begin{aligned} &\kappa_1 > 2(1+r), \\ &\kappa_2 > \kappa_1 + 1 + r \quad \text{and} \quad 2^{\frac{p-1}{p-N}}(\kappa_2+1) + 1 + r < \frac{\kappa_2(\kappa_2+1)}{\kappa_1}, \\ &2^{\frac{p-1}{p-N}}(\kappa_2+1) + r < \kappa_3 \leq \frac{\kappa_2(\kappa_2+1)}{\kappa_1} - 1. \end{aligned} \tag{5.39}$$

Indeed, fixing first $\kappa_1 > 2(1+r)$ and then κ_2 sufficiently large, the first three conditions above are easily achieved. The last condition follows from: $2^{\frac{p-1}{p-N}}(\kappa_2 + 1) + r < \frac{\kappa_2(\kappa_2+1)}{\kappa_1} - 1$, valid in view of the third inequality in (5.39).
Consider the annulus walk as in Theorem 5.11, with the radii $R_1 = \kappa_1$, $R_2 = \kappa_2 + 1$, $R_3 = \kappa_3 - r$ and the boundary thickness $\tilde{\epsilon}_0 = \frac{\kappa_1}{3} < \frac{R_1}{2}$. Observe that:

$$\frac{v(R_2) - v(R_1)}{v(R_3) - v(R_1)} < \frac{1}{2},$$

because of the last condition in (5.39):

$$\Big(2(\kappa_2+1)^{\frac{p-N}{p-1}} - \kappa_1^{\frac{p-N}{p-1}}\Big)^{\frac{p-1}{p-N}} < 2^{\frac{p-1}{p-N}}(\kappa_2+1) < \kappa_3 - r.$$

It now follows from (5.25), that there exists $\hat{\zeta} < 1$ with the property that for every $z_0 \in B_{R_2}(y_0) \setminus \bar{B}_{R_1}(y_0)$ and every $\zeta \leq \hat{\zeta}$, Player II has a strategy so that regardless of the strategy of Player I, there holds:

$$\mathbb{P}\Big(\tilde{Z}_{\tilde{\tau}} \notin \bar{B}_{R_3 - \zeta}(y_0)\Big) \leq \bar{p} < \frac{1}{2}. \tag{5.40}$$

Here, $\bar{p}$ depends again only on p and N, whereas $\{\tilde{Z}_n\}_{n=0}^\infty$ and $\tilde{\tau}$ correspond to the token positions and the stopping time in the boundary aware Tug-of-War game played on the annulus $\tilde{\mathcal{D}} = B_{R_3}(y_0) \setminus \bar{B}_{R_1}(y_0)$, thickened by the boundary of size $\tilde{\epsilon}_0$.
2. Let $\rho = \frac{1}{\hat{\zeta}} > 1$ and define k_1, k_2, k_3 (again depending only on p and N) by:

$$k_1 = \rho\kappa_1, \qquad k_2 + 1 = \rho(\kappa_2+1), \qquad k_3 - r = \rho(\kappa_3 - r). \tag{5.41}$$

By Corollary 5.13, used with the rescaling ρ, it follows that for every $\zeta \le 1$ and every starting position $z_0 \in B_{k_2+1}(y_0) \setminus \bar{B}_{k_1}(y_0)$ in the ζ-walk $\{Z_n\}_{n=0}^{\infty}$, Player II has a strategy so that for every strategy of Player I there holds:

$$\mathbb{P}\Big(\min\{n \ge 0;\ Z_n \notin \bar{B}_{k_3-r}(y_0)\} < \min\{n \ge 0;\ Z_n \in B_{k_1+\zeta}(y_0)\}\Big) \le \bar{p}. \tag{5.42}$$

Indeed, the event in the left-hand side above is a subset of the event $\{\tilde{Z}^{\rho}_{\tilde{\tau}^{\rho}} \notin \bar{B}_{k_3-r}(y_0) = \bar{B}_{\rho R_3}(y_0)\}$ where now $\{\tilde{Z}^{\rho}_n\}_{n=0}^{\infty}$ and $\tilde{\tau}^{\rho}$ correspond, as in the proof of Corollary 5.13, to the game on $\tilde{\mathcal{D}}^{\rho} = B_{\rho R_3}(y_0) \setminus \bar{B}_{\rho R_1}(y_0)$ with the step size $\zeta = \rho$. The probability of $\{\tilde{Z}^{\rho}_{\tilde{\tau}^{\rho}} \notin \bar{B}_{\rho R_3}(y_0)\}$ is clearly bounded by $\bar{p}$, according to (5.40).

Observe that k_1, k_2, k_3 in (5.41) satisfy (5.36) (we propose to check this statement in Exercise 5.21). For every $i \ge 4$, we define:

$$k_i \doteq \begin{cases} \alpha^{m-1}k_2 + \dfrac{\alpha^{m-1}-1}{\alpha-1}r & \text{for } i = 2m \\ \alpha^{m-1}k_3 + \dfrac{\alpha^{m-1}-1}{\alpha-1}r & \text{for } i = 2m+1 \end{cases}$$

where:

$$\alpha = \frac{k_3 - r}{k_1} > 1. \tag{5.43}$$

We, likewise, leave it as an exercise to prove (5.37).

For any fixed $i \ge 1$ and $\epsilon > 0$ we presently use the scaling property in Corollary 5.13, with the scaling factor $\rho = \frac{k_i}{k_1}\epsilon$. The claimed property (5.38) is now established in view of (5.42) and of the third condition in (5.37). Indeed, it suffices to write $\epsilon = \rho\zeta$ and apply (5.42) to $\zeta = \frac{k_1}{k_i} \le 1$ (Fig. 5.6). □

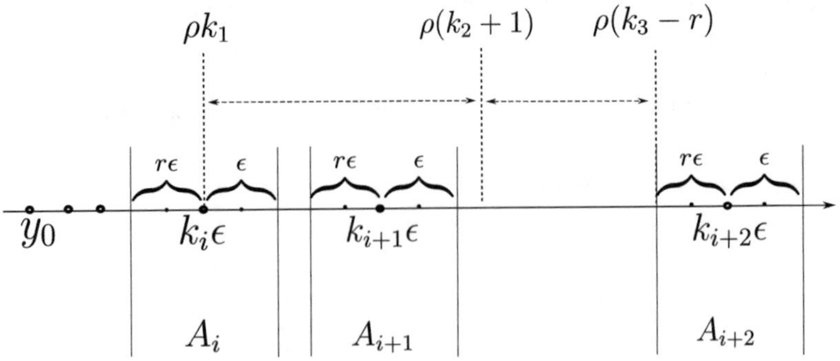

Fig. 5.6 Application of Theorem 5.11 in the proof of (5.38)

Fix $\delta \ll 1$ and, recalling the definition of α in (5.43), set:

$$\hat{\epsilon} = \frac{\delta}{k_3 - r} < \frac{\delta}{2+r} \qquad \text{and} \qquad \hat{\delta} = \frac{\delta}{\alpha}. \tag{5.44}$$

For a given $\epsilon \in (0, \hat{\epsilon})$, $z_0 \in B_{\hat{\delta}}(y_0)$, we now define the strategy $\sigma_{0,II}$ of Player II, and prove that (5.34) holds with some $\eta_1 > 0$ (depending only on p and N).

Firstly, using the fact that the sequence $\{k_i - r\}_{i=1}^{\infty}$ increases to $+\infty$, we set:

$$m \doteq \max\big\{i \geq 2;\ \delta \geq (k_{i+1} - r)\epsilon\big\}.$$

Note that in view of the third equality in (5.37) there holds:

$$|z_0 - y_0| < \frac{\delta}{\alpha} = \delta\frac{k_m}{k_{m+2} - r} < k_m\epsilon,$$

The strategy $\sigma_{0,II}$:

$$\sigma_{0,II}^n = \sigma_{0,II}^n\big(z_0, (z_1, w_1, a_1, b_1), \ldots, (z_n, w_n, a_n, b_n)\big) \qquad \text{for all } n \geq 0$$

is constructed separately in the following three cases (Fig. 5.7).

Case 1. If $|z_i - y_0| \geq (k_{m-1} + 1)\epsilon$ for all $i \leq n$, then we set:

$$\sigma_{0,II}^n = \sigma_{0,II,m}^n\big(z_0, (z_1, w_1, a_1, b_1), \ldots, (z_n, w_n, a_n, b_n)\big).$$

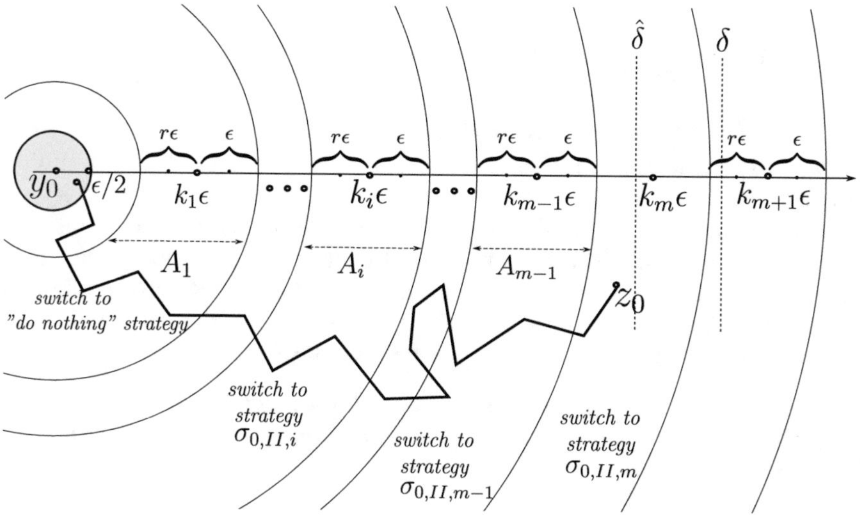

Fig. 5.7 Construction of the strategy $\sigma_{0,II}$ in the proof of Theorem 5.16

Case 2. If $|z_i - y_0| < (k_1 + 1)\epsilon$ for some $i \leq n$, then we set:

$$\sigma_{0,II}^n = 0.$$

Case 3. Otherwise, define:

$$i = i(z_0, z_1, \dots, z_n) \doteq \max\Big\{1 \leq i < m-1; \ \ \forall 0 \leq j \leq n \quad |z_j - y_0| \geq (k_i + 1)\epsilon\Big\},$$

$$j \doteq \min\Big\{0 \leq j \leq n; \ \ |z_j - y_0| < (k_{i+1} + 1)\epsilon\Big\},$$

and set:

$$\sigma_{0,II}^n = \sigma_{0,II,i+1}^{n-j}\Big(z_j, (z_{j+1}, w_{j+1}, a_{j+1}, b_{j+1}), \dots, (z_n, w_n, a_n, b_n)\Big).$$

In other words, denoting by $\{A_i\}_{i=1}^\infty$ the disjoint open rings in:

$$A_i \doteq B_{(k_i+1)\epsilon}(y_0) \setminus \bar{B}_{(k_i - r)\epsilon}(y_0) \qquad \text{for all } \ i \geq 1,$$

the strategy $\sigma_{0,II}$ follows the strategy $\sigma_{0,II,m}$ before entering the ring A_{m-1} (Case 1), then switches to $\sigma_{0,II,i}$ as soon as the token enters A_i, for any $i > 1$ (Case 3). Upon entering A_1, the strategy switches to the "do nothing" strategy (Case 2). Since:

$$|z_0 - y_0| < k_m \epsilon < (k_m + 1)\epsilon \leq \frac{k_{m-1}}{k_1}(k_2 + 1)\epsilon,$$

by the second inequality in (5.37), and since:

$$A_i \subset B_{\frac{k_{i-1}}{k_1}(k_2+1)\epsilon}(y_0) \setminus \bar{B}_{(k_{i-1}+1)\epsilon}(y_0) \qquad \text{for all } \ i \geq 2, \tag{5.45}$$

Cases 1 through 3 cover all possible scenarios.

It is also not hard to check that each $\sigma_{0,II}^n : H_n \to B_r(0) \subset \mathbb{R}^N$ is (Borel) measurable, as required.

Before completing the proof of Theorem 5.16, we make another simple observation, whose proof is similar to that of Lemma 3.13, to the effect that the "do nothing" strategy of Case 2 above advances the token, once in the ring A_1, to a position in the final ball $B_{\epsilon/2}(y_0)$, with positive probability that is bounded below by a constant depending only on p and N. In the application below, we will take $R = k_1 + 1$.

Lemma 5.19 *Let $R > 1$. There exists a probability bound $\bar{Q}_0 < 1$, depending only on p, N, R, such that for every $z_0 \in B_{R\epsilon}(0)$ and any strategies of Player I and Player II, there holds:*

$$\mathbb{P}\Big(\min\{n \geq 0; \ Z_n \in B_{\epsilon/2}(0)\} > \min\{n \geq 0; \ Z_n \notin B_{R\epsilon}(0)\}\Big) \leq \bar{Q}_0.$$

Proof Without loss of generality, we may assume that $z_0 = |z_0|e_1$. Consider the following set D_{adv} of "advancing" random outcomes (Fig. 5.8):

$$\epsilon D_{adv} \doteq \Big(-\frac{\epsilon}{2}, -\frac{\epsilon}{4}\Big) \times B^{N-1}_{\frac{\epsilon}{16R}}(0) \subset B_\epsilon(0),$$

where $B^{N-1}_{\frac{\epsilon}{16R}}(0)$ is the $(N-1)$-dimensional ball with centre 0 and radius $\frac{\epsilon}{16R}$.

Note that since $|z_0| < R\epsilon$, it takes $n \leq \lfloor 4R \rfloor$ moves by any vectors $\{w_i\}_{i=1}^n$ in D_{adv}, to achieve $|\langle Z_n, e_1\rangle| < \frac{\epsilon}{4}$. On the other hand, the sum of lengths of these advancements in the direction orthogonal to e_1 is smaller than $\lfloor 4R \rfloor \frac{\epsilon}{16R} \leq \frac{\epsilon}{4}$. Hence, for all $w_1, \ldots, w_{4R} \in \epsilon D_{adv}$ we have (Fig. 5.9):

$$\exists n \leq \lfloor 4R \rfloor \qquad z_0 + \epsilon \sum_{i=1}^n w_i \in \Big(-\frac{\epsilon}{4}, \frac{\epsilon}{4}\Big) \times B^{N-1}_{\frac{\epsilon}{4}}(0) \subset B_{\frac{\epsilon}{2}}(0).$$

Consequently:

$$\begin{aligned} &\mathbb{P}\Big(\min\{n \geq 0;\ Z_n \in B_{\epsilon/2}(0)\} < \min\{n \geq 0;\ Z_n \notin B_{R\epsilon}(0)\}\Big) \\ &\qquad \geq \mathbb{P}\Big(\big(D_{adv} \times \{3\} \times (0,1)\big)^{\lfloor 4R \rfloor} \times \prod_{i=\lfloor 4R \rfloor+1}^{\infty} \Omega_1\Big) \end{aligned}$$

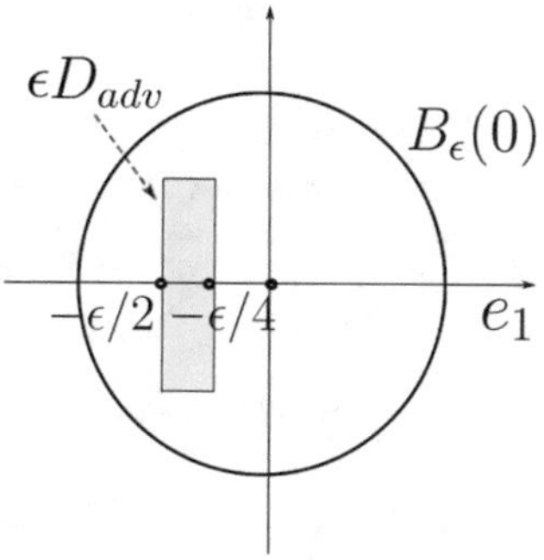

Fig. 5.8 The set of "advancing" outcomes in the proof of Lemma 5.19

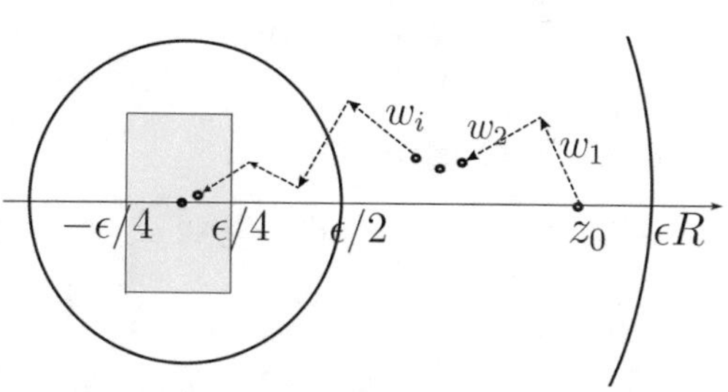

Fig. 5.9 The path of consecutive token positions advancing to $B_{\epsilon/2}(0)$

$$= \Big(\mathbb{P}_1\big(D_{adv} \times \{3\} \times (0,1)\big)\Big)^{\lfloor 4R \rfloor} = \left(\frac{V_{N-1}}{12(16R)^{N-1}V_N}\right)^{\lfloor 4R \rfloor},$$

proving the claim with: $\bar{Q}_0 = 1 - \left(\frac{V_{N-1}}{12(16R)^{N-1}V_N}\right)^{\lfloor 4R \rfloor} < 1.$ □

Applying Lemma 5.19 to $R = k_1 + 1$, we obtain the following bound:

Corollary 5.20 *For any $z_0 \in B_{(k_1+1)\epsilon}(y_0)$ we have:*

$$\begin{aligned} &\mathbb{P}\Big(\min\{n \geq 0;\ Z_n \in B_{\epsilon/2}(y_0)\} < \min\{n \geq 0;\ Z_n \notin B_{(k_1+1)\epsilon}(y_0)\}\Big) \\ &\qquad \geq 1 - \bar{Q}_0 = \left(\frac{V_{N-1}}{12(16(k_1+1))^{N-1}V_N}\right)^{4(k_1+1)}, \end{aligned} \tag{5.46}$$

where the expression in the right-hand side above depends only on p and N.

Proof of Theorem 5.16

1. In virtue of Lemma 5.17, it suffices to prove the estimate (5.34), where we fixed $\delta < \epsilon_0$ together with $\epsilon \in (0, \hat{\epsilon})$ and $z_0 \in B_{\hat{\delta}}(y_0)$ (according to $\hat{\epsilon}$ and $\hat{\delta}$ chosen as in (5.44) and for the strategy $\sigma_{0,II}$ defined above).

   To this end, we first remark that (5.34) trivially holds when $z_0 \in B_{(k_1-r)\epsilon}(y_0)$, because by (5.46) we get:

$$\begin{aligned} &\mathbb{P}\big(\min\{n \geq 0;\ Z_n \in B_{\frac{\epsilon}{2}}(y_0)\} < \min\{n \geq 0;\ Z_n \notin B_{\delta}(y_0)\}\big) \\ &\quad \geq \mathbb{P}\big(\min\{n \geq 0;\ Z_n \in B_{\frac{\epsilon}{2}}(y_0)\} < \min\{n \geq 0;\ Z_n \notin B_{(k_1+1)\epsilon}(y_0)\}\big) \geq 1 - \bar{Q}_0. \end{aligned}$$

   Otherwise, let $1 \leq i \leq m-2$ be such that $|z_0 - y_0| \in ((k_i - r)\epsilon, (k_{i+1} - r)\epsilon]$, or set $i = m-1$ in case of $|z_0 - y_0| > (k_{m-1} - r)\epsilon$. In this step we will show that:

$$\begin{aligned} &\mathbb{P}\big(\min\{n \geq 0;\ Z_n \in B_{\frac{\epsilon}{2}}(y_0)\} < \min\{n \geq 0;\ Z_n \notin B_{\delta}(y_0)\}\big) \\ &\quad \geq \mathbb{P}\big(j \doteq \min\{n \geq 0;\ Z_n \in A_i\} < \min\{n \geq 0;\ Z_n \in A_{i+2}\} \\ &\qquad\qquad \text{and } \min\{n \geq j;\ Z_n \in B_{\frac{\epsilon}{2}}(y_0)\} < \min\{n \geq j;\ Z_n \in A_{i+1}\}\big) \\ &\quad \geq (1 - \bar{p})(1 - \bar{Q}_i), \end{aligned} \tag{5.47}$$

   where we define, for all $i \geq 1$:

$$\bar{Q}_i \doteq \sup_{z_0 \in A_i} \sup_{\sigma_I} \mathbb{P}\big(\min\{n \geq 0;\ Z_n \in B_{\frac{\epsilon}{2}}(y_0)\} > \min\{n \geq 0;\ Z_n \in A_{i+1}\big) \tag{5.48}$$

   The first inequality in (5.47) is straightforward from the inclusion between the events in the left and right-hand sides.

To prove the second inequality, we proceed as in the proof of Theorem 5.9. Firstly, observe that if $j = 0$ then $z_0 \in A_i$ and (5.47) is simply a consequence of the definition in (5.48). To treat the case $j > 0$, denote:

$$\tilde{\Omega} = \Omega \cap \Big\{ \min\{n \geq 0;\ Z_n \in A_i\} < \min\{n \geq 0;\ Z_n \in A_{i+2}\} \Big\}.$$

and consider the probability space $(\tilde{\Omega}, \tilde{\mathcal{F}}, \tilde{\mathbb{P}})$ in:

$$\tilde{\mathcal{F}} = \{C \cap \tilde{\Omega};\ C \in \mathcal{F}\} \quad \text{and} \quad \tilde{\mathbb{P}}(C) = \frac{\mathbb{P}(C)}{\mathbb{P}(\tilde{\Omega})} \quad \text{for all } C \in \tilde{\mathcal{F}},$$

where by (5.38) we assure that:

$$\mathbb{P}(\tilde{\Omega}) \geq 1 - \bar{p} > 0. \tag{5.49}$$

Define also the measurable space $(\Omega_{fin}, \mathcal{F}_{fin})$ by setting $\Omega_{fin} = \bigcup_{n=1}^{\infty} \Omega_n$ and by taking $\mathcal{F}_{fin}$ to be the smallest σ-algebra containing $\bigcup_{n=1}^{\infty} \mathcal{F}_n$. There, $j = \min\{n \geq 1;\ Z_n \in A_i\}$ defined as in (5.47) is a stopping time on $\tilde{\Omega}$ with respect to the filtration $\{C \cap \tilde{\Omega};\ C \in \mathcal{F}_n\}_{n=0}^{\infty}$. We further consider the random variables:

$$\begin{aligned} &Y_1 : \tilde{\Omega} \to \Omega_{fin} \qquad Y_1\big(\{(w_n, a_n, b_n)\}_{n=1}^{\infty}\big) \doteq \{(w_n, a_n, b_n)\}_{n=1}^{j}, \\ &Y_2 : \tilde{\Omega} \to \Omega \qquad\quad Y_2\big(\{(w_n, a_n, b_n)\}_{n=1}^{\infty}\big) \doteq \{(w_n, a_n, b_n)\}_{n=j+1}^{\infty}, \end{aligned} \tag{5.50}$$

and note that they are independent (see Exercise 5.21 (iii)).

Applying Lemma A.21 to Y_1, Y_2 and to the measurable indicator function $F : \Omega_{fin} \times \Omega \to \mathbb{R}$ (see Exercise 5.21 (iv)):

$$F\big(\{(w_n, a_n, b_n)\}_{n=1}^{s}, \{(w_n, a_n, b_n)\}_{n=s+1}^{\infty}\big) = \mathbb{1}_{\big\{\min\{n>s;\ Z_n \in B_{\frac{\epsilon}{2}}(y_0)\} < \min\{n>s;\ Z_n \in A_{i+1}\}\big\}},$$

we obtain:

$$\begin{aligned} &\mathbb{P}\big(j \doteq \min\{n \geq 0;\ Z_n \in A_i\} < \min\{n \geq 0;\ Z_n \in A_{i+2}\} \\ &\qquad\qquad \text{and } \min\{n \geq j;\ Z_n \in B_{\frac{\epsilon}{2}}(y_0)\} < \min\{n \geq j;\ Z_n \in A_{i+1}\}\big) \\ &= \mathbb{P}(\tilde{\Omega}) \cdot \int_{\tilde{\Omega}} F\big(Y_1(\omega), Y_2(\omega)\big)\, \mathrm{d}\tilde{\mathbb{P}}(\omega) = \mathbb{P}(\tilde{\Omega}) \cdot \int_{\tilde{\Omega}} f(\omega_1)\, \mathrm{d}\tilde{\mathbb{P}}(\omega_1), \end{aligned} \tag{5.51}$$

where for a given $\omega_1 = \{(w_n, a_n, b_n)\}_{n=1}^\infty \in \tilde{\Omega}$ the integrand function f returns:

$$f(\omega_1) = \tilde{\mathbb{P}}\Big(\omega_2 = \{(\bar{w}_n, \bar{a}_n, \bar{b}_n)\}_{n=1}^\infty \in \tilde{\Omega};\\ \min\Big\{n > j(\omega_1);\ Z_n\big(\{(w_s, a_s, b_s)\}_{s=1}^{j(\omega_1)}, \{(\bar{w}_s, \bar{a}_s, \bar{b}_s)\}_{s=j(\omega_2)+1}^\infty\big) \in B_{\frac{\epsilon}{2}}(y_0)\Big\}\\ < \min\Big\{n > j(\omega_1);\ Z_n \in A_{i+1}\Big\}\Big).$$

Consequently, for every $\omega_1 \in \tilde{\Omega}$ as above, we have:

$$\mathbb{P}(\tilde{\Omega}) \cdot f(\omega_1) = \mathbb{P}\Big(\omega_2 = \{(\bar{w}_n, \bar{a}_n, \bar{b}_n)\}_{n=1}^\infty \in \tilde{\Omega};\\ \min\Big\{n > 0;\ Z_n^{Z_{j(\omega_1)}, \sigma_I, \sigma_{0,II}}\big(\{(\bar{w}_s, \bar{a}_s, \bar{b}_s)\}_{s=j(\omega_2)+1}^\infty\big) \in B_{\frac{\epsilon}{2}}(y_0)\Big\}\\ < \min\Big\{n > 0;\ Z_n \in A_{i+1}\Big\}\Big)\\ \geq \mathbb{P}(\tilde{\Omega}) \cdot \big(1 - \bar{Q}_i\big) \geq \big(1 - \bar{p}\big)\big(1 - \bar{Q}_i\big),$$

in view of the definition in (5.48) and the estimate (5.49). Thus, (5.51) implies (5.47), as requested.

2. Define now, for all $i \geq 1$, all $z_0 \in A_i$ and all strategies σ_I of Player I:

$$Q_i(z_0, \sigma_I) = \mathbb{P}\big(\min\{n \geq 0;\ Z_n^{z_0, \sigma_I, \sigma_{0,II}} \in A_1\}\\ > \min\{n \geq 0;\ Z_n \in A_{i+1}\}\big), \tag{5.52}$$

$$Q_i = \sup_{z_0 \in A_i} \sup_{\sigma_I} Q_i(z_0, \sigma_I).$$

We then have, for all i, z_0 and σ_I as specified above:

$$\mathbb{P}\big(\min\{n \geq 0;\ Z_n^{z_0, \sigma_I, \sigma_{0,II}} \in A_{i+1}\} < \min\{n \geq 0;\ Z_n \in B_{\frac{\epsilon}{2}}(y_0)\}\big)\\ = \mathbb{P}\big(\min\{n \geq 0;\ Z_n \in A_{i+1}\} < \min\{n \geq 0;\ Z_n \in A_1\}\big)\\ + \mathbb{P}\big(j \doteq \min\{n \geq 0;\ Z_n \in A_1\} < \min\{n \geq 0;\ Z_n \in A_{i+1}\}\\ \text{and } \min\{n \geq j;\ Z_n \in A_{i+1}\} < \min\{n \geq j;\ Z_n \in B_{\frac{\epsilon}{2}}(y_0)\}\big)\\ \leq Q_i(z_0, \sigma_I)\\ + \big(1 - Q_i(z_0, \sigma_I)\big) \cdot\\ \cdot \sup_{z_0 \in A_1} \sup_{\sigma_I} \mathbb{P}\big(\min\{n \geq 0;\ Z_n \in A_{i+1}\} < \min\{n \geq 0;\ Z_n \in B_{\frac{\epsilon}{2}}(y_0)\}\big)\\ \leq Q_i(z_0, \sigma_I) + \big(1 - Q_i(z_0, \sigma_I)\big)\bar{Q}_0\\ = \bar{Q}_0 + \big(1 - \bar{Q}_0\big) Q_i(z_0, \sigma_I) \leq \bar{Q}_0 + \big(1 - \bar{Q}_0\big) Q_i. \tag{5.53}$$

The bound of the second term in the left-hand side of first inequality above by the product of probabilities:

$$\begin{aligned}&\big(1 - Q_i(z_0, \sigma_I)\big) \cdot \\ &\cdot \sup_{z_0\in A_1} \sup_{\sigma_I} \mathbb{P}\big(\min\{n \geq 0;\ Z_n \in A_{i+1}\} < \min\{n \geq 0;\ Z_n \in B_{\frac{\epsilon}{2}}(y_0)\}\big)\end{aligned}$$

is obtained in the same manner as in the proof of the bound in (5.47), so we suppress the details. Further, the second inequality in (5.53) follows from (5.46) and the last inequality is a consequence of the definition of Q_i. Therefore, we arrive at:

$$\bar{Q}_i \leq \bar{Q}_0 + (1 - \bar{Q}_0)Q_i \qquad \text{for all } i \geq 1. \tag{5.54}$$

3. We now estimate the probability bound Q_i in (5.52). For $i = 1$, we have: $Q_1 = 0$. Fix $i \geq 2$ and $z_0 \in A_i$, together with a strategy σ_I of Player I. Then:

$$\begin{aligned}1 - Q_i(z_0, \sigma_I) &= \mathbb{P}\big(j \doteq \min\{n \geq 0;\ Z_n \in A_{i-1}\} < \min\{n \geq 0;\ Z_n \in A_{i+1}\} \\ &\qquad\qquad \text{and } \min\{n \geq j;\ Z_n \in A_1\} < \min\{n \geq j;\ Z_n \in A_{i+1}\}\big) \\ &\geq (1 - \bar{p}) \cdot \inf_{z_0\in A_{i-1}} \inf_{\sigma_I} \mathbb{P}\big(\min\{n \geq 0;\ Z_n \in A_1\} < \min\{n \geq 0;\ Z_n \in A_{i+1}\}\big),\end{aligned}$$

in view of (5.38) and reasoning again as in the proof of (5.47). On the other hand, for every $z_0 \in A_{i-1}$ and every σ_I we have:

$$\begin{aligned}&\mathbb{P}\big(\min\{n \geq 0;\ Z_n \in A_1\} < \min\{n \geq 0;\ Z_n \in A_{i+1}\}\big) \\ &\quad = \mathbb{P}\big(\min\{n \geq 0;\ Z_n \in A_1\} < \min\{n \geq 0;\ Z_n \in A_i\}\big) \\ &\qquad + \mathbb{P}\big(j \doteq \min\{n \geq 0;\ Z_n \in A_i\} < \min\{n \geq 0;\ Z_n \in A_1\} \\ &\qquad\qquad \text{and } \min\{n \geq j;\ Z_n \in A_1\} < \min\{n \geq j;\ Z_n \in A_{i+1}\}\big) \\ &\quad = \big(1 - Q_{i-1}(z_0, \sigma_I)\big) + \big(1 - Q_i\big)Q_{i-1}(z_0, \sigma_I) \\ &\quad = 1 - Q_i Q_{i-1}(z_0, \sigma_I) \geq 1 - Q_i Q_{i-1},\end{aligned}$$

by the definition (5.52) and reasoning as in the proof of (5.47).

Concluding, we obtain:

$$1 - Q_i(z_0, \sigma_I) \geq (1 - \bar{p})\big(1 - Q_i Q_{i-1}\big) \qquad \text{for all } \ i \geq 2,\ z_0 \in A_i,\ \sigma_I,$$

which results in: $Q_i \leq 1 - (1 - \bar{p})\big(1 - Q_i Q_{i-1}\big)$ or equivalently:

$$\begin{cases} 0 \leq Q_i \leq \dfrac{\bar{p}}{1 - (1 - \bar{p})Q_{i-1}} & \text{for all } \ i \geq 2 \\ Q_1 = 0. \end{cases} \tag{5.55}$$

It is now an immediate observation that:

$$0 \le Q_i \le \frac{\bar{p}}{1 - (1 - \bar{p})} \qquad \text{for all } i \ge 1.$$

Indeed, the above inequality holds for $i = 1$ and if it is valid for Q_{i-1} then $0 \le (1 - \bar{p})Q_{i-1} \le \bar{p}$ so (5.55) implies the same bound for Q_i.

By (5.54) and the above, we get:

$$1 - \bar{Q}_i \ge \big(1 - \bar{Q}_0\big) \cdot \big(1 - Q_i\big) \ge \big(1 - \bar{Q}_0\big)\frac{1 - 2\bar{p}}{1 - \bar{p}}.$$

Finally, (5.47) yields:

$$\mathbb{P}\big(\min\{n \ge 0;\ Z_n \in B_{\frac{\epsilon}{2}}(y_0)\} < \min\{n \ge 0;\ Z_n \notin B_\delta(y_0)\}\big) \ge \big(1 - \bar{Q}_0\big)\big(1 - 2\bar{p}\big).$$

The constant in the right-hand side above is clearly positive, as $\bar{p} < \frac{1}{2}$ by (5.40), and it depends only on the exponent p and the dimension N. This ends the proof of (5.34) and achieves Theorem 5.16. □

Exercise 5.21

(i) Verify conditions in (5.36).
(ii) Verify the properties (5.37).
(iii) Prove that the random variables Y_1 and Y_2 in (5.50) are independent.
(iv) Show that the function $F : \Omega_{fin} \times \Omega \to \Omega$ used in (5.51) is measurable.

5.7 Bibliographical Notes

The game-theoretical constructions in this chapter are based on the seminal analysis in Peres and Sheffield (2008), here adapted to the dynamic programming principle (4.1), and expanded to contain all the probability-related details. Definition 5.5 is a natural extension of the simpler notion of walk-regularity in Definition 2.12 that was valid for the random walk process (the ball walk discussed in Chap. 2) and related to the linear case $p = 2$. We recall that this notion has been, in turn, derived from the classical Doob's regularity condition of boundary points for the Brownian motion, discussed in Sect. 3.7*.

Chapter 6
Mixed Tug-of-War with Noise: Case $p \in (1, \infty)$

Our discussion was started in Chap. 2 by analysing the linear case $p = 2$. There, we derived the probabilistic interpretation of harmonic functions via their mean value property, used as the dynamic programming principle of the *ball walk*. In Chaps. 3 and 4 we developed a parallel description for p-harmonic functions, via the asymptotic mean value expansions (3.13), (3.14) viewed as the *game-theoretical dynamic programming principles*, valid for exponent ranges $p \geq 2$ and $p > 2$, respectively. In the present chapter, we will follow the same program, based on a family of asymptotic expansions that are viable in the entire range $1 < p < \infty$.

In Sect. 6.1 below, we introduce a new *finite difference approximation* to the Dirichlet problem for the homogeneous p-Laplace equation posed on a domain $\mathcal{D} \subset \mathbb{R}^N$, under the boundary data F on $\partial\mathcal{D}$. The approximation is based on superposing the "deterministic averages $\frac{1}{2}(\inf + \sup)$" taken over balls, with the "stochastic averages $\fint$", taken over N-dimensional ellipsoids whose aspect ratio depends on N, p and whose orientations span all directions while determining inf / sup. In Sect. 6.2 we prove the well-posedness of the induced *mean value equations*, utilizing the interpolation to the boundary as in Chap. 4. The probabilistic interpretation u^ϵ of their solutions, at each averaging scale $\epsilon > 0$, is put forward in Sect. 6.3, where we describe the *mixed Tug-of-War game with noise*. In this game, the token is initially placed at $x_0 \in \mathcal{D}$, and at each step it is further advanced according to the following rule. First, either of the two players (each acting with probability $\frac{1}{2}$) shifts the token by a chosen vector y of length at most ϵ; second, the token is further shifted within an ellipsoid centred at the current position, with radius $\gamma_p\epsilon$, oriented along $|y|$, and with aspect ratio that varies quadratically between 1 and another parameter ρ_p as the magnitude of the shift $|y|$ increases from 0 to ϵ. The scaling factors γ_p and ρ_p depend on p and N. Whenever the token reaches the 2ϵ-neighbourhood of $\partial\mathcal{D}$, the game is stopped with probability proportional to: 1 minus the distance of the token's location from $\mathbb{R}^N \setminus (\mathcal{D} + B_\epsilon(0))$. Then, $u^\epsilon(x_0)$ is set to be the expectation of F at the stopping position $x_\tau \in \mathcal{D}$, subject to both

M. Lewicka, *A Course on Tug-of-War Games with Random Noise*, Universitext,
https://doi.org/10.1007/978-3-030-46209-3_6

players playing optimally. As in Chap. 3, the optimality criterion is based on Player II disbursing the payoff $F(x_\tau)$ to Player I at the termination of the game; due to the min-max property, the order of supremizing the outcomes over strategies of one player and infimizing over strategies of the opponent, is immaterial.

In Sect. 6.4, we define the related *game-regularity* of the boundary points. As in Chap. 5, this notion is essentially equivalent to the local equicontinuity of the family $\{u^\epsilon\}_{\epsilon\to 0}$, and ultimately to its uniform convergence to the unique viscosity solution of the studied Dirichlet problem. It is expected that game-regularity is equivalent to the Wiener p-regularity condition in Definition C.47 and Theorem C.48, as discussed in Appendix C. In the absence of such result at the time of preparing these Course Notes, we prove in Sect. 6.4 that a sufficient condition for game-regularity is provided by the *exterior corkscrew condition*. Another sufficient condition $p > N$ can be achieved along the same lines as in Chap. 5, and is thus left to the reader. In Sect. 6.5 we show that in dimension $N = 2$, every *simply connected* domain is game-regular.

6.1 The Third Averaging Principle

For $\gamma, \rho > 0$ and a unit vector $\nu \in \mathbb{R}^N$, we denote by $E(0, \gamma; \rho, \nu) \subset \mathbb{R}^N$ the ellipsoid centred at 0, with radius γ, and with aspect ratio ρ oriented along ν (Fig. 6.1):

$$E(0, \gamma; \rho, \nu) = \Big\{y \in \mathbb{R}^N;\ \frac{\langle y, \nu\rangle^2}{\rho^2} + |y - \langle y, \nu\rangle\nu|^2 < \gamma^2\Big\}.$$

For $x \in \mathbb{R}^N$, we have the translated ellipsoid:

$$E(x, \gamma; \rho, \nu) = x + E(0, \gamma; \rho, \nu).$$

Note that, when $\nu = 0$, this formula also makes sense and returns the ball $E(x, \gamma; \rho, 0) = B_\gamma(x)$. Given a function $u : \mathbb{R}^N \to \mathbb{R}$, its average will be denoted:

$$\mathcal{A}\big(u; \gamma, \rho, \nu\big)(x) = \fint_{E(x,\gamma;\rho,\nu)} u(y)\,\mathrm{d}y = \fint_{B_1(0)} u\big(x + \gamma y + \gamma(\rho - 1)\langle y, \nu\rangle\nu\big)\,\mathrm{d}y.$$

In particular, $\mathcal{A}(u; \gamma, 1, \nu)(x) = \mathcal{A}(u; \gamma, \rho, 0)(x) = \mathcal{A}_\gamma u(x)$ is consistent with the notation $\mathcal{A}_\gamma u(x) = \fint_{B_\gamma(x)} u(y)\,\mathrm{d}y$ from Chap. 2. We also observe the following linear change of variables:

$$B_1(0) \ni y \mapsto \gamma\rho\langle y, \nu\rangle\nu + \gamma\big(y - \langle y, \nu\rangle\nu\big) \in E(0, \gamma; \rho, \nu)$$

that will be often used in the sequel.

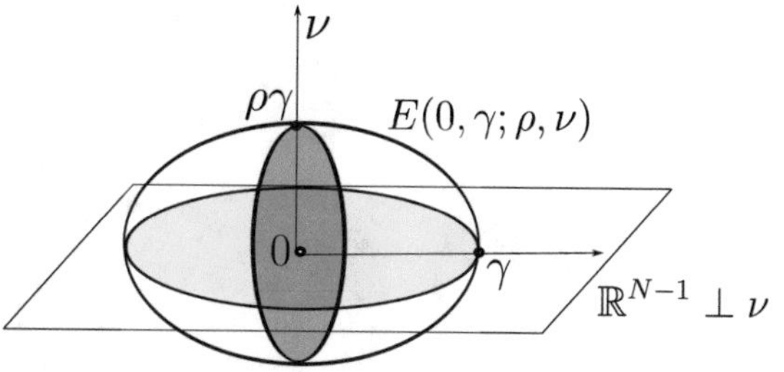

Fig. 6.1 The averaging ellipsoid in $\mathcal{A}(u;\gamma,\rho,\nu)(x)$

Theorem 6.1 *Given $p \in (1,\infty)$, fix any scaling factors $\gamma_p, \rho_p > 0$ with:*

$$\frac{N+2}{\gamma_p^2} + \rho_p^2 = p - 1. \tag{6.1}$$

Let $\mathcal{D} \subset \mathbb{R}^N$ be an open set and assume that $u \in C^2(\mathcal{D})$. Then, for every $x_0 \in \mathcal{D}$ such that $\nabla u(x_0) \neq 0$, and $\epsilon > 0$ satisfying $\bar{B}_{\epsilon(1+\gamma_p(1\vee\rho_p))}(x_0) \subset \mathcal{D}$, we have (Fig. 6.2):

$$\begin{aligned} \frac{1}{2}\Big(\inf_{x\in B_\epsilon(x_0)} + \sup_{x\in B_\epsilon(x_0)}\Big)\mathcal{A}\Big(u;\gamma_p\epsilon, 1+(\rho_p-1)\frac{|x-x_0|^2}{\epsilon^2}, \frac{x-x_0}{|x-x_0|}\Big)(x) \\ = u(x_0) + \frac{\gamma_p^2\epsilon^2}{2(N+2)}|\nabla u(x_0)|^{2-p}\Delta_p u(x_0) + o(\epsilon^2). \end{aligned} \tag{6.2}$$

The rate of convergence $o(\epsilon^2)$ depends on: p, N, γ_p and (in increasing manner) on $|\nabla u(x_0)|$, $|\nabla^2 u(x_0)|$ and the modulus of continuity of $\nabla^2 u$ at x_0.

It is convenient to write the expression in the left-hand side of the formula (6.2) as the average $\frac{1}{2}(\inf + \sup)$ on the ball $B_\epsilon(x_0)$, of the function $x \mapsto f_u(x;x_0,\epsilon)$ in:

$$\begin{aligned} f_u(x;x_0,\epsilon) &= \mathcal{A}\Big(u;\gamma\epsilon, 1+(\rho-1)\frac{|x-x_0|^2}{\epsilon^2}, \frac{x-x_0}{|x-x_0|}\Big)(x) \\ &= \fint_{B_1(0)} u\Big(x+\gamma\epsilon y + \frac{\gamma(\rho-1)}{\epsilon}\langle y, x-x_0\rangle(x-x_0)\Big)\,\mathrm{d}y, \end{aligned} \tag{6.3}$$

where $\gamma = \gamma_p$ and $\rho = \rho_p$. We will frequently use the notation:

$$S_\epsilon u(x_0) = \frac{1}{2}\Big(\inf_{x\in B_\epsilon(x_0)} + \sup_{x\in B_\epsilon(x_0)}\Big) f_u(x;x_0,\epsilon). \tag{6.4}$$

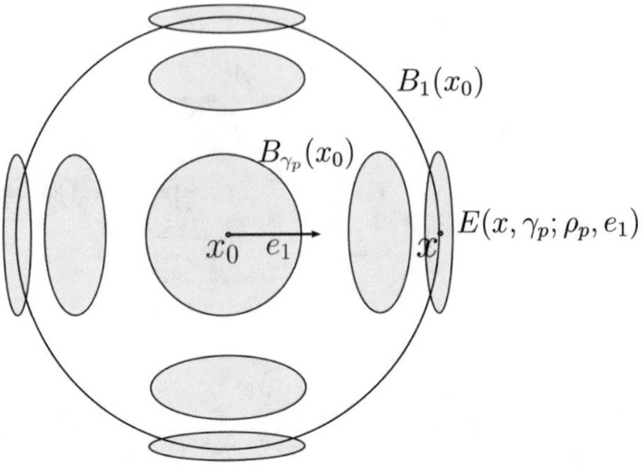

Fig. 6.2 The sampling ellipsoids in the expansion (6.2), when $\epsilon = 1$

For each $x \in B_\epsilon(x_0)$ the integral quantity in (6.3) returns the average of u on the N-dimensional ellipse centred at x, with radius $\gamma\epsilon$, and with aspect ratio $\frac{\epsilon^2+(\rho-1)|x-x_0|^2}{\epsilon^2}$ along the orientation vector $\frac{x-x_0}{|x-x_0|}$. Equivalently, writing $x = x_0 + \epsilon y$, the value $f_u(x; x_0, \epsilon)$ is the average of u on the scaled ellipse:

$$x_0 + \epsilon E\big(y, \gamma; 1 + (\rho - 1)|y|^2, \frac{y}{|y|}\big).$$

Since the aspect ratio changes smoothly from 1 to ρ as $|x - x_0|$ increases from 0 to ϵ, the said ellipse coincides with the ball $B(x_0, \gamma\epsilon)$ at $x = x_0$ and it interpolates as $|x - x_0| \to \epsilon-$, to $E\big(x, \gamma\epsilon; \rho, \frac{x-x_0}{|x-x_0|}\big)$.

Proof of Theorem 6.1

1. We fix $\gamma, \rho > 0$ and consider the Taylor expansion of u at a given $x \in B_\epsilon(x_0)$ under the integral in (6.3). Observe that the first order increments are linear in y, hence they integrate to 0 on $B_1(0)$. These increments are of order ϵ, so:

$$\begin{aligned}
&f_u(x;\ x_0, \epsilon)\\
&= u(x) + \frac{1}{2}\Big\langle \nabla^2 u(x) : \fint_{B_1(0)} \Big(\gamma\epsilon y + \frac{\gamma(\rho-1)}{\epsilon}\langle y, x - x_0\rangle (x - x_0)\Big)^{\otimes 2}\,\mathrm{d}y\Big\rangle + o(\epsilon^2)\\
&= u(x) + \frac{\gamma^2}{2}\Big\langle \nabla^2 u(x) : \epsilon^2 \fint_{B_1(0)} y^{\otimes 2}\,\mathrm{d}y\\
&\qquad\qquad + 2(\rho - 1)\fint_{B_1(0)} \langle y, x - x_0\rangle y\,\mathrm{d}y \otimes (x - x_0)\\
&\qquad\qquad + \frac{(\rho-1)^2}{\epsilon^2}\Big(\fint_{B_1(0)} \langle y, x - x_0\rangle^2\,\mathrm{d}y\Big)(x - x_0)^{\otimes 2}\Big\rangle + o(\epsilon^2).
\end{aligned} \tag{6.5}$$

Recall that: $\mathcal{A}_\epsilon y^{\otimes 2}(0) = \frac{1}{N+2} Id_N$ by Exercise 3.7 (i). Hence (6.5) becomes:

$$\begin{aligned} f_u(x;x_0,\epsilon) &= u(x) + \frac{\gamma^2\epsilon^2}{2(N+2)}\Delta u(x) \\ &+ \frac{\gamma^2(\rho-1)}{2}\Big(\frac{2}{N+2} + \frac{(\rho-1)|x-x_0|^2}{\epsilon^2(N+2)}\Big)\big\langle\nabla^2 u(x) : (x-x_0)^{\otimes 2}\big\rangle + o(\epsilon^2) \\ &= \bar{f}_u(x;x_0,\epsilon) + o(\epsilon^2), \end{aligned}$$

where a further Taylor expansion of u at x_0 gives:

$$\begin{aligned} \bar{f}_u(x;x_0,\epsilon) &= u(x_0) + \langle\nabla u(x_0), x-x_0\rangle + \frac{\gamma^2\epsilon^2}{2(N+2)}\Delta u(x_0) \\ &+ \Big(\frac{1}{2} + \frac{\gamma^2(\rho-1)}{2}\Big(\frac{2}{N+2} + \frac{(\rho-1)|x-x_0|^2}{\epsilon^2(N+2)}\Big)\Big)\big\langle\nabla^2 u(x_0) : (x-x_0)^{\otimes 2}\big\rangle. \end{aligned}$$

The left-hand side of (6.2) thus satisfies:

$$\begin{aligned} &\frac{1}{2}\Big(\inf_{x\in B_\epsilon(x_0)} f_u(x;x_0,\epsilon) + \sup_{x\in B_\epsilon(x_0)} f_u(x;x_0,\epsilon)\Big) \\ &\qquad = \frac{1}{2}\Big(\inf_{x\in B_\epsilon(x_0)} \bar{f}_u(x;x_0,\epsilon) + \sup_{x\in B_\epsilon(x_0)} \bar{f}_u(x;x_0,\epsilon)\Big) + o(\epsilon^2), \end{aligned} \tag{6.6}$$

Since on $B_\epsilon(x_0)$ we have: $\bar{f}_u(x;x_0,\epsilon) = u(x_0) + \langle\nabla u(x_0), x-x_0\rangle + O(\epsilon^2)$, the assumption $\nabla u(x_0)\neq 0$ implies that the continuous function $\bar{f}_u(\cdot\,;x_0,\epsilon)$ attains its extrema on the boundary $\partial B_\epsilon(x_0)$, provided that ϵ is sufficiently small. This reasoning justifies that $\bar{f}_u$ in (6.6) may be replaced by the quadratic polynomial:

$$\begin{aligned} \bar{\bar{f}}_u(x;x_0,\epsilon) &= u(x_0) + \frac{\gamma^2\epsilon^2}{2(N+2)}\Delta u(x_0) + \langle\nabla u(x_0), x-x_0\rangle \\ &\qquad + \Big(\frac{1}{2} + \frac{\gamma^2(\rho^2-1)}{2(N+2)}\Big)\big\langle\nabla^2 u(x_0) : (x-x_0)^{\otimes 2}\big\rangle. \end{aligned}$$

2. We now recall that $\bar{\bar{f}}_u$ attains its extrema on $\bar{B}_\epsilon(x_0)$, up to error $O(\epsilon^3)$ whenever r is sufficiently small, precisely at the opposite boundary points $x_0 + \epsilon\frac{\nabla u(x_0)}{|\nabla u(x_0)|}$ and $x_0 - \epsilon\frac{\nabla u(x_0)}{|\nabla u(x_0)|}$, as shown in Step 3 of the proof of Theorem 3.4. Consequently,

for $\gamma = \gamma_p, \rho = \rho_p$ satisfying (6.1), there holds:

$$\begin{aligned}
&\frac{1}{2}\Big(\inf_{x\in B_\epsilon(x_0)} \bar{\bar{f}}_u(x;x_0,\epsilon) + \sup_{x\in B_\epsilon(x_0)} \bar{\bar{f}}_u(x;x_0,\epsilon)\Big)\\
&\quad = u(x_0) + \frac{\gamma^2\epsilon^2}{2(N+2)}\Delta u(x_0) + \epsilon^2\Big(\frac{1}{2} + \frac{\gamma^2(\rho^2-1)}{2(N+2)}\Big)\Delta_\infty u(x_0) + O(\epsilon^3)\\
&\quad = u(x_0) + \frac{\gamma^2\epsilon^2}{2(N+2)}\Big(\Delta u(x_0) + \Big(\frac{N+2}{\gamma^2} + \rho^2 - 1\Big)\Delta_\infty u(x_0)\Big) + O(\epsilon^3),
\end{aligned} \tag{6.7}$$

further equal to:

$$\begin{aligned}
u(x_0) &+ \frac{\gamma_p^2\epsilon^2}{2(N+2)}\Big(\Delta u(x_0) + (p-2)\Delta_\infty u(x_0)\Big) + O(\epsilon^3)\\
&= u(x_0) + \gamma_p^2\epsilon^2\frac{|\nabla u(x_0)|^{2-p}}{2(N+2)}\Delta_p u(x_0) + O(\epsilon^3).
\end{aligned}$$

This completes the proof of (6.2) in view of (6.6).

Remark 6.2

(i) When $p \to \infty$, one can take $\rho_p = 1$ and $\gamma_p \sim 0$ in (6.1), whereas for $\Delta_\infty u(x_0) = 0$, the asymptotic formula (6.2) formally becomes: $u(x_0) = \frac{1}{2}\big(\inf_{x\in B_\epsilon(x_0)} u(x) + \sup_{x\in B_\epsilon(x_0)} u(x)\big)$, consistently with the AMLE property of the ∞-harmonic functions (see Remark 3.6 (iii)).

(ii) When $p = 2$, then choosing $\rho_p = 1$ and $\gamma_p \sim \infty$ corresponds to taking both types of averages on balls whose radii have ratio $\sim \infty$. Equivalently, one may take the integral average on $B_\epsilon(x_0)$ and the external one on $B_0(x_0) \sim \{x_0\}$, consistently with the mean value property of harmonic functions: $u(x_0) = \mathcal{A}_\epsilon u(x_0)$.

(iii) On the other hand, when $p \to 1+$, then there must be $\rho_p \to 0+$ and the critical choice $\rho_p = 0$ is the only one valid for every $p \in (1,\infty)$. It corresponds to varying the aspect ratio along the radius of $B_\epsilon(x_0)$ from 1 to 0 rather than to $\rho_p > 0$, and taking the sampling domains to be the ellipsoids: $E\big(x, \gamma\epsilon; 1 - \frac{|x-x_0|^2}{\epsilon^2}, \frac{x-x_0}{|x-x_0|}\big)$, with the radius $\gamma\epsilon$ scaled by the factor $\gamma = \sqrt{\frac{N+2}{p-1}}$. At $x = x_0$, the aforementioned ellipsoid coincides with the ball $B_{\gamma\epsilon}(x_0)$, whereas as $|x - x_0| \to \epsilon-$ it degenerates to the $(N-1)$-dimensional ball:

$$E\Big(x, \gamma\epsilon; 0, \frac{x-x_0}{|x-x_0|}\Big) = x + \Big\{y \in \mathbb{R}^N;\ \langle y, x-x_0\rangle = 0 \text{ and } |y| < \epsilon\sqrt{\frac{N+2}{p-1}}\Big\}.$$

The resulting mean value expansion is then:

$$\begin{aligned}\frac{1}{2}\Big(\inf_{x\in B_\epsilon(x_0)} + \sup_{x\in B_\epsilon(x_0)}\Big)\mathcal{A}\Big(u;\gamma\epsilon, 1-\frac{|x-x_0|^2}{\epsilon^2}, \frac{x-x_0}{|x-x_0|}\Big)(x)\\ = u(x_0) + \frac{\epsilon^2}{2(p-1)}|\nabla u(x_0)|^{2-p}\Delta_p u(x_0) + o(\epsilon^2).\end{aligned} \tag{6.8}$$

Exercise 6.3 In Peres and Sheffield (2008), instead of averaging on an N-dimensional ellipsoid, the average is taken on the $(N-2)$-dimensional sphere centred at x, with some radius $\gamma|x-x_0|$, and contained within the hyperplane perpendicular to $x-x_0$. The radius thus increases linearly from 0 to $\gamma\epsilon$ with a factor $\gamma > 0$, as $|x-x_0|$ varies from 0 to ϵ. This corresponds to evaluating on $B_\epsilon(x_0)$ the averages $\frac{1}{2}(\sup + \inf)$ of:

$$f_u^\gamma(x;x_0,\epsilon) = \fint_{\partial B^{N-1}} u\big(x+\gamma|x-x_0|R(x)y\big)\,\mathrm{d}y.$$

Here, $R(x) \in SO(N)$ is such that $R(x)e_N = \frac{x-x_0}{|x-x_0|}$, and ∂B^{N-1} stands for the $(N-2)$-dimensional sphere of unit radius, viewed as a subset of $\mathbb{R}^N$ contained in the subspace $\mathbb{R}^{N-1}$ orthogonal to e_N (note that $x \mapsto R(x)$ can be only locally defined as a C^2 function). Apply the argument as in the proof of Theorem 6.1 to show the following expansion of any $u \in C^2(\mathcal{D})$ at $x_0 \in \mathcal{D}$ satisfying $\nabla u(x_0) \neq 0$:

$$\begin{aligned}\frac{1}{2}\Big(\inf_{x\in B_\epsilon(x_0)} + \sup_{x\in B_\epsilon(x_0)}\Big) f_u^{\sqrt{\frac{N-1}{p-1}}}(x;x_0,\epsilon)\\ = u(x_0) + \frac{\epsilon^2}{2(p-1)}|\nabla u(x_0)|^{2-p}\Delta_p u(x_0) + o(\epsilon^2).\end{aligned} \tag{6.9}$$

Exercise 6.4 Prove the following expansion that has been put forward in Arroyo et al. (2017). Let $u \in C^2(\mathcal{D})$ satisfy $\Delta_p u = 0$ in $\mathcal{D}$ and let $x_0 \in \mathcal{D}$ be such that $\nabla u(x_0) \neq 0$. Then we have:

$$u(x_0) = \frac{1}{2}\Big(\inf_{|\nu|=\epsilon} + \sup_{|\nu|=\epsilon}\Big)\Big(\frac{p-1}{p+N}u(x_0+\nu) + \frac{N+1}{p+N}\fint_{B_\epsilon^\nu} u(x_0+y)\,\mathrm{d}\sigma(y)\Big), \tag{6.10}$$

where B_ϵ^ν denotes the $(N-1)$-dimensional ball, centred at 0, having radius ϵ, and orthogonal to the given vector ν.

6.2 The Dynamic Programming Principle and the Basic Convergence Theorem

As in Chap. 3, we begin by proving the well-posedness of the dynamic programming principle obtained by truncating error terms in (6.2) and continuously interpolating to the boundary data on $\partial\mathcal{D}$. In order to avoid the stopping of the induced Tug-of-War game outside of $\mathcal{D}$ (which was the case in the boundary aware game discussed in Chap. 4), the interpolation occurs in the $(\epsilon, 2\epsilon)$ neighbourhood of the boundary, rather than in the $(0, \epsilon)$ neighbourhood (Fig. 6.3).

Also, instead of dealing with the thickened and outer boundaries Γ, Γ_{out} as in Definition 3.8, we will assume that the boundary data F is defined on the whole $\mathbb{R}^N$. In the limit $\epsilon \to 0$ and when $F \in C(\mathbb{R}^N)$, the statements below will depend only on the trace $F_{|\partial\mathcal{D}}$ and not on its particular extension on $\mathbb{R}^N$.

Theorem 6.5 *Let $\mathcal{D} \subset \mathbb{R}^N$ be open, bounded, and connected. Given $\gamma_p, \rho_p > 0$ as in (6.1), for every $\epsilon \in (0, 1)$ and every bounded, Borel $F : \mathbb{R}^N \to \mathbb{R}$, there exists a unique bounded, Borel $u_\epsilon : \mathbb{R}^N \to \mathbb{R}$, satisfying:*

$$u_\epsilon(x) = d_\epsilon(x) S_\epsilon u_\epsilon(x) + \big(1 - d_\epsilon(x)\big) F(x) \qquad \text{for all } x \in \mathbb{R}^N. \tag{6.11}$$

Here, S_ϵ is defined in (6.4) and we set:

$$d_\epsilon(x) = \frac{1}{\epsilon} \min\big\{\epsilon, \operatorname{dist}\big(x, (\mathbb{R}^N \setminus \mathcal{D}) + \bar{B}_\epsilon(0)\big)\big\}. \tag{6.12}$$

The solution operator to (6.11) is monotone, i.e. if $F \le \bar{F}$, then the corresponding solutions satisfy: $u_\epsilon \le \bar{u}_\epsilon$. Moreover: $\|u_\epsilon\|_{L^\infty(\mathbb{R}^N)} \le \|F\|_{L^\infty(\mathbb{R}^N)}$.

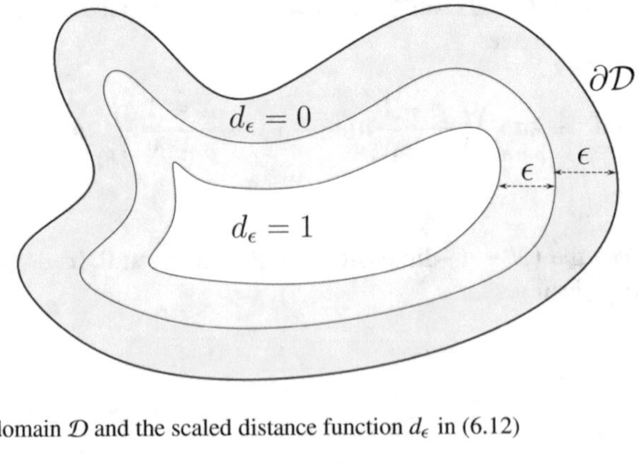

Fig. 6.3 The domain $\mathcal{D}$ and the scaled distance function d_ϵ in (6.12)

Proof

1. The proof follows steps of the proof of Theorem 4.1. To ease the notation, we drop the subscript ϵ and write u instead of u_ϵ. Clearly, the solution u of (6.11) is a fixed point of $T_\epsilon v = d_\epsilon S_\epsilon v + (1 - d_\epsilon)F$. Recall that:

$$\begin{aligned} (S_\epsilon v)(x) &= \frac{1}{2}\Big(\inf_{z\in B_1(0)} + \sup_{z\in B_1(0)}\Big) f_v(x+\epsilon z; x, \epsilon) \\ \text{where:} \quad f_v(x+\epsilon z; x,\epsilon) &= \fint_{x+\epsilon E(z,\gamma_p;1+(\rho_p-1)|z|^2,\frac{z}{|z|})} v(w)\,dw. \end{aligned} \tag{6.13}$$

For a fixed ϵ and x, and given a bounded, Borel function $v : \mathbb{R}^N \to \mathbb{R}$, the average f_v is continuous in $z \in \bar{B}_1(0)$. In view of continuity of d_ϵ, we conclude that both T_ϵ, S_ϵ return a bounded, Borel function. We further note that S_ϵ and T_ϵ are monotone: $S_\epsilon v \le S_\epsilon \bar{v}$ and $T_\epsilon v \le T_\epsilon \bar{v}$ if $v \le \bar{v}$.

The solution u of (6.11) is obtained as the limit of iterations $u_{n+1} = T_\epsilon u_n$, where we set $u_0 \equiv \text{const} \le \inf F$. Since $u_1 = T_\epsilon u_0 \ge u_0$ on $\mathbb{R}^N$, by monotonicity of T_ϵ, the sequence $\{u_n\}_{n=0}^\infty$ is nondecreasing. It is also bounded (by $\|F\|_{L^\infty(\mathbb{R}^N)}$) and thus it converges pointwise to a (bounded, Borel) limit $u : \mathbb{R}^N \to \mathbb{R}$. Observe now that:

$$\begin{aligned} |T_\epsilon u_n(x) - T_\epsilon u(x)| &\le |S_\epsilon u_n(x) - S_\epsilon u(x)| \\ &\le \sup_{z\in B_1(0)} \fint_{x+\epsilon E(z,\gamma_p;1+(\rho_p-1)|z|^2,\frac{z}{|z|})} |u_n - u|(w)\,dw \\ &\le C_\epsilon \int_{\mathcal{D}} |u_n - u|(w)\,dw, \end{aligned} \tag{6.14}$$

where C_ϵ is the lower bound on the volume of the sampling ellipses. By the monotone convergence theorem, it follows that the right-hand side in (6.14) converges to 0 as $n \to \infty$. Consequently, $u = T_\epsilon u$, proving existence of solutions to (6.11).

2. We now show uniqueness. If $u, \bar{u}$ both solve (6.11), then define:

$$M = \sup_{x\in\mathbb{R}^N} |u(x) - \bar{u}(x)| = \sup_{x\in\mathcal{D}} |u(x) - \bar{u}(x)|.$$

The proof is the same as in Step 2 of the proof of Theorem 4.1; we now indicate its simplified content for the case $u, \bar{u} \in C(\mathbb{R}^N)$ (which holds automatically for continuous F). Consider any maximizer $x_0 \in \mathcal{D}$, where $|u(x_0) - \bar{u}(x_0)| = M$.

By (6.14) we get:

$$M = |u(x_0) - \bar{u}(x_0)| = d_\epsilon(x_0)|S_\epsilon u(x_0) - S_\epsilon \bar{u}(x_0)|$$
$$\leq \sup_{z \in B_1(0)} f_{|u-\bar{u}|}(x + \epsilon z; x, \epsilon) \leq M,$$

yielding in particular $\fint_{B(x_0,\gamma_p\epsilon)} |u-\bar{u}|(w)\,dw = M$. Consequently, $B(x_0, \gamma_p\epsilon) \subset D_M = \{|u-\bar{u}| = M\}$ and hence the set D_M is open in $\mathbb{R}^N$. Since D_M is obviously closed and nonempty, there must be $D_M = \mathbb{R}^N$ and since $u - \bar{u} = 0$ on $\mathbb{R}^N \setminus \mathcal{D}$, it follows that $M = 0$. Thus $u = \bar{u}$, proving the claim. Finally, monotonicity of S_ϵ yields the monotonicity of the solution operator to (6.11). □

Remark 6.6 It follows from (6.14) that the sequence $\{u_n\}_{n=1}^\infty$ in the proof of Theorem 6.5 converges to $u = u_\epsilon$ uniformly. In fact, the iteration procedure $u_{n+1} = Tu_n$ started by any bounded, Borel u_0 converges uniformly to the uniquely given u_ϵ. We further remark that if F is continuous, then u_ϵ is likewise continuous, and if F is Lipschitz, then u_ϵ is Lipschitz, with Lipschitz constant depending (in nondecreasing manner) on: $1/\epsilon$, $\|F\|_{C(\partial\mathcal{D})}$ and the Lipschitz constant of $F_{|\partial\mathcal{D}}$.

We conclude this section by showing the basic convergence result, which is parallel to Theorem 4.6 stated for the averaging principle (3.14).

Theorem 6.7 *Let $F \in C^2(\mathbb{R}^N)$ be a bounded data function that satisfies on some open set U, compactly containing $\mathcal{D}$:*

$$\Delta_p F = 0 \quad \text{and} \quad \nabla F \neq 0 \quad \text{in } U. \tag{6.15}$$

Then the solutions u_ϵ of (6.11) converge to F uniformly in $\mathbb{R}^N$, namely:

$$\|u_\epsilon - F\|_{C(\mathcal{D})} \leq C\epsilon \qquad \text{as } \epsilon \to 0, \tag{6.16}$$

with a constant C depending on F, U, $\mathcal{D}$ and p, but not on ϵ.

Proof

1. Since $u_\epsilon = F$ on $\mathbb{R}^N \setminus \mathcal{D}$ by construction, (6.16) indeed implies the uniform convergence of u_ϵ in $\mathbb{R}^N$. Also, by translating $\mathcal{D}$ if necessary, we may assume that $B_1(0) \cap U = \emptyset$. Let v_ϵ be the function resulting in Lemma 4.8, namely $v_\epsilon(x) = F(x) + \epsilon|x|^s$ for some $s \geq 2$, satisfying for every $\epsilon \in (0, \hat{\epsilon})$:

$$\nabla v_\epsilon \neq 0 \quad \text{and} \quad \Delta_p v_\epsilon \geq \epsilon s \cdot |\nabla v_\epsilon|^{p-2} \quad \text{in } \bar{\mathcal{D}}. \tag{6.17}$$

Parameters s and $\hat{\epsilon}$ can be further chosen in a way that for all $\epsilon \in (0, \hat{\epsilon})$:

$$v_\epsilon \leq S_\epsilon v_\epsilon \quad \text{in } \bar{\mathcal{D}}. \tag{6.18}$$

Indeed, analysis of the remainder terms in the expansion (6.2) reveals that:

$$v_\epsilon(x) - S_\epsilon v_\epsilon(x) = -\frac{\epsilon^2}{p-1}|\nabla v_\epsilon(x)|^{2-p}\Delta_p v_\epsilon(x) + R_2(\epsilon, s), \tag{6.19}$$

where: $|R_2(\epsilon, s)| \leq C_p\epsilon^2 \mathrm{osc}_{B(x,(1+\gamma_p)\epsilon)}|\nabla^2 v_\epsilon| + C\epsilon^3$. We denoted by C_p a constant depending only on p, whereas C is a constant depending only $|\nabla v_\epsilon|$ and $|\nabla^2 v_\epsilon|$, that remain uniformly bounded for small ϵ. Since v_ϵ is the sum of the smooth on U function $x \mapsto \epsilon|x|^s$, and a p-harmonic function F that is also smooth in virtue of its nonvanishing gradient, we obtain that (6.19) and (6.17) imply (6.18) for s sufficiently large and taking ϵ appropriately small.

2. Let A be a compact set in: $\mathcal{D} \subset A \subset U$. For each $x \in A$ and $\epsilon \in (0, \hat{\epsilon})$, let:

$$\phi_\epsilon(x) = v_\epsilon(x) - u_\epsilon(x) = F(x) - u_\epsilon(x) + \epsilon|x|^s.$$

By (6.18) and (6.11) we get:

$$\begin{aligned}\phi_\epsilon(x) &= d_\epsilon(x)(v_\epsilon(x) - S_\epsilon u_\epsilon(x)) + (1 - d_\epsilon(x))(v_\epsilon(x) - F(x))\\ &\leq d_\epsilon(x)(S_\epsilon v_\epsilon(x) - S_\epsilon u_\epsilon(x)) + (1 - d_\epsilon(x))(v_\epsilon(x) - F(x))\\ &\leq d_\epsilon(x) \sup_{y\in B_1(0)} f_{\phi_\epsilon}\big(x + \epsilon y, x, \epsilon\big) + (1 - d_\epsilon(x))\big(v_\epsilon(x) - F(x)\big).\end{aligned} \tag{6.20}$$

Define: $M_\epsilon = \max_A \phi_\epsilon$. We claim that there exists $x_0 \in A$ with $d_\epsilon(x_0) < 1$ and such that $\phi_\epsilon(x_0) = M_\epsilon$. To prove the claim, let $\mathcal{D}^\epsilon = \big\{x \in \mathcal{D};\ \mathrm{dist}(x, \partial\mathcal{D}) \geq 2\epsilon\}$. We can assume that the closed set $\mathcal{D}^\epsilon \cap \{\phi_\epsilon = M_\epsilon\}$ is nonempty; otherwise the claim would be obvious. Let $\mathcal{D}^\epsilon_0$ be a nonempty connected component of $\mathcal{D}^\epsilon$ and denote $\mathcal{D}^\epsilon_M = \mathcal{D}^\epsilon_0 \cap \{\phi_\epsilon = M_\epsilon\}$. Clearly, $\mathcal{D}^\epsilon_M$ is closed in $\mathcal{D}^\epsilon_0$; we now show that it is also open. Let $x \in \mathcal{D}^\epsilon_M$. Since $d_\epsilon(x) = 1$ from (6.20) it follows that:

$$M_\epsilon = \phi_\epsilon(x) \leq \sup_{y\in B(x,\epsilon)} \mathcal{A}\Big(\phi_\epsilon; \gamma_p\epsilon + (\rho_p - 1)\frac{|y-x|^2}{\epsilon^2}, \frac{y-x}{|y-x|}\Big)(y) \leq M_\epsilon.$$

Consequently, $\phi_\epsilon \equiv M_\epsilon$ in $B(x, \gamma_p\epsilon)$ and thus we obtain the openness of $\mathcal{D}^\epsilon_M$ in $\mathcal{D}^\epsilon_0$. In particular, $\mathcal{D}^\epsilon_M$ contains a point $\bar{x} \in \partial\mathcal{D}^\epsilon$. Repeating the previous argument for $\bar{x}$ results in $\phi_\epsilon \equiv M_\epsilon$ in $B(\bar{x}, \gamma_p\epsilon)$, proving the claim.

3. We now complete the proof of Theorem 6.7 by deducing a bound on M_ϵ. If $M_\epsilon = \phi_\epsilon(x_0)$ for some $x_0 \in \bar{\mathcal{D}}$ with $d_\epsilon(x_0) < 1$, then (6.20) yields: $M_\epsilon =$

$\phi_\epsilon(x_0) \leq d_\epsilon(x_0) M_\epsilon + (1 - d_\epsilon(x_0))\big(v_\epsilon(x_0) - F(x_0)\big)$, which implies:

$$M_\epsilon \leq v_\epsilon(x_0) - F(x_0) = \epsilon |x_0|^s.$$

On the other hand, if $M_\epsilon = \phi_\epsilon(x_0)$ for some $x_0 \in A \setminus \mathcal{D}$, then $d_\epsilon(x_0) = 0$, hence likewise: $M_\epsilon = \phi_\epsilon(x_0) = v_\epsilon(x_0) - F(x_0) = \epsilon|x_0|^s$. In either case:

$$\max_{\bar{\mathcal{D}}}(u - u_\epsilon) \leq \max_{\bar{\mathcal{D}}} \phi_\epsilon + C\epsilon \leq 2C\epsilon,$$

where $C = \max_{x \in V} |x|^s$ is independent of ϵ. A symmetric argument applied to $-u$ after noting that $(-u)_\epsilon = -u_\epsilon$ gives: $\min_{\bar{\mathcal{D}}}(u - u_\epsilon) \geq -2C\epsilon$. The proof is done. □

6.3 Mixed Tug-of-War with Noise

Below, we develop the probability setting similar to that of Sects. 3.4 and 4.3, but related to the expansion (6.2). We fix $p \in (1, \infty)$, the scaling factors γ_p, ρ_p as in (6.1), and the sampling radius parameter $\epsilon \in (0, 1)$.

1. Let $\Omega_1 = B_1(0) \times \{1, 2\} \times (0, 1)$ and define:

$$\Omega = (\Omega_1)^{\mathbb{N}} = \Big\{\omega = \{(w_i, a_i, b_i)\}_{i=1}^\infty; \\ w_i \in B_1(0),\ a_i \in \{1, 2\},\ b_i \in (0, 1) \quad \text{for all } i \in \mathbb{N}\Big\}.$$

The probability space $(\Omega, \mathcal{F}, \mathbb{P})$ is given as the countable product of $(\Omega_1, \mathcal{F}_1, \mathbb{P}_1)$. Here, $\mathcal{F}_1$ is the smallest σ-algebra containing all products $D \times S \times B$ where $D \subset B_1(0) \subset \mathbb{R}^N$ and $B \subset (0, 1)$ are Borel, and $A \subset \{1, 2\}$. The measure $\mathbb{P}_1$ is the product of: the normalized Lebesgue measure on $B_1(0)$, the uniform counting measure on $\{1, 2\}$ and the Lebesgue measure on $(0, 1)$:

$$\mathbb{P}_1(D \times S \times B) = \frac{|D|}{|B_1(0)|} \cdot \frac{|A|}{2} \cdot |B|.$$

For each $n \in \mathbb{N}$, the probability space $(\Omega_n, \mathcal{F}_n, \mathbb{P}_n)$ is the product of n copies of $(\Omega_1, \mathcal{F}_1, \mathbb{P}_1)$. The σ-algebra $\mathcal{F}_n$ is always identified with the sub-σ-algebra of $\mathcal{F}$, consisting of sets $F \times \prod_{i=n+1}^\infty \Omega_1$ for all $F \in \mathcal{F}_n$. The sequence $\{\mathcal{F}_n\}_{n=0}^\infty$ where $\mathcal{F}_0 = \{\emptyset, \Omega\}$, is a filtration of $\mathcal{F}$.
2. Given are two families of functions $\sigma_I = \{\sigma_I^n\}_{n=0}^\infty$ and $\sigma_{II} = \{\sigma_{II}^n\}_{n=0}^\infty$, defined on the corresponding spaces of "finite histories" $H_n = \mathbb{R}^N \times (\mathbb{R}^N \times \Omega_1)^n$:

$$\sigma_I^n, \sigma_{II}^n : H_n \to B_1(0) \subset \mathbb{R}^N,$$

assumed to be measurable with respect to the (target) Borel σ-algebra in $B_1(0)$ and the (domain) product σ-algebra on H_n. For every $x_0 \in \mathbb{R}^N$ we now recursively define the sequence of random variables: As usual, we often suppress some of the superscripts $x_0, \sigma_I, \sigma_{II}$ and write X_n (or $X_n^{x_0}$, or $X_n^{\sigma_I,\sigma_{II}}$, etc.) instead of $X_n^{x_0,\sigma_I,\sigma_{II}}$, if no ambiguity arises. Let:

$$\begin{aligned}
&X_0 \equiv x_0,\\
&X_n\big((w_1,a_1,b_1),\ldots,(w_n,a_n,b_n)\big)\\
&= x_{n-1} + \begin{cases} \epsilon\Big(\sigma_I^{n-1}(h_{n-1}) + \gamma_p w_n \\ \quad +\gamma_p(\rho_p-1)\langle w_n, \sigma_I^{n-1}(h_{n-1})\rangle\sigma_I^{n-1}(h_{n-1})\Big) & \text{for } a_n = 1\\ \epsilon\Big(\sigma_{II}^{n-1}(h_{n-1}) + \gamma_p w_n \\ \quad +\gamma_p(\rho_p-1)\langle w_n, \sigma_{II}^{n-1}(h_{n-1})\rangle\sigma_{II}^{n-1}(h_{n-1})\Big) & \text{for } a_n = 2,\end{cases}\\
&\text{where}\quad x_{n-1} = X_{n-1}\big((w_1,a_1,b_1),\ldots,(w_{n-1},a_{n-1},b_{n-1})\big)\\
&\text{and}\quad h_{n-1} = \big(x_0,(x_1,w_1,a_1,b_1),\ldots,(x_{n-1},w_{n-1},a_{n-1},b_{n-1})\big) \in H_{n-1}.
\end{aligned} \tag{6.21}$$

In this "game", the position x_{n-1} is first advanced (deterministically) according to the two players' "strategies" σ_I and σ_{II} by a shift $\epsilon y \in B_\epsilon(0)$, and then (randomly) uniformly by a further shift in the ellipsoid $\epsilon E\big(0, \gamma_p; 1 + (\rho_p - 1)|y|^2, \frac{y}{|y|}\big)$. The deterministic shifts are activated by the value of the equally probable outcomes: $a_n = 1$ activates σ_I and $a_n = 2$ activates σ_{II} (Fig. 6.4).

3. The auxiliary variables $b_n \in (0,1)$ serve as thresholds for reading the eventual value from the prescribed boundary data. Let $\mathcal{D} \subset \mathbb{R}^N$ be an open, bounded and

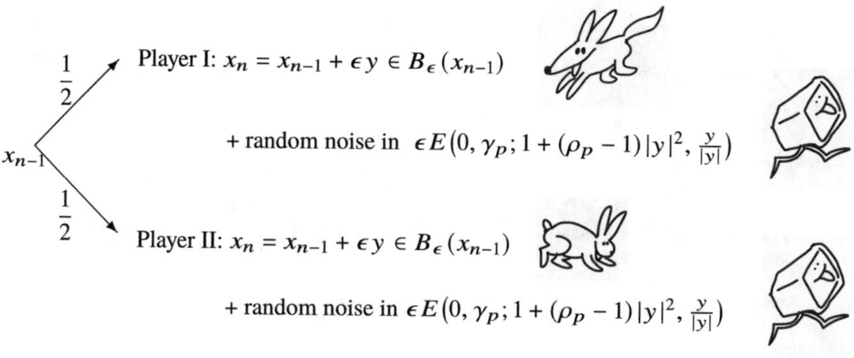

Fig. 6.4 Player I, Player II and random noise in the mixed Tug-of-War

connected set. Define $\tau^{\epsilon, x_0, \sigma_I, \sigma_{II}} : \Omega \to \mathbb{N} \cup \{\infty\}$ in:

$$\tau^{x_0, \sigma_I, \sigma_{II}}\big((w_1, a_1, b_1), (w_2, a_2, b_2), \dots\big) \\ = \min\big\{n \geq 1;\ b_n > d_\epsilon(x_{n-1})\big\}. \tag{6.22}$$

As before, we drop the superscripts and write τ instead of $\tau^{x_0, \sigma_I, \sigma_{II}}$ if there is no ambiguity. Our game is thus terminated, with probability $1 - d_\epsilon(x_{n-1})$, whenever the position x_{n-1} reaches the ϵ-neighbourhood of $\partial\mathcal{D}$.

Lemma 6.8 *If the scaling factors $\rho_p, \gamma_p > 0$ in (6.1) satisfy:*

$$\rho_p \leq 1 \ \text{ and } \ \gamma_p\rho_p > 1 \qquad \text{or} \qquad \rho_p \geq 1 \ \text{ and } \ \gamma_p > 1, \tag{6.23}$$

then τ is a stopping time relative to the filtration $\{\mathcal{F}_n\}_{n=0}^\infty$, namely:

$$\mathbb{P}(\tau < \infty) = 1.$$

Further, for any $p \in (1, \infty)$ there exist positive ρ_p, γ_p with (6.1) and (6.23).

Proof Let $\rho_p \leq 1$ and $\gamma_p\rho_p > 1$. Then, for some $\beta > 0$, there also holds: $\gamma_p(\rho_p - \beta) > 1$. Define an open set of "advancing random shifts":

$$D_{adv} = \big\{w \in B_1(0);\ \langle w, e_1\rangle > 1 - \beta\big\}.$$

For every $\sigma \in B_1(0)$ and every $w \in D_{adv}$ we have:

$$\gamma_p\big\langle w + (\rho_p - 1)\langle w, \sigma\rangle\sigma, e_1\big\rangle \geq \gamma_p\big(\langle w, e_1\rangle + \rho_p - 1\big) > \gamma_p(\rho_p - \beta).$$

Since $\mathcal{D}$ is bounded, the above estimate implies existence of $n \geq 1$ (depending on ϵ) such that for all $x_0 \in \mathcal{D}$ and all deterministic shifts $\{\sigma^i \in B_1(0)\}_{i=1}^n$ there holds:

$$x_0 + \epsilon\sum_{i=1}^n \big(\sigma^i + \gamma_p w_i + \gamma_p(\rho_p - 1)\langle w_i, \sigma^i\rangle\sigma^i\big) \notin \mathcal{D} \qquad \text{for all } \ \{w_i \in D_{adv}\}_{i=1}^n.$$

In conclusion:

$$\mathbb{P}(\tau \leq n) \geq \mathbb{P}_n\Big(\big(D_{adv} \times \{1, 2\} \times (0, 1)\big)^n\Big) = \Big(\frac{|D_{adv}|}{|B_1(0)|}\Big)^n = \eta > 0,$$

and so: $\mathbb{P}(\tau > kn) \leq (1 - \eta)^k$ for all $k \in \mathbb{N}$, yielding:

$$\mathbb{P}(\tau = \infty) = \lim_{k\to\infty} \mathbb{P}(\tau > kn) = 0.$$

The proof proceeds similarly when $\rho_p \geq 1$ and $\gamma_p > 1$. Fix $\bar{\beta} > 0$ such that $\gamma_p(1-\bar{\beta}) > 1$ and define D_{adv} as before, for a small $0 < \beta \ll \bar{\beta}$, ensuring that:

$$\gamma_p\big\langle w + (\rho_p - 1)\langle w, \sigma\rangle\sigma, e_1\big\rangle \geq \gamma_p\big(\langle w, e_1\rangle - (\rho_p - 1)\sqrt{2\beta}\big) > \gamma_p(1-\bar{\beta})$$

for every $\sigma \in B_1(0)$ and every $w \in D_{adv}$. Again, after at most $\left\lceil \frac{\operatorname{diam} \mathcal{D}}{\epsilon(\gamma_p(1-\bar{\beta})-1)} \right\rceil$ shifts, the token will leave $\mathcal{D}$ (unless it is stopped earlier) and the game will be terminated.

It remains to prove existence of $\gamma_p, \rho_p > 0$ satisfying (6.1) and (6.23). We observe that the viability of $\rho_p \leq 1$, $\gamma_p\rho_p > 1$ is equivalent to: $\frac{1}{\gamma_p^2} < p - 1 - \frac{N+2}{\gamma_p^2} \leq 1$ and further to: $\frac{p-2}{N+2} \leq \frac{1}{\gamma_p^2} < \frac{p-1}{N+3}$, which allows for choosing γ_p (and ρ_p) for $p < N+4$. On the other hand, viability of $\rho_p \geq 1$, $\gamma_p > 1$ is equivalent to: $\gamma_p^2 > 1$ and $p - 1 - \frac{N+2}{\gamma_p^2} \geq 1$, that is:

$$\frac{1}{\gamma_p^2} < \min\Big\{1, \frac{p-2}{N+2}\Big\},$$

implying existence of γ_p, ρ_p for $p > 2$. □

1. From now on, we will work under the additional requirement (6.23). In our "game", the first "player" collects from his opponent the payoff given by the data F at the stopping position. The incentive of the collecting "player" to maximize the outcome and of the disbursing "player" to minimize it, leads to the definition of the two game values below.

   Let $F : \mathbb{R}^N \to \mathbb{R}$ be a bounded, Borel function. Then we have:

$$\begin{aligned} u_I^\epsilon(x_0) &= \sup_{\sigma_I} \inf_{\sigma_{II}} \mathbb{E}\Big[F \circ \big(X^{x_0,\sigma_I,\sigma_{II}}\big)_{\tau^{x_0,\sigma_I,\sigma_{II}}-1}\Big], \\ u_{II}^\epsilon(x_0) &= \inf_{\sigma_{II}} \sup_{\sigma_I} \mathbb{E}\Big[F \circ \big(X^{x_0,\sigma_I,\sigma_{II}}\big)_{\tau^{x_0,\sigma_I,\sigma_{II}}-1}\Big]. \end{aligned} \tag{6.24}$$

The main result in Theorem 6.9 below will show that $u_I^\epsilon = u_{II}^\epsilon$ coincide with the unique solution to the dynamic programming principle in Sect. 6.2, modelled on the expansion (6.2). It is also clear that $u_{I,II}^\epsilon$ depend only on the values of F in the ϵ-neighbourhood of $\partial\mathcal{D}$. In Sect. 6.4 we will prove that as $\epsilon \to 0$, the uniform limit of $u_{I,II}^\epsilon$ that depends only on the continuous $F_{|\partial\mathcal{D}}$, is p-harmonic in $\mathcal{D}$ and attains F on $\partial\mathcal{D}$, provided that $\partial\mathcal{D}$ is regular.

Theorem 6.9 *For every $\epsilon \in (0,1)$, let u_I^ϵ, u_{II}^ϵ be as in (6.24) and u_ϵ as in Theorem 6.5. Then:*

$$u_I^\epsilon = u_\epsilon = u_{II}^\epsilon.$$

Proof

1. The proof is as that of Theorem 4.12, with only technical modifications. We drop the sub/superscript ϵ to ease the notation. To show that $u_{II} \leq u$, fix $x_0 \in \mathbb{R}^N$ and $\eta > 0$. We first observe that there exists a strategy $\sigma_{0,II}$ where $\sigma_{0,II}^n(h_n) = \sigma_{0,II}^n(x_n)$ satisfies for every $n \geq 0$ and $h_n \in H_n$:

$$f_u(x_n + \epsilon\sigma_{0,II}^n(x_n); x_n, \epsilon) \leq \inf_{z \in B_1(0)} f_u(x_n + \epsilon z; x_n, \epsilon) + \frac{\eta}{2^{n+1}}. \tag{6.25}$$

Indeed, using continuity of (6.3), we note that there exists $\delta > 0$ such that:

$$\Big|\inf_{z \in B_1(0)} f_u(y+\epsilon z; y, \epsilon) - \inf_{z \in B_1(0)} f_u(\bar{y}+\epsilon z; \bar{y}, \epsilon)\Big| < \frac{\eta}{2^{n+2}} \quad \text{for all } |y-\bar{y}| < \delta.$$

Let $\{B_\delta(y_i)\}_{i=1}^\infty$ be a locally finite covering of $\mathbb{R}^N$. For each $i = 1\ldots\infty$, choose $z_i \in B_1(0)$ satisfying:

$$\big|\inf_{z \in B_1(0)} f_u(y_i + \epsilon z; y_i, \epsilon) - f_u(y_i + \epsilon z_i; y_i, \epsilon)\big| < \frac{\eta}{2^{n+2}}.$$

Finally, set:

$$\sigma_{0,II}^n(y) = z_i \quad \text{for } y \in B_\delta(y_i) \setminus \bigcup_{j=1}^{i-1} B_\delta(y_j).$$

The piecewise constant $\sigma_{0,II}^n$ is obviously Borel and it satisfies (6.25). An alternative argument to show (6.25) is to invoke Lemma 3.15.
2. Fix a strategy σ_I and consider the random variables $M_n : \Omega \to \mathbb{R}$:

$$M_n = (u \circ X_n)\mathbb{1}_{\tau > n} + (F \circ X_{\tau-1})\mathbb{1}_{\tau \leq n} + \frac{\eta}{2^n}.$$

Then $\{M_n\}_{n=0}^\infty$ is a supermartingale with respect to the filtration $\{\mathcal{F}_n\}_{n=0}^\infty$. Clearly:

$$\begin{aligned}\mathbb{E}\big(M_n \mid \mathcal{F}_{n-1}\big) = {} & \mathbb{E}\big((u \circ X_n)\mathbb{1}_{\tau > n} \mid \mathcal{F}_{n-1}\big) + \mathbb{E}\big((F \circ X_{n-1})\mathbb{1}_{\tau = n} \mid \mathcal{F}_{n-1}\big) \\ & + \mathbb{E}\big((F \circ X_{\tau-1})\mathbb{1}_{\tau < n} \mid \mathcal{F}_{n-1}\big) + \frac{\eta}{2^n} \qquad \text{a.s.}\end{aligned} \tag{6.26}$$

We readily observe that: $\mathbb{E}\big((F\circ X_{\tau-1})\mathbb{1}_{\tau<n} \mid \mathcal{F}_{n-1}\big) = (F\circ X_{\tau-1})\mathbb{1}_{\tau<n}$. Further, writing $\mathbb{1}_{\tau=n} = \mathbb{1}_{\tau\geq n}\mathbb{1}_{b_n>d_\epsilon(x_{n-1})}$, it follows that:

$$\begin{aligned}\mathbb{E}\big((F\circ X_{n-1})\mathbb{1}_{\tau=n} \mid \mathcal{F}_{n-1}\big) &= \mathbb{E}\big(\mathbb{1}_{b_n>d_\epsilon(x_{n-1})} \mid \mathcal{F}_{n-1}\big)\cdot (F\circ X_{n-1})\mathbb{1}_{\tau\geq n}\\ &= \big(1-d_\epsilon(x_{n-1})\big)(F\circ X_{n-1})\mathbb{1}_{\tau\geq n} \qquad \text{a.s.}\end{aligned}$$

Similarly, since $\mathbb{1}_{\tau>n} = \mathbb{1}_{\tau\geq n}\mathbb{1}_{b_n\leq d_\epsilon(x_{n-1})}$, we get in view of (6.25):

$$\begin{aligned}&\mathbb{E}\big((u\circ X_n)\mathbb{1}_{\tau>n} \mid \mathcal{F}_{n-1}\big) = \mathbb{E}\big(u\circ X_n \mid \mathcal{F}_{n-1}\big)\cdot d_\epsilon(x_{n-1})\mathbb{1}_{\tau\geq n}\\ &= \int_{\Omega_1}(u\circ X_n)\mathbb{1}_{b_n\leq d_\epsilon(x_{n-1})}\,\mathrm{d}\mathbb{P}_1\cdot\mathbb{1}_{\tau\geq n} \qquad \text{a.s.}\\ &= \frac{1}{2}\Big(\mathcal{A}\Big(u;\gamma_p\epsilon, 1+(\rho_p-1)|\sigma_I^{n-1}|^2, \frac{\sigma_I^{n-1}}{|\sigma_I^{n-1}|}\Big)(x_{n-1}+\epsilon\sigma_I^{n-1})\\ &\qquad + \mathcal{A}\Big(u;\gamma_p\epsilon, 1+(\rho_p-1)|\sigma_{0,II}^{n-1}|^2, \frac{\sigma_{0,II}^{n-1}}{|\sigma_{0,II}^{n-1}|}\Big)(x_{n-1}+\epsilon\sigma_{0,II}^{n-1})\Big)\cdot d_\epsilon(x_{n-1})\mathbb{1}_{\tau\geq n}\end{aligned}$$

so that:

$$\mathbb{E}\big((u\circ X_n)\mathbb{1}_{\tau>n} \mid \mathcal{F}_{n-1}\big) \leq \big(S_\epsilon u\circ X_{n-1} + \frac{\eta}{2^n}\big)d_\epsilon(x_{n-1})\mathbb{1}_{\tau\geq n} \qquad \text{a.s.}$$

Concluding, by (6.11) the decomposition (6.26) yields:

$$\begin{aligned}\mathbb{E}\big(M_n \mid \mathcal{F}_{n-1}\big) &\leq \Big(d_\epsilon(x_{n-1})\big(S_\epsilon u\circ X_{n-1}\big) + (1-d_\epsilon(x_{n-1}))\big(F\circ X_{n-1}\big)\Big)\mathbb{1}_{\tau\geq n}\\ &\quad + (F\circ X_{\tau-1})\mathbb{1}_{\tau\leq n-1} + \frac{\eta}{2^{n-1}} = M_{n-1} \qquad \text{a.s.}\end{aligned}$$

3. The supermartingale property of $\{M_n\}_{n=0}^\infty$ being established, we get:

$$u(x_0)+\eta = \mathbb{E}\big[M_0\big] \geq \mathbb{E}\big[M_\tau\big] = \mathbb{E}\big[F\circ X_{\tau-1}\big] + \frac{\eta}{2^\tau}.$$

Thus: $u_{II}(x_0) \leq \sup_{\sigma_I}\mathbb{E}\big[F\circ (X^{\sigma_I,\sigma_{II,0}})_{\tau-1}\big] \leq u(x_0)+\eta$. As $\eta>0$ was arbitrary, we obtain the claimed comparison $u_{II}(x_0)\leq u(x_0)$. For the reverse inequality $u(x_0)\leq u_I(x_0)$ we use a symmetric argument, with an almost-maximizing strategy $\sigma_{0,I}$ and the submartingale $\bar{M}_n = (u\circ X_n)\mathbb{1}_{\tau>n} + (F\circ X_{\tau-1})\mathbb{1}_{\tau\leq n} - \frac{\eta}{2^n}$ along a given yet arbitrary strategy σ_{II}. The obvious estimate $u_I(x_0)\leq u_{II}(x_0)$ ends the proof. □

6.4 Sufficient Conditions for Game-Regularity: Exterior Corkscrew Condition

In this section we carry out the program described in Chap. 5. Most of the results and proofs are the same as in the context of the averaging principle (3.14) with $p > 2$, hence we only indicate a few necessary modifications. The reader familiar with our previous discussions should have no problem in filling out the details.

Towards checking convergence of the family $\{u_\epsilon\}_{\epsilon\to 0}$, we first observe that its equicontinuity is implied by the equicontinuity "at $\partial\mathcal{D}$". This last property will be, in turn, implied by the "game-regularity" condition (6.28) below.

Lemma 6.10 *Let $\mathcal{D} \subset \mathbb{R}^N$ be an open, bounded, connected domain and let $F \in C(\mathbb{R}^N)$ be a bounded, continuous data function. Assume that for every $\eta > 0$ there exists $\delta > 0$ and $\hat{\epsilon} \in (0, 1)$ such that for all $\epsilon \in (0, \hat{\epsilon})$ there holds:*

$$|u_\epsilon(y_0) - u_\epsilon(x_0)| \leq \eta \qquad \textit{for all } \ y_0 \in \mathcal{D}, \ x_0 \in \partial\mathcal{D} \quad \textit{satisfying } \ |x_0 - y_0| \leq \delta. \tag{6.27}$$

Then the family $\{u_\epsilon\}_{\epsilon\to 0}$ of solutions to (6.11) is equicontinuous in $\bar{\mathcal{D}}$.

The proof is left as an exercise (Exercise 6.19) since it essentially follows the analytical proof of Theorem 5.3.

As in Chap. 5, we say that a point $y_0 \in \partial\mathcal{D}$ is game-regular if, whenever the game starts near x_0, one of the players has a strategy for making the game terminate still near x_0, with high probability.

Definition 6.11 Fix $p \in (1, \infty)$, $\epsilon \in (0, 1)$ and γ_p, ρ_p as in (6.1), (6.23). Consider the Tug-of-War game with noise (6.21), as defined in Sect. 6.3. Let $\mathcal{D} \subset \mathbb{R}^N$ be open, bounded and connected.

(a) We say that a point $y_0 \in \partial\mathcal{D}$ is *game-regular* if for every $\eta, \delta > 0$ there exist $\hat{\delta} \in (0, \delta)$ and $\hat{\epsilon} \in (0, 1)$ such that the following holds. Fix $\epsilon \in (0, \hat{\epsilon})$ and $x_0 \in B_{\hat{\delta}}(y_0)$; there exists then a strategy $\sigma_{0,I}$ with the property that for every strategy σ_{II} we have:

$$\mathbb{P}\big((X^{x_0,\sigma_{0,I},\sigma_{II}})_{\tau-1} \in B_\delta(y_0)\big) \geq 1 - \eta, \tag{6.28}$$

where $\tau = \tau^{x_0,\sigma_{0,I},\sigma_{II}}$ is the stopping time in (6.22).

(b) We say that $\mathcal{D}$ is game-regular if every boundary point $y_0 \in \partial\mathcal{D}$ is game-regular.

Observe that if condition (b) above holds, then $\hat{\delta}$ and $\hat{\epsilon}$ in part (a) can be chosen independently of y_0. Also, game-regularity is symmetric in σ_I and σ_{II}.

Theorem 6.12 *Let $\mathcal{D}$, p, γ_p, δ_p be as in Definition 6.11.*

(i) Assume that for every bounded data $F \in C(\mathbb{R}^N)$, the family of solutions $\{u_\epsilon\}_{\epsilon\to 0}$ to (6.11) is equicontinuous in $\bar{\mathcal{D}}$. Then $\mathcal{D}$ is game-regular.
(ii) Conversely, if $\mathcal{D}$ is game-regular then $\{u_\epsilon\}_{\epsilon\to 0}$ satisfies (6.27), and hence it is equicontinuous in virtue of Lemma 6.10, for every bounded and continuous data $F \in C(\mathbb{R}^N)$.

The proof is again verbatim the same as the proof of Theorems 5.7 and 5.8. As a consequence, we easily get:

Theorem 6.13 *Let $F \in C(\mathbb{R}^N)$ be a bounded data function and let $\mathcal{D}$ be open, bounded and game-regular with respect to p, γ_p, δ_p be as in Definition 6.11. Then the family $\{u_\epsilon\}_{\epsilon\to 0}$ of solutions to (6.11) converges uniformly in $\bar{\mathcal{D}}$ to the unique viscosity solution of (6.29).*

Proof By Theorem 6.12 (ii) and the Ascoli–Arzelà theorem, every subsequence of $\{u_\epsilon\}_{\epsilon\to 0}$ contains a further subsequence $\{u_\epsilon\}_{\epsilon\in J}$ that converges uniformly as $\epsilon \to 0$, $\epsilon \in J$ to some $u \in C(\bar{\mathcal{D}})$. As in the proof of Theorem 5.2, it follows that u is a viscosity solution to:

$$\Delta_p u = 0 \quad \text{in } \mathcal{D}, \qquad u = F \quad \text{on } \partial\mathcal{D}, \tag{6.29}$$

according to Definition 5.1. Indeed, fix $x_0 \in \mathcal{D}$ and let ϕ be a test function as in (5.2). As in the previous case, there exists a sequence $\{x_\epsilon\}_{\epsilon\in J} \in \mathcal{D}$, such that:

$$\lim_{\epsilon\to 0, \epsilon\in J} x_\epsilon = x_0 \quad \text{and} \quad u_\epsilon(x_\epsilon) - \phi(x_\epsilon) = \min_{\bar{\mathcal{D}}} (u_\epsilon - \phi).$$

Consequently: $\phi(x) \leq u_\epsilon(x) + \big(\phi(x_\epsilon) - u_\epsilon(x_\epsilon)\big)$ for all $x \in \bar{\mathcal{D}}$, and further:

$$S_\epsilon\phi(x_\epsilon) - \phi(x_\epsilon) \leq S_\epsilon u(x_\epsilon) + \big(\phi(x_\epsilon) - u_\epsilon(x_\epsilon)\big) - \phi(x_\epsilon) = 0, \tag{6.30}$$

for all ϵ sufficiently small. On the other hand, (6.2) yields:

$$S_\epsilon\phi(x_\epsilon) - \phi(x_\epsilon) = \frac{\epsilon^2}{p-1}|\nabla\phi(x_\epsilon)|^{2-p}\Delta_p\phi(x_\epsilon) + o(\epsilon^2),$$

for ϵ small enough to get $\nabla\phi(x_\epsilon) \neq 0$. Combining the above with (6.30) gives:

$$\Delta_p\phi(x_\epsilon) \leq o(1).$$

Passing to the limit with $\epsilon \to 0$, $\epsilon \in J$ establishes the desired inequality $\Delta_p \phi(x_0) \leq 0$ and proves part (i) of Definition 5.1. The verification of part (ii) is done along the same lines. The proof is now done, in virtue of the uniqueness of viscosity solutions in Corollary C.63. □

Definition 6.14 We say that a given boundary point $y_0 \in \partial\mathcal{D}$ satisfies the *exterior corkscrew condition* provided that there exists $\mu \in (0, 1)$ such that for all sufficiently small $r > 0$ there exists a ball $B_{\mu r}(x)$ such that:

$$B_{\mu r}(x) \subset B_r(y_0) \setminus \bar{\mathcal{D}}.$$

The main result of this section is an improvement of the sufficiency, stated in Theorem 5.15, of the exterior cone condition for game-regularity:

Theorem 6.15 *Assume that $\mathcal{D}, p, \gamma_p, \delta_p$ are as in Definition 6.11. If $y_0 \in \partial\mathcal{D}$ satisfies the exterior corkscrew condition, then y_0 is game-regular.*

The proof is based on the concatenating strategies technique, the analysis of the annulus walk and the equivalent notion of game regularity stated below. The indicated results are the counterparts of Theorems 5.9, 5.11 and Corollary 5.12, in the present setting of the mixed averaging principle (6.2).

Theorem 6.16 *In the context of Definition 6.11, for a given $y_0 \in \partial\mathcal{D}$, assume that there exists $\theta_0 \in (0, 1)$ such that for every $\delta > 0$ there exists $\hat{\delta} \in (0, \delta)$ and $\hat{\epsilon} \in (0, 1)$ with the following property. Fix $\epsilon \in (0, \hat{\epsilon})$ and choose an initial position $x_0 \in B_{\hat{\delta}}(y_0) \cap \mathcal{D}$; there is a strategy $\sigma_{0,II}$ such that for every σ_I we have:*

$$\mathbb{P}\big(\exists n < \tau^{x_0,\sigma_I,\sigma_{0,II}} \quad X_n^{x_0,\sigma_I,\sigma_{0,II}} \notin B_\delta(y_0)\big) \leq \theta_0. \tag{6.31}$$

Then y_0 is game-regular.

Theorem 6.17 *For given radii $0 < R_1 < R_2 < R_3$, consider the annulus $\tilde{\mathcal{D}} = B_{R_3}(0) \setminus \bar{B}_{R_1}(0) \subset \mathbb{R}^N$. For every $\xi > 0$, there exists $\hat{\epsilon} \in (0, 1)$ depending on R_1, R_2, R_3 and ξ, p, N, such that for every $x_0 \in \tilde{\mathcal{D}} \cap B_{R_2}(0)$ and every $\epsilon \in (0, \hat{\epsilon})$, there exists a strategy $\tilde{\sigma}_{0,II}$ with the property that for every $\tilde{\sigma}_I$ there holds:*

$$\mathbb{P}\Big(\tilde{X}_{\tilde{\tau}-1} \notin \bar{B}_{R_3-2\epsilon}(0)\Big) \leq \frac{v(R_2) - v(R_1)}{v(R_3) - v(R_1)} + \xi. \tag{6.32}$$

Here, $v : (0, \infty) \to \mathbb{R}$ is given by:

$$v(t) = \begin{cases} \operatorname{sgn}(p - N)\, t^{\frac{p-N}{p-1}} & \text{for } p \neq N \\ \log t & \text{for } p = N, \end{cases} \tag{6.33}$$

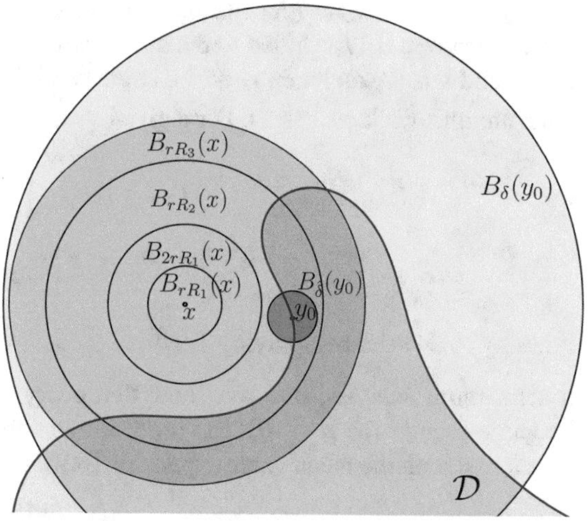

Fig. 6.5 Positions of the concentric balls $B(y_0)$ and $B(x_0)$ in the proof of Theorem 6.15

and $\{\tilde{X}_n = \tilde{X}_n^{x_0,\tilde{\sigma}_I,\tilde{\sigma}_{0,II}}\}_{n=0}^\infty$ *and* $\tilde{\tau} = \tilde{\tau}^{x_0,\tilde{\sigma}_I,\tilde{\sigma}_{0,II}}$ *denote, as before, the random variables corresponding to positions and stopping time in the random Tug-of-War game on* $\tilde{\mathcal{D}}$*. The estimate (6.32) can be replaced by:*

$$\mathbb{P}\big(\tilde{X}_{\tilde{\tau}-1} \notin \bar{B}_{R_3-2\epsilon}(0)\big) \le \theta_0, \tag{6.34}$$

valid for any $\theta_0 > 1 - \big(\frac{R_2}{R_1}\big)^{\frac{p-N}{p-1}}$ *in* $p \in (1, N)$ *and any* $\theta_0 > 0$ *if* $p \ge N$*, upon choosing* R_3 *sufficiently large with respect to* R_1 *and* R_2*. The bounds (6.32) and (6.34) remain true if we replace* R_1, R_2, R_3 *by* rR_1, rR_2, rR_3*, the domain* $\tilde{\mathcal{D}}$ *by* $r\tilde{\mathcal{D}}$ *and* $\hat{\epsilon}$ *by* $r\hat{\epsilon}$*, for any* $r > 0$.

Proof of Theorem 6.15

With the help of Theorem 6.17, we will show that the assumption of Theorem 6.16 is satisfied, with probability $\theta_0 < 1$ depending only on p, N and $\mu \in (0, 1)$ in Definition 6.14. Namely, set $R_1 = 1$, $R_2 = \frac{2}{\mu}$ and $R_3 > R_2$ according to (6.34) in order to have $\theta_0 = \theta_0(p, N, R_1, R_2) < 1$. Further, set $r = \frac{\delta}{2R_3}$ so that $rR_2 = \frac{\delta}{\mu R_3}$. Using the corkscrew condition, we obtain:

$$B_{2rR_1}(x) \subset B_{\delta/(\mu R_3)}(y_0) \setminus \bar{\mathcal{D}},$$

for some $x \in \mathbb{R}^N$ (Fig. 6.5). In particular: $|x - y_0| < rR_2$, so $y_0 \in B_{rR_2}(x) \setminus \bar{B}_{2rR_1}(x)$. It now easily follows that there exists $\hat{\delta} \in (0, \delta)$ with the property that:

$$B_{\hat{\delta}}(y) \subset B_{rR_2}(x) \setminus \bar{B}_{2rR_1}(x).$$

Finally, observe that $B_{rR_3}(x) \subset B_\delta(y_0)$ as $rR_3+|x-y_0| < rR_3+rR_2 < 2rR_3 = \delta$.

Let $\hat{\epsilon}/r > 0$ be as in Theorem 6.17, applied to the annuli with radii R_1, R_2, R_3. For a given $x_0 \in B_{\hat{\delta}}(y_0)$ and $\epsilon \in (0, \hat{\epsilon})$, let $\tilde{\sigma}_{0,II}$ be the strategy ensuring validity of the bound (6.34) in the annulus walk on $x + \tilde{\mathcal{D}}$. For a given σ_I there holds:

$$\Big\{\omega \in \Omega;\ \exists n < \tau^{x_0,\sigma_I,\sigma_{0,II}}(\omega) \qquad X_n^{x_0,\sigma_I,\sigma_{0,II}}(\omega) \notin B_\delta(y_0)\Big\}$$

$$\subset \Big\{\omega \in \Omega;\ \tilde{X}_{\tilde{\tau}-1}^{x_0,\tilde{\sigma}_I,\tilde{\sigma}_{0,II}}(\omega) \notin B_{rR_3-2\epsilon}(x)\Big\}.$$

The final claim follows by (6.34) and by applying Theorem 6.16.

Remark 6.18 With a bit more analysis, one can show that every open, bounded domain $\mathcal{D} \subset \mathbb{R}^N$ is game-regular for $p > N$. The proof mimics the argument of Sect. 5.6 for the process based on the mean value expansion (6.9), so we omit it.

Exercise 6.19

(i) Work out the details of the proof of Lemma 6.10 based on the proof of Theorem 5.3.
(ii) Work out the proofs of Theorems 6.12, 6.16, 6.17, adjusting the appropriate arguments in Chap. 5.
(iii) Give an example of an open, bounded set in $\mathbb{R}^N$ that satisfies the exterior corkscrew condition but does not satisfy the exterior cone condition.

Exercise 6.20 * Follow the outline of proof of Theorem 3.21 to show that at $p = 2$, condition of game-regularity in Definition 6.11 (a) is equivalent to Doob's regularity (3.44).

6.5 Sufficient Conditions for Game-Regularity: Simply Connectedness in Dimension $N = 2$

In this section we derive a new sufficient condition for game-regularity, namely: all simply connected domains are game-regular when $N = 2$. The idea of the proof below is based on the observation that each player has a strategy to keep the token trajectory close to a given (polygonal) curve in $\mathcal{D}$, with positive probability, and regardless of the oponent's strategy. Given a boundary point y_0, we then consider a surrounding path as in Fig. 6.6, where the origin plays the role and y_0 and where the whole diagram gets scaled by a chosen $\delta > 0$. We use a topological argument: since such line must intersect $\partial\mathcal{D}$ at a point different than y_0, it follows that the token remains in the ball $B_\delta(y_0)$ until the stopping time, as requested in condition (6.31) which ensures game-regularity of y_0.

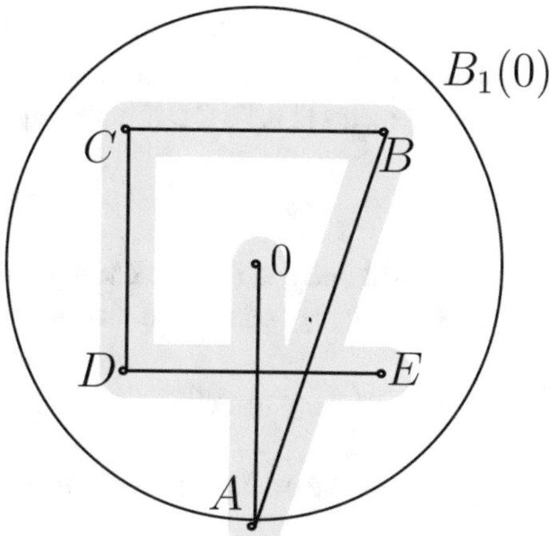

Fig. 6.6 The directed polygonal line $0ABCDE$ and the area of location of the consecutive game positions X_n in the proof of Theorem 6.21

Theorem 6.21 *Let $\mathcal{D} \subset \mathbb{R}^2$ be open, bounded, connected and simply connected. Assume that $p \in (1, \infty)$ and γ_p, ρ_p satisfy (6.1) and (6.23) with respect to $N = 2$. Then $\mathcal{D}$ is game-regular.*

Proof

1. By applying Theorem 6.17 to $R_1 = \frac{1}{2}$ and $R_2 = 2$, we conclude that there are $R_3 > 2$, $\eta > 0$ and $\bar{\epsilon} > 0$ such that for every $x_0 \in B_1(-e_1)$ and every $\epsilon \in (0, \bar{\epsilon})$ there exists $\tilde{\sigma}_{0,II}$ with the property that for every $\tilde{\sigma}_I$ there holds:

$$\mathbb{P}\Big(\min\big\{n \geq 0;\ X_n^{x_0,\tilde{\sigma}_I,\tilde{\sigma}_{0,II}} \in B_{\frac{1}{2}+2\epsilon}(0)\big\} < \min\big\{n \geq 0;\ X_n^{x_0,\tilde{\sigma}_I,\tilde{\sigma}_{0,II}} \notin \bar{B}_{R_3-2\epsilon}(0)\big\}\Big) \geq \eta.$$

Recall that the planar process $\{X_n\}_{n=0}^{\infty}$ above is defined in (6.21). Given $r > 0$, this implies that for every $x_0 \in B_r(-re_1)$ and every $\epsilon \in (0, r\bar{\epsilon})$:

$$\mathbb{P}\Big(\min\big\{n \geq 0;\ X_n^{x_0,\tilde{\sigma}_I,\tilde{\sigma}_{0,II}} \in B_r(0)\big\} < \min\big\{n \geq 0;\ X_n^{x_0,\tilde{\sigma}_I,\tilde{\sigma}_{0,II}} \notin B_{R_3 r}(0)\big\}\Big) \geq \eta. \tag{6.35}$$

2. Consider the polygonal line $0ABCDE$ with vertexes: $0 = (0, 0)$, $A = (0, -1)$, $B = (\frac{1}{2}, \sqrt{2} - 1)$, $C = -\frac{1}{2}, \sqrt{2} - 1)$, $D = (-\frac{1}{2}, \sqrt{2} - 2)$, $E = (\frac{1}{2}, \sqrt{2} - 2)$, depicted in Fig. 6.6. Let n be such that:

$$\frac{R_3}{2^n} < \frac{1}{2} \min_{z \in ABCDE} |z|.$$

Call $r = \frac{1}{2^n}$ and denote $\{z_i\}_{i=0}^{11\cdot 2^{n-1}}$ the consecutive points along the oriented polygonal $\overrightarrow{0ABCDE}$, started at $z_0 = 0$ and distanced r apart from each other. Since segments in $0ABCDE$ have respective lengths $1, \frac{3}{2}, 1, 1, 1$, it follows that:

$$z_{2^n} = A, \quad z_{5\cdot 2^{n-1}} = B, \quad z_{7\cdot 2^{n-1}} = C, \quad z_{9\cdot 2^{n-1}} = D, \quad z_{11\cdot 2^{n-1}} = E.$$

Fix $x_0 \in \overline{0A}$ and $\epsilon \in (0, r\bar{\epsilon})$. Clearly $x_0 \in B_r(z_i)$ for some $i = 0 \dots 2^n$. We now define the strategy $\sigma_{0,II}$ such that for every σ_I there holds:

$$\begin{split}\bar{p} \doteq \mathbb{P}\Big(\exists n \geq 0;\ X_n^{x_0,\sigma_I,\sigma_{0,II}} \in B_r(E) \text{ and}\\ \text{the oriented polygonal line } \overrightarrow{x_0X_1X_2\dots X_n} \text{ passes through all}\\ \text{balls } \{B_r(z_j)\}_{j=i}^{11\cdot 2^{n-1}}, \text{ while staying in the } R_3r \text{ neighbourhood}\\ \text{of } \overrightarrow{0ABCDE}\Big) \geq \eta^{11\cdot 2^{n-1}}.\end{split} \tag{6.36}$$

Condition on the oriented polygonal $\overrightarrow{x_0X_1X_2\dots X_n}$ in (6.36) implies that it surrounds 0 and intersects itself, while remaining within $B_{\frac{3}{2}}(0)$.

Define the stopping times:

$$\tau_0 = 0, \quad \tau_j = \min\{k \geq j-1;\ X_k \in B_r(z_j)\} \quad \text{for all } j = 1 \dots 11\cdot 2^{n-1}.$$

Also, for all j as above, let $f_j : \mathbb{R}^2 \to \mathbb{R}^2$ denote the rigid motions $f_j(x) = rR_jx + z_j$ where $R_j \in SO(2)$ satisfies $R_je_1 = \frac{1}{r}(z_j - z_{j-1})$. Then, we put:

$$\begin{split}&\sigma_{0,II}^k\big(x_0, (x_1, w_1, a_1, b_1), \dots, (x_k, w_k, a_k, b_k)\big)\\ &= \big(f_j \circ \tilde{\sigma}_{0,II}^{k-\tau_{j-1}}\big)\Big(f_j^{-1}(x_{\tau_{j-1}}), (f_j^{-1}(x_{\tau_{j-1}+1}),\\ &\qquad R_j^{-1}w_1, a_1, b_1), \dots, (f_j^{-1}(x_k), R_j^{-1}w_k, a_k, b_k)\Big)\\ &\qquad \text{for all } k \in [\tau_{j-1}, \tau_j) \text{ and all } j = i+1, \dots, 11\cdot 2^{n-1},\end{split}$$

where the strategy $\tilde{\sigma}_{0,II}$ satisfies (6.35). Consequently:

$$\begin{split}\mathbb{P}\Big(&\min\big\{n \geq 0;\ X_n^{x_0,\sigma_I,\sigma_{0,II}} \in B_r(z_{i+1})\big\}\\ &< \min\big\{n \geq 0;\ X_n^{x_0,\sigma_I,\sigma_{0,II}} \notin \bar{B}_{R_3r}(z_{i+1})\big\}\Big) \geq \eta,\end{split}$$

and further, for all $j = i+1, \ldots, 11 \cdot 2^{n-1} - 1$:

$$\begin{aligned}
&\mathbb{P}\Big(\exists n \geq \tau_j;\ X_n^{x_0,\sigma_I,\sigma_{0,II}} \in B_r(z_{j+1}) \\
&\qquad\qquad \text{and } \forall k < n\ \exists i \leq s \leq j+1\ X_k^{x_0,\sigma_I,\sigma_{0,II}} \in B_{R_3 r}(z_s)\Big) \\
&\geq \eta \cdot \mathbb{P}\Big(\exists n \geq \tau_{j-1};\ X_n^{x_0,\sigma_I,\sigma_{0,II}} \in B_r(z_j) \\
&\qquad\qquad \text{and } \forall k < n\ \exists i \leq s \leq j\ X_k^{x_0,\sigma_I,\sigma_{0,II}} \in B_{R_3 r}(z_s)\Big),
\end{aligned} \tag{6.37}$$

which follows by an application of Lemma A.21 as in the proof of Theorem 5.9 (see Exercise 6.24). By induction, we hence derive (6.36):

$$\bar{p} \geq \mathbb{P}\Big(\forall 0 \leq k \leq \tau_{11\cdot 2^{n-1}}\ \exists i \leq s \leq 11 \cdot 2^{n-1}\quad X_k^{x_0,\sigma_I,\sigma_{0,II}} \in B_{R_3 r}(z_s)\Big) \geq \eta^{11\cdot 2^{n-1}}.$$

3. Call $\bar{\theta}_0 = \eta^{11\cdot 2^{n-1}}$. Let $y_0 \in \partial\mathcal{D}$ and let $\delta > 0$. We claim that (6.31) is valid with some universal threshold $\theta_0 < 1$ and :

$$\hat{\delta} = \frac{\delta}{2} \quad \text{and} \quad \hat{\epsilon} = \frac{\delta}{2}\bar{\epsilon} \wedge \frac{\delta}{8(1+\gamma_p(1 \vee \delta_p))}.$$

Indeed, fix $\epsilon \in (0, \hat{\epsilon})$ and choose an initial position $x_0 \in B_{\hat{\delta}}(y_0) \cap \mathcal{D}$. By performing an orthogonal change of variables, we may without loss of generality assume that $y_0 = 0$ and $x_0 \in \frac{2}{\delta}\overline{0A}$. Let $\sigma_{0,II}$ be as constructed in Step 1. Then:

$$\begin{aligned}
&\mathbb{P}\Big(\exists n \geq 0;\ X_n^{x_0,\sigma_I,\sigma_{0,II}} \in B_{\frac{\delta}{2}r}(\frac{\delta}{2}E) \text{ and the oriented polygonal line} \\
&\overrightarrow{x_0X_1X_2\ldots X_n} \text{ surrounds } 0,\ \text{intersects itself, and stays in } B_{\frac{3}{4}\delta}(0)\Big) \geq 1 - \bar{\theta}_0.
\end{aligned}$$

in virtue of (6.36). Since $\mathcal{D}$ is simply connected, the aforementioned path must then cross $\partial\mathcal{D}$, which yields:

$$\begin{aligned}
&\mathbb{P}\Big(\min\big\{n \geq 0;\ X_n^{x_0,\sigma_I,\sigma_{0,II}} \in \partial\mathcal{D} + B_{2\epsilon(1+\gamma_p(1\vee\rho_p))}(0)\big\} \\
&\qquad\qquad < \min\big\{n \geq 0;\ X_n^{x_0,\sigma_I,\sigma_{0,II}} \not\in \bar{B}_{\frac{3}{4}\delta}(0)\big\}\Big) \geq 1 - \bar{\theta}_0.
\end{aligned} \tag{6.38}$$

We now note the following simple counterpart of Lemma 5.19: □

Lemma 6.22 *Let $R > 1$. There exists $\bar{p}_0 > 0$, depending on p, γ_p, δ_p, R, such that for every $x_0 \in B_{R\epsilon}(0)$ there exists $\tilde{\sigma}_{0,II}$ with the property that for every $\tilde{\sigma}_I$ we have:*

$$\mathbb{P}\Big(\min\big\{n \geq 0;\ X_n^{x_0,\tilde{\sigma}_I,\tilde{\sigma}_{0,II}} \in B_{\frac{\epsilon}{2}}\big\} < \min\big\{n \geq 0;\ X_n^{x_0,\tilde{\sigma}_I,\tilde{\sigma}_{0,II}} \notin \bar{B}_{R\epsilon}(0)\big\}\Big) \geq \bar{p}_0.$$

Proof Without loss of generality, we may assume that $x_0 = |x_0|e_1$. Define the strategy $\tilde{\sigma}_{0,II} = \sigma = -\frac{1}{3}e_1$ and the set of "neutral" random outcomes:

$$D_{neu} = B_s(0) \subset B_1(0) \qquad s \leq \frac{1}{12\lfloor 4R\rfloor \gamma_p(\rho_p + 1)}.$$

Observe that for all $w \in D_{neu}$ there holds:

$$\begin{aligned}&\big\langle \sigma + \gamma_p(w + (\rho_p - 1)\langle w, \sigma\rangle\sigma), e_1\big\rangle \in \Big(-\frac{1}{4}, \frac{1}{2}\Big),\\ &\Big|\big\langle \sigma + \gamma_p(w + (\rho_p - 1)\langle w, \sigma\rangle\sigma), e_1\big\rangle\Big| < \frac{1}{12\lfloor 4R\rfloor}.\end{aligned}$$

Thus, it takes some $n \leq \lfloor 4R\rfloor$ moves by any vectors as above, to achieve: $|\langle X_n, e_1\rangle| < \frac{\epsilon}{4}$ and also: $n \cdot \frac{\epsilon}{\lfloor 4R\rfloor} < \frac{\epsilon}{4}$. Consequently:

$$\exists n \leq \lfloor 4r\rfloor \quad x_0 + \sum_{i=1}^{n}\Big(\sigma + \gamma_p(w + (\rho_p - 1)\langle w, \sigma\rangle\sigma)\Big) \in \Big(-\frac{\epsilon}{2}, \frac{\epsilon}{2}\Big)^2 \subset B_{\frac{\epsilon}{2}}(0)$$

for any $w_1, \ldots, w_n \in D_{neu}$.

Hence, we arrive at:

$$\begin{aligned}&\mathbb{P}\Big(\min\big\{n \geq 0;\ X_n^{x_0,\tilde{\sigma}_I,\tilde{\sigma}_{0,II}} \in B_{\frac{\epsilon}{2}}\big\} < \min\big\{n \geq 0;\ X_n^{x_0,\tilde{\sigma}_I,\tilde{\sigma}_{0,II}} \notin \bar{B}_{R\epsilon}(0)\big\}\Big)\\ &\qquad \geq \mathbb{P}\Big(\big(D_{neu} \times \{2\} \times (0,1)\big)^{\lfloor 4R\rfloor} \times \prod_{i=\lfloor 4r\rfloor}^{\infty} \Omega_1\Big) = \Big(\frac{\pi s^2}{2}\Big)^{\lfloor 4R\rfloor} = \bar{p}_0,\end{aligned}$$

as claimed.

We now complete the proof of Theorem 6.21. Applying Lemma 6.22 to $R = 1 + \gamma_p \max\{1, \rho_p\}$, we see that for all $x_0 \in \mathcal{D}$ with $\operatorname{dist}(x_0, \partial\mathcal{D}) < \epsilon R$ there exists $\tilde{\sigma}_{0,II}$ such that for every $\tilde{\sigma}_I$ there holds:

$$\mathbb{P}\Big(\forall n \leq \tau^{x_0,\tilde{\sigma}_I,\tilde{\sigma}_{0,II}} \quad X_n^{x_0,\tilde{\sigma}_I,\tilde{\sigma}_{0,II}} \in B_{4\epsilon(1+\gamma_p\max\{1,\rho_p\})}(x_0)\Big) \geq \frac{\bar{p}_0}{2},$$

with some $\bar{p}_0 > 0$ that depends only on p, γ_p, ρ_p. Recalling (6.38) we obtain:

$$\mathbb{P}\Big(\forall n \leq \tau^{x_0,\sigma_I,\sigma_{0,II}} \quad X_n^{x_0,\bar{\sigma}_I,\bar{\sigma}_{0,II}} \in B_\delta(0)\Big) \geq (1-\bar{\theta}_0)\cdot\frac{\bar{p}_0}{2}.$$

This yields (6.31) in Theorem 6.16 with $\theta_0 = 1 - \frac{1}{2}(1-\bar{\theta}_0)\bar{p}_0$. □

Remark 6.23 By the same argument as in the proof above, one can see the following. Let $\mathcal{D} \subset \mathbb{R}^2$ be open, bounded, connected and let $y_0 \in \partial\mathcal{D}$ satisfy the *exterior curve condition*, i.e. there exists a continuous curve contained in $\mathbb{R}^2 \setminus \mathcal{D}$, of positive length and with one endpoint at y_0. Then y_0 is walk-regular.

Exercise 6.24 Provide all details of the proof of the inductive formula (6.37).

Exercise 6.25 An open, bounded, simply connected set $\mathcal{D} \subset \mathbb{R}^2$ does not have to satisfy the exterior curve condition as in Remark 6.23. Consider $K_0 \subset \mathbb{R}^2$ that is the union of the vertical segment $\{1\} \times [-1, 1]$ with the graph of the function: $\sin\frac{\pi}{x-1}$ over the interval $(1, 2]$. Define:

$$K = \{(0,0)\} \cup \bigcup_{n=2}^{\infty} \frac{1}{2^n}K_0, \qquad \mathcal{D} = (-1,1)^2 \setminus K.$$

Show that the open set $\mathcal{D} \subset \mathbb{R}^2$ is bounded and simply connected, but $(0, 0)$ is a boundary point of $\mathcal{D}$ without an exterior curve.

6.6 Bibliographical Notes

The construction in this chapter is mostly taken from Lewicka (2018). The fundamental paper by Peres and Sheffield (2008) developed the dynamic programming principle and the Tug-of-War game valid in the whole exponent range $p \in (1, \infty)$, that was based on the averaging expansion different than our (6.2). In particular, since it involved averaging on the codimension 2 sets, the regularity of the game values as well as the coincidence of the upper and lower values were not clear. In Arroyo et al. (2017); Hartikainen (2016), yet another dynamic programming principle (6.10), still sampling on sets of measure zero, has been analysed and convergence of its solutions proved in domains satisfying the exterior corkscrew condition has been shown in Arroyo et al. (2018).

The proof in Sect. 6.5 is an adaptation of the argument in Peres and Sheffield (2008) to the mixed averaging principle and the resulting Tug of War game in Sect. 6.3. The present set-up has the advantage of utilizing the full N-dimensional sampling on ellipses, rather than on spheres. This implies that the solutions of the dynamic programming principle at each scale $\epsilon > 0$ are unique, continuous and coincide with the well-defined game values; much like in the linear $p = 2$ case where the N-dimensional averaging guaranteed smoothness of harmonic functions.

Appendix A
Background in Probability

In this chapter we recall definitions and statements on the following chosen topics in probability: probability and measurable spaces, random variables, product spaces, conditional expectation, independence, martingales in discrete times, stopping times, Doob's optional stopping theorem and convergence of martingales.

In this chapter we recall definitions and preliminary facts on the chosen topics in probability: probability spaces, conditional expectation and martingales in discrete times. We limit ourselves to the material that is necessary to carry out the constructions of the ball walk and the tug of war games discrete processes. This material may be found in any textbook on probability, for example in: Williams (1991); Durrett (2010); Kallenberg (2002); Dudley (2004).

A.1 Probability and Measurable Spaces

Definition A.1 Let Ω be a given set endowed with a σ-algebra $\mathcal{F}$ of its subsets and a probability measure $\mathbb{P}$ on $\mathcal{F}$. We call $(\Omega, \mathcal{F})$ a *measurable space* and $(\Omega, \mathcal{F}, \mathbb{P})$ a *probability space*.

Recall that the following are required for $\mathcal{F} \subset 2^{\Omega}$ to be a σ*-algebra*:

(i) $\emptyset \in \mathcal{F}$.
(ii) If $A \in \mathcal{F}$, then $\Omega \setminus A \in \mathcal{F}$.
(iii) If $\{A_i\}_{i=1}^{\infty}$ is a sequence of sets $A_i \in \mathcal{F}$, then $\bigcup_{i=1}^{\infty} A_i \in \mathcal{F}$.

M. Lewicka, *A Course on Tug-of-War Games with Random Noise*, Universitext,
https://doi.org/10.1007/978-3-030-46209-3

A *measure* on a measurable space $(\Omega, \mathcal{F})$ is, by definition, a nonnegative and countably additive set function $\mu : \mathcal{F} \to \bar{\mathbb{R}}_+ = [0, +\infty]$, that is:

(i) $0 = \mu(\emptyset) \leq \mu(A)$, for all $A \in \mathcal{F}$.
(ii) If the sets in $\{A_i \in \mathcal{F}\}_{i=1}^{\infty}$ are pairwise disjoint, then: $\mu\big(\bigcup_{i=1}^{\infty} A_i\big) = \sum_{i=1}^{\infty} \mu(A_i)$.

A *probability measure* μ (usually denoted by $\mathbb{P}$) is a measure such that:

(iii) $\mu(\Omega) = 1$.

More generally, we shall deal with σ-*finite* measures, which means that:

(iii)' $\Omega = \bigcup_{i=1}^{\infty} A_i$, such that $A_i \in \mathcal{F}$ and $\mu(A_i) < +\infty$ for each $i \in \mathbb{N}$.

One primary example of a measurable space consists of a Borel set $\Omega \subset \mathbb{R}^N$ and the σ-algebra $\mathcal{F} = \mathcal{B}(\Omega)$ of all Borel subsets of Ω. Recall that the *Borel σ-algebra* $\mathcal{B}(\mathbb{R}^N)$ is, by definition, the smallest σ-algebra containing all the open N-dimensional sets, whereas $\mathcal{B}(\Omega) = \{A \cap \Omega;\ A \in \mathcal{B}(\mathbb{R}^N)\}$. We further have the following examples of a probability space $(\Omega, \mathcal{F}, \mathbb{P})$:

Example A.2

(i) *(Uniform probability measure on a finite set).* We take: $\Omega = \{1, 2, \ldots, n\}$, $\mathcal{F} = 2^{\Omega}$ and $\mathbb{P}(A) = \frac{|A|}{n}$, where $|A|$ denotes the number of elements in the finite set $A \subset \Omega$.
(ii) *(Discrete probability measure).* Let $\Omega = \mathbb{N}$, $\mathcal{F} = 2^{\Omega}$ and let $\sum_{i=1}^{\infty} p_i$ be a nonnegative series, summable to 1. Then we define: $\mathbb{P}(A) = \sum_{i \in A} p_i$.
(iii) *(Normalized Lebesgue measure on a ball).* We set $\Omega = B_\epsilon(0) \subset \mathbb{R}^N$, $\mathcal{F} = \mathcal{B}(\Omega)$ and $\mathbb{P}(A) = \frac{|A|}{\epsilon^N V_N}$, where $|A|$ denotes the N-dimensional Lebesgue measure (i.e. the volume) of $A \in \mathcal{B}(\Omega)$. We write $V_N = |B_1(0)|$ for the volume of the unit ball $B_1(0)$ in $\mathbb{R}^N$.
(iv) *(Probability measure with a density).* Here, $\Omega \subset \mathbb{R}^N$ is a given Lebesgue (or Borel) measurable set and $\mathcal{F}$ consists of all its Lebesgue (or Borel) measurable subsets. We define:

$$\mathbb{P}(A) = \int_A f(x)\, \mathrm{d}x, \tag{A.1}$$

where $f : \Omega \to \mathbb{R}$ is a given nonnegative Lebesgue (or Borel) measurable function with $\int_\Omega f(x)\, \mathrm{d}x = 1$. We call f the *density* of $\mathbb{P}$; observe that f with property (A.1) is unique up to modifications on sets of $\mathbb{P}$-measure 0.

An element $\omega \in \Omega$ is called an *outcome* and a set $A \in \mathcal{F}$ is called an *event*. Throughout the book, we use probability theory convention and suppress the notion of outcomes, when no ambiguity arises. Accordingly, for a function $X : \Omega \to \mathbb{R}$ and $r \in \mathbb{R}$, we write $\{X \leq r\}$ instead of $\{\omega \in \Omega;\ X(\omega) \leq r\}$, or $A \cap \{X = 0\}$ instead of $\{\omega \in A;\ X(\omega) = 0\}$, etc.

We also adopt another convention: when a certain property, for example $X(\omega) \leq r$, is satisfied on a set of full measure: $\mathbb{P}(\{X \leq r\}) = 1$, we write: $X \leq r$ *a.s.* (to be read: *almost surely*), or $X \leq r$ $\mathbb{P}$-a.s. in case there is need to specify the probability measure $\mathbb{P}$. Only in case of possible ambiguity, we write: $Y(\omega) \leq r$ for $\mathbb{P}$-a.e. ω (to be read: $\mathbb{P}$-*almost every* ω). To avoid repeated parentheses, we often replace $\mathbb{P}(\{X \leq r\})$ by $\mathbb{P}(X \leq r)$, unless a further clarification is needed.

One can construct a probability measure on a σ-algebra as an extension of a given premeasure on an algebra. This is an important technique, used to introduce the Lebesgue measure on $\Omega = \mathbb{R}^N$, as well as the countable product of probability spaces, explained in Sect. A.3. Recall that $\mathcal{A} \subset 2^\Omega$ is an *algebra* if:

(i) $\emptyset \in \mathcal{A}$,
(ii) If $A \in \mathcal{A}$, then $\Omega \setminus A \in \mathcal{A}$.
(iii) If $A_1, A_2 \in \mathcal{A}$, then $A_1 \cup A_2 \in \mathcal{A}$.

A *premeasure* on an algebra $\mathcal{A}$ is a set function $\mu_0 : \mathcal{A} \to \bar{\mathbb{R}}_+$, such that:

(i) $0 = \mu_0(\emptyset) \leq \mu_0(A)$ for all $A \in \mathcal{A}$.
(ii) If $\{A_i\}_{i=1}^\infty$ is a sequence of pairwise disjoint sets $A_i \in \mathcal{A}$ such that $\bigcup_{i=1}^\infty A_i \in \mathcal{A}$, then $\mu_0\big(\bigcup_{i=1}^\infty A_i\big) = \sum_{i=1}^\infty \mu_0(A_i)$.

The fact that every premeasure generates a measure is known as the *Caratheodory extension theorem*:

Theorem A.3 *Let μ_0 be a premeasure on an algebra $\mathcal{A} \subset 2^\Omega$. Let $\mathcal{F}$ be the smallest σ-algebra $\mathcal{F}$ of subsets of Ω, that contains $\mathcal{A}$. Then, there exists a measure μ on $\mathcal{F}$, satisfying:*

$$\mu(A) = \mu_0(A) \qquad \textit{for all } A \in \mathcal{A}.$$

If $\mu_0(\Omega) < +\infty$, then such measure μ is unique.

A.2 Random Variables and Expectation

Definition A.4 A *random variable* on a measurable space $(\Omega, \mathcal{F})$ is a $\mathcal{F}$-measurable function $X : \Omega \to \bar{\mathbb{R}} \doteq \mathbb{R} \cup \{-\infty, +\infty\}$. Namely, we have:

$$\{X \leq r\} \in \mathcal{F} \quad \text{for all } r \in \mathbb{R}.$$

The simplest example of a random variable is the *indicator* of a set $A \in \mathcal{F}$:

$$\mathbb{1}_A(\omega) = \begin{cases} 1 \text{ if } \omega \in A \\ 0 \text{ if } \omega \in \Omega \setminus A. \end{cases}$$

More generally, we also consider random variables $X : \Omega_1 \to \Omega_2$ between two measurable spaces $(\Omega_1, \mathcal{F}_1)$ and $(\Omega_2, \mathcal{F}_2)$. The measurability condition on X is then:

$$\{X \in A_2\} \in \mathcal{F}_1 \quad \text{for all } A_2 \in \mathcal{F}_2.$$

Clearly, taking $(\Omega_1, \mathcal{F}_1) = (\Omega, \mathcal{F})$ and $(\Omega_2, \mathcal{F}_2) = (\bar{\mathbb{R}}, \mathcal{B}(\bar{\mathbb{R}}))$ yields the scalar valued case of Definition A.4 where, with a slight abuse of notation, $\mathcal{B}(\bar{\mathbb{R}})$ stands for the smallest σ-algebra of subsets of $\bar{\mathbb{R}}$ containing all the sets $[r, \infty) \cup \{+\infty\}$ for $r \in \mathbb{R}$. Unless stated otherwise, by "random variable" we always mean the $\bar{\mathbb{R}}$-valued measurable function as in Definition A.4, that is a.s. finite.

Given a random variable $X : \Omega \to \bar{\mathbb{R}}$ on a probability space $(\Omega, \mathcal{F}, \mathbb{P})$, there is a standard construction of the *integral* (also called *expectation* in this setting) which proceeds in three steps:

(i) If $X = \sum_{i=1}^n \alpha_i \mathbb{1}_{A_i}$ with $A_i \in \mathcal{F}$ and $\alpha_i \in \mathbb{R}$ for $i = 1 \dots n$ and some $n \in \mathbb{N}$, then we call X a *simple function* and define:

$$\int_\Omega X \, \mathrm{d}\mathbb{P} \doteq \sum_{i=1}^n \alpha_i \mathbb{P}(A_i) \in \mathbb{R}.$$

(ii) Every nonnegative random variable X is a nondecreasing pointwise limit of simple functions $\{X_i\}_{i=1}^\infty$. For such X we define:

$$\int_\Omega X \, \mathrm{d}\mathbb{P} \doteq \lim_{i \to \infty} \int_\Omega X_i \, \mathrm{d}\mathbb{P} \in \bar{\mathbb{R}}_+.$$

(iii) When $\int_\Omega |X| \, \mathrm{d}\mathbb{P} < \infty$ we say that X is $\mathbb{P}$-*integrable*. We then define X_+, X_- to be the positive and negative parts of X, namely: $X = X_+ - X_-$ and $X_+, X_- \geq 0$ a.s., and we set:

$$\int_\Omega X \, \mathrm{d}\mathbb{P} \doteq \int_\Omega X_+ \, \mathrm{d}\mathbb{P} - \int_\Omega X_- \, \mathrm{d}\mathbb{P} \in \mathbb{R}.$$

Definition A.5 Given a probability space $(\Omega, \mathcal{F}, \mathbb{P})$, we denote by $L^1(\Omega, \mathcal{F}, \mathbb{P})$ the linear space of all $\mathbb{P}$-integrable random variables X on Ω. The *expectation* of $X \in L^1(\Omega, \mathcal{F}, \mathbb{P})$ is then:

$$\mathbb{E}[X] \doteq \int_\Omega X \, \mathrm{d}\mathbb{P}.$$

The following fundamental results allow for passing to the limit under the integral sign. Theorem A.6 is usually referred to as *Fatou's lemma*, part (i) of Theorem A.7 as the *monotone convergence theorem* and part (ii) as the *Lebesgue dominated convergence theorem*.

Theorem A.6 *Let $\{X_i\}_{i=1}^\infty$ be a sequence of a.s. nonnegative random variables on a probability space $(\Omega, \mathcal{F}, \mathbb{P})$. Then we have:*

$$\int_\Omega \liminf_{i\to\infty} X_i \, d\mathbb{P} \leq \liminf_{i\to\infty} \int_\Omega X_i \, d\mathbb{P}.$$

Theorem A.7 *Let $\{X_i\}_{i=1}^\infty$ be a sequence of random variables on a probability space $(\Omega, \mathcal{F}, \mathbb{P})$, converging pointwise a.s. to a random variable X. Assume that one of the following properties holds:*

(i) For each $i \in \mathbb{N}$ we have: $X_i \geq 0$ and $X_i \leq X_{i+1}$ a.s.
(ii) There exists an integrable random variable $Z \in L^1(\Omega, \mathcal{F}, \mathbb{P})$ such that $|X_i| \leq Z$ a.s. for every $i \in \mathbb{N}$.

Then one can pass to the limit under the integral: $\lim_{i\to\infty} \int_\Omega X_i \, d\mathbb{P} = \int_\Omega X \, d\mathbb{P}$.

The useful construction of the *push-forward measures* yields the following *change of variable formula*:

Exercise A.8 Let $(\Omega_1, \mathcal{F}_1, \mathbb{P}_1)$ and $(\Omega_2, \mathcal{F}_2)$ be a probability space and a measurable space, respectively. Given a random variable $X : \Omega_1 \to \Omega_2$, define:

$$\mathbb{P}_2(A) \doteq \mathbb{P}_1(X \in A) \qquad \text{for all } A \in \mathcal{F}_2.$$

Show that $(\Omega_2, \mathcal{F}_2, \mathbb{P}_2)$ is a probability space and that:

$$\int_{\Omega_1} Z \circ X \, \mathrm{d}\mathbb{P}_1 = \int_{\Omega_2} Z \, \mathrm{d}\mathbb{P}_2,$$

for every nonnegative or integrable random variable $Z : \Omega_2 \to \bar{\mathbb{R}}$.

Example A.9 Let $(\Omega_1, \mathcal{F}_1, \mathbb{P}_1)$ be the normalized Lebesgue measure on the Borel σ-algebra $\mathcal{F}_1$ of subsets of $\Omega_1 = B_r(0) \subset \mathbb{R}^N$. Let $\Omega_2 = \partial B_r(0)$ and let $X : \Omega_1 \to \Omega_2$ be the projection given by $X(\omega) = r\frac{\omega}{|\omega|}$ for all $\omega \in \Omega_1 \setminus \{0\}$. The push-forward construction of Exercise A.8 results then in the probability measure $\mathbb{P}_2$ on the σ-algebra $\mathcal{F}_2$ of Borel subsets of Ω_2, namely: $\mathcal{F}_2 = \{A \subset \Omega_2;\ X^{-1}(A) \in \mathcal{F}_1\}$. We call $\mathbb{P}_2$ the *normalized spherical measure*, whereas the following measure will be called the *spherical measure* on $\partial B_r(0)$:

$$\sigma^{N-1} \doteq \frac{2\pi^{\frac{N}{2}}}{\Gamma(\frac{N}{2})} r^{N-1} \mathbb{P}_2,$$

so that $\sigma^{N-1}(\partial B_r(0)) = |\partial B_r(0)|$ equals the surface area of $\partial B_r(0)$. Since $|\partial B_r(0)| = \frac{N}{r}|B_r(0)|$, we obtain the "polar coordinates" integration formula for all $u \in C(\bar{B}_r(0))$:

$$\int_{B_r(0)} u(y) \, \mathrm{d}y = \int_0^r \int_{\partial B_s(0)} u(y) \, \mathrm{d}\sigma^{N-1}(y) \, \mathrm{d}s. \tag{A.2}$$

Finally, we state the result on the (weak-$*$) convergence of probability measures, which is a version of the fundamental *Prohorov's theorem*:

Theorem A.10 *Let (Ω, d) be a compact metric space and let $\{\mathbb{P}_i\}_{i=1}^{\infty}$ be a sequence of probability measures on Ω equipped with the Borel σ-algebra $\mathcal{F}$. Then there exists a subsequence $\{\mathbb{P}_{i_k}\}_{k=1}^{\infty}$ converging weakly-$*$ to some probability measure $\mathbb{P}$ on $(\Omega, \mathcal{F})$, which means that:*

$$\lim_{k\to\infty} \int_\Omega X \, d\mathbb{P}_{i_k} = \int_\Omega X \, d\mathbb{P} \qquad \textit{for all } X \in C(\Omega).$$

A.3 Product Measures

We now recall the notion and construction of a *product measurable space* and a *product measure*. Given a finite number of probability spaces $\{(\Omega_i, \mathcal{F}_i, \mathbb{P}_i)\}_{i=1}^n$, we form a Cartesian product $\Omega = \prod_{i=1}^n \Omega_i$ and endow it with the product σ-algebra $\mathcal{F}$, defined as the smallest σ-algebra containing all the product sets: $A = \prod_{i=1}^n A_i$, where $A_i \in \mathcal{F}_i$ for each $i = 1, \dots, n$. The product measure $\mathbb{P}$ on $\mathcal{F}$ is then the unique measure such that:

$$\mathbb{P}(A) = \prod_{i=1}^n \mathbb{P}(A_i).$$

A fundamental result in this context is the *Fubini–Tonelli theorem*. For its statement below and also in the sequel, we adopt the following convention. Let $A \in \mathcal{F}$ be a set of full measure in the probability space $(\Omega, \mathcal{F}, \mathbb{P})$, i.e. $\mathbb{P}(A) = 1$. Let $Z : A \to \bar{\mathbb{R}}$ be a random variable on the induced probability space $(A, \mathcal{F}_A, \mathbb{P}_{|\mathcal{F}_A})$ with $\mathcal{F}_A = \{B \cap A;\ B \in \mathcal{F}\}$. Then we write:

$$\int_\Omega Z \, \mathrm{d}\mathbb{P} \doteq \int_A Z \, \mathrm{d}\mathbb{P}_{|\mathcal{F}_A},$$

whenever the integral in the right hand side above is well defined.

Theorem A.11 *Let $(\Omega, \mathcal{F}, \mathbb{P})$ be the product of two probability spaces, denoted $(\Omega_1, \mathcal{F}_1, \mathbb{P}_1)$ and $(\Omega_2, \mathcal{F}_2, \mathbb{P}_2)$. Let $X : \Omega \to \bar{\mathbb{R}}$ be a $\mathcal{F}$-measurable random variable which is either $\mathbb{P}$-integrable or nonnegative. Then the function $Y : \Omega_1 \to \bar{\mathbb{R}}$ given by $Y(\omega_1) = \int_{\Omega_2} X(\omega_1, \omega_2) \, d\mathbb{P}_2(\omega_2)$ is well defined $\mathbb{P}_1$-a.s., it is $\mathcal{F}_1$-measurable, and:*

$$\int_\Omega X \, \mathrm{d}\mathbb{P} = \int_{\Omega_1} Y \, \mathrm{d}\mathbb{P}_1.$$

In the same manner, when $(\Omega, \mathcal{F}, \mathbb{P})$ is the product of $n \geq 2$ probability spaces, the integral of any $\mathbb{P}$-integrable or nonnegative $\mathcal{F}$-measurable random variable $X : \Omega \to \bar{\mathbb{R}}$ can be expressed by means of iterated integrals:

$$\int_{\Omega} X(\omega_1, \ldots, \omega_n) \, d\mathbb{P}$$
$$= \int_{\Omega_1} \cdots \int_{\Omega_n} X(\omega_1, \ldots, \omega_n) \, d\mathbb{P}_n(\omega_n) \ldots d\mathbb{P}_1(\omega_1). \tag{A.3}$$

Given now a sequence of probability spaces $\{(\Omega_i, \mathcal{F}_i, \mathbb{P}_i)\}_{i=1}^{\infty}$, the parallel construction of the product $(\Omega, \mathcal{F}, \mathbb{P})$ requires a bit more care. For the measurable space $(\Omega, \mathcal{F})$, we let $\Omega = \prod_{i=1}^{\infty} \Omega_i$ to consist of all sequences $\{\omega_i\}_{i=1}^{\infty}$ with $\omega_i \in \Omega_i$ for all $i \in \mathbb{N}$, while the σ-algebra $\mathcal{F}$ is the smallest σ-algebra of subsets of Ω, containing all sets of the form $\prod_{i=1}^{\infty} A_i$, where $A_n \in \mathcal{F}_n$ for some $n \in \mathbb{N}$ and $A_i = \Omega_i$ for all other indices $i \neq n$. We then have:

Theorem A.12 *Let $\{(\Omega_i, \mathcal{F}_i, \mathbb{P}_i)\}_{i=1}^{\infty}$ be a sequence of probability spaces. There exists a unique probability measure $\mathbb{P}$ on the product $(\Omega, \mathcal{F})$, such that:*

$$\mathbb{P}\Big(\prod_{i=1}^{\infty} A_i\Big) = \prod_{i=1}^{\infty} \mathbb{P}_i(A_i), \tag{A.4}$$

for every sequence of sets $\{A_i\}_{i=1}^{\infty}$ such that $A_i \in \mathcal{F}_i$ for each $i \in \mathbb{N}$ and $A_i = \Omega_i$ for all but finitely many indices i.

Proof

1. Consider the family generated by the finite "cylinders" of the form:

$$\mathcal{A} \doteq \Big\{ \Big(\bigcup_{k=1}^{m} \prod_{i=1}^{n} A_{ik}\Big) \times \Big(\prod_{i=n+1}^{\infty} \Omega_i\Big); \tag{A.5}$$
$$A_{ik} \in \mathcal{F}_i \quad \text{for all } i = 1, \ldots, n, \; k = 1, \ldots, m, \; n, m \in \mathbb{N} \Big\}.$$

It is easy to observe that $\mathcal{A}$ is an algebra of subsets of Ω and that the smallest σ-algebra containing $\mathcal{A}$ coincides with $\mathcal{F}$. Define the set function μ_0 on $\mathcal{A}$ by:

$$\mu_0\Big(F \times \prod_{i=n+1}^{\infty} \Omega_i\Big) = \bar{\mathbb{P}}_n(F), \tag{A.6}$$

for all sets F of the form: $F = \bigcup_{k=1}^{m} \prod_{i=1}^{n} A_{ik}$ as in (A.5), and where $\bar{\mathbb{P}}_n$ is the product measure on the product of n probability spaces $\{(\Omega_i, \mathcal{F}_i, \mathbb{P}_i)\}_{i=1}^{n}$. We will now show that μ_0 is a premeasure on $\mathcal{A}$; the existence of its unique extension $\mathbb{P}$ on $\mathcal{F}$ will then follow from Theorem A.3.

To this end, consider a sequence $\{A_i\}_{i=1}^{\infty}$ of sets $A_i \in \mathcal{A}$ that are pairwise disjoint and with $\bigcup_{i=1}^{\infty} A_i \in \mathcal{A}$. Clearly, for every $n \geq 1$ we have:

$$\sum_{i=1}^{n} \mu_0(A_i) = \mu_0\Big(\bigcup_{i=1}^{n} A_i\Big) \leq \mu_0\Big(\bigcup_{i=1}^{\infty} A_i\Big),$$

where we used (A.6) and the resulting monotonicity of μ_0. To show that $\sum_{i=1}^{\infty} \mu_0(A_i) = \lim_{n\to\infty} \sum_{i=1}^{n} \mu_0(A_1) = \mu_0\big(\bigcup_{i=1}^{\infty} A_i\big)$, it suffices to check:

$$\lim_{n\to\infty} \mu_0\Big(\bigcup_{i=n+1}^{\infty} A_i\Big) = 0.$$

This property will follow from the general fact below, applied to the decreasing family of sets $B_n = \bigcup_{i=n+1}^{\infty} A_i$ with empty intersection.

2. Let $\{B_n\}_{n=1}^{\infty}$ be a decreasing family of sets $B_n \in \mathcal{A}$, such that:

$$\bigcap_{n=1}^{\infty} B_n = \emptyset. \tag{A.7}$$

We will show that $\lim_{n\to\infty} \mu_0(B_n) = 0$. Assume, by contradiction, that:

$$\mu_0(B_n) \geq \epsilon > 0 \qquad \text{for every } n \geq 1. \tag{A.8}$$

Without loss of generality, we may take each B_n to be of the form as in (A.5), i.e. $B_n = F_n \times (\prod_{i=n+1}^{\infty} \Omega_i)$, where F_n is a finite union of Cartesian products of measurable subsets of $\Omega_1, \ldots, \Omega_n$. Given a k-tuple $(x_1, \ldots x_k) \in \prod_{i=1}^{k} \Omega_i$, we denote: $B_n(x_1, \ldots x_k) = \{(x_{k+1}, \ldots) \in \prod_{i=k+1}^{\infty} \Omega_i;\ (x_1, x_2, \ldots) \in B_n\}$. Clearly then: $(\prod_{i=1}^{k} \Omega_i) \times B_n(x_1, \ldots, x_k) \in \mathcal{A}$.

In virtue of Theorem A.11 we observe that:

$$\begin{aligned}\mu_0(B_n) &= \int_{\Omega_1} \mu_0\big(\Omega_1 \times B_n(x_1)\big)\, \mathrm{d}\mathbb{P}_1(x_1)\\ &\leq \mathbb{P}_1\Big(x_1 \in \Omega_1;\ \mu_0(\Omega_1 \times B_n(x_1)) \geq \frac{\epsilon}{2}\Big) + \frac{\epsilon}{2}.\end{aligned}$$

Together with (A.8), this yields:

$$\mathbb{P}_1\Big(x_1 \in \Omega_1;\ \mu_0(\Omega_1 \times B_n(x_1)) \geq \frac{\epsilon}{2}\Big) \geq \frac{\epsilon}{2} \qquad \text{for all } n \geq 1.$$

Since the given above subsets of Ω_1 are decreasing as $n \to \infty$, it follows that their intersection must be nonempty, namely:

$$\exists \tilde{x}_1 \in \Omega_1 \quad \forall n \geq 1 \qquad \mu_0(\Omega_1 \times B_n(\tilde{x}_1)) \geq \frac{\epsilon}{2}. \tag{A.9}$$

In a similar manner, we obtain:

$$\begin{aligned}\mu_0(\Omega_1 \times B_n(\tilde{x}_1)) &= \int_{\Omega_2} \mu_0\big(\Omega_1 \times \Omega_2 \times B_n(\tilde{x}_1, x_2)\big)\, \mathrm{d}\mathbb{P}_2(x_2) \\ &\leq \mathbb{P}_2\Big(x_2 \in \Omega_2;\ \mu_0(\Omega_1 \times \Omega_2 \times B_n(\tilde{x}_1, x_2)) \geq \frac{\epsilon}{4}\Big) + \frac{\epsilon}{4},\end{aligned}$$

which combined with (A.9) yields:

$$\mathbb{P}_2\Big(x_2 \in \Omega_2;\ \mu_0(\Omega_1 \times \Omega_2 \times B_n(\tilde{x}_1, x_2)) \geq \frac{\epsilon}{4}\Big) \geq \frac{\epsilon}{4} \quad \text{for all } n \geq 1,$$

and thus:

$$\exists \tilde{x}_2 \in \Omega_2 \quad \forall n \geq 1 \qquad \mu_0(\Omega_1 \times \Omega_2 \times B_n(\tilde{x}_1, \tilde{x}_2)) \geq \frac{\epsilon}{4}.$$

Repeating this procedure, we inductively obtain a sequence $\{\tilde{x}_i \in \Omega_i\}_{i=1}^\infty$ so that:

$$\mu_0\Big(\big(\prod_{i=1}^k \Omega_k\big) \times B_n(\tilde{x}_1, \ldots, \tilde{x}_k)\Big) \geq \frac{\epsilon}{2^k} \quad \text{for all } n, k \geq 1.$$

In particular, taking $k = n$ we observe that each set $B_n(\tilde{x}_1, \ldots, \tilde{x}_n)$ is nonempty, and so $(\tilde{x}_1, \ldots, \tilde{x}_n) \in F_n$ for all $n \geq 1$. Thus, there must be: $(\tilde{x}_1, \tilde{x}_2, \ldots) \in \bigcap_{n=1}^\infty B_n$, contradicting (A.7). The proof is done. □

Example A.13

(i) The countable product of the probability spaces in Example A.2 (iii). Namely, let $B_\epsilon(0) \subset \mathbb{R}^N$ be endowed with the Borel σ-algebra $\mathcal{B}(B_\epsilon(0))$ and the normalized Lebesgue measure $\frac{|\cdot|}{|B_\epsilon(0)|}$. Theorem A.12 defines the probability space on $\Omega = (B_\epsilon(0))^{\mathbb{N}} \doteq \prod_{i=1}^\infty B_\epsilon(0)$.

(ii) In Chap. 3 we will be working with the probability space $(\Omega, \mathcal{F}, \mathbb{P})$, where $\Omega = \big(B_1(0) \times \{1, 2, 3\} \times (0, 1)\big)^{\mathbb{N}}$, constructed as the infinite product of countably many copies of the product $B_1(0) \times \{1, 2, 3\} \times [0, 1]$. This last probability space is the product of: the normalized Lebesgue measure space as in Example A.2 (iii), the uniform measure space as in Example A.2 (i) with $n = 3$, and the Lebesgue measure space on $(0, 1)$.

In fact, Theorem A.12 holds for arbitrary (not necessarily countable) products of probability spaces. The same proof works in the general case as well, since every set

A belonging to the product algebra $\mathcal{A}$ depends only on finitely many coordinates and each set in the product σ-algebra $\mathcal{F}$ depends only on countably many coordinates.

A.4 Conditional Expectation

Definition A.14 Let $X \in L^1(\Omega, \mathcal{F}, \mathbb{P})$ be an integrable random variable on a probability space $(\Omega, \mathcal{F}, \mathbb{P})$ and let $\mathcal{G} \subset \mathcal{F}$ be a sub- σ-algebra. The *conditional expectation* of X relative to $\mathcal{G}$, denoted by $\mathbb{E}(X \mid \mathcal{G})$, is a random variable $Y = \mathbb{E}(X \mid \mathcal{G})$ such that $Y \in L^1(\Omega, \mathcal{G}, \mathbb{P}_{|\mathcal{G}})$ and:

$$\int_A Y \, \mathrm{d}\mathbb{P} = \int_A X \, \mathrm{d}\mathbb{P} \qquad \text{for all } A \in \mathcal{G}.$$

Observe that, automatically, there holds: $\mathbb{E}[\mathbb{E}(X \mid \mathcal{G})] = \mathbb{E}[X]$. Existence of conditional expectation follows from the fundamental *Radon–Nikodym theorem*:

Theorem A.15 *Let μ, ν be two σ-finite measures on a measurable space $(\Omega, \mathcal{F})$. If ν is* absolutely continuous *with respect to μ (we write $\nu \ll \mu$), i.e.*

$$\nu(A) = 0 \quad \textit{for all} \quad A \in \mathcal{F} \textit{ such that } \mu(A) = 0,$$

then there exists a $\mathcal{F}$-measurable random variable Y on Ω, such that:

$$\nu(A) = \int_A Y \, \mathrm{d}\mu \qquad \textit{for all} \quad A \in \mathcal{F}.$$

Any Y as above is nonnegative μ-a.s. and if Y_1, Y_2 are such, then: $Y_1 = Y_2$ μ-a.s. We call Y the Radon–Nikodym derivative *of ν with respect to μ.*

Write now $X = X_+ - X_-$ as the difference of the positive and negative parts of an integrable random variable X on a probability space $(\Omega, \mathcal{F}, \mathbb{P})$. Given a sub-$\sigma$-algebra $\mathcal{G} \subset \mathcal{F}$, define $\mu = \mathbb{P}_{|\mathcal{G}}$ and:

$$\nu(A) = \int_A X_+ \, \mathrm{d}\mathbb{P} \quad \text{for all } A \in \mathcal{G}.$$

Clearly, μ and ν are two σ-finite measures on the measurable space $(\Omega, \mathcal{G})$ and $\nu \ll \mu$, so ν has its Radon–Nikodym derivative Y_+ with respect to μ. The same construction for X_- results in the $\mathcal{G}$-measurable random variable Y_-. We now set $Y = Y_+ - Y_-$. We leave it to the reader to check that the conditions of Definition A.14 are satisfied to the effect that $Y = \mathbb{E}(X \mid \mathcal{G})$.

Exercise A.16

(i) If Y_1, Y_2 are two random variables satisfying the conditions of Definition A.14, then $Y_1 = Y_2$ a.s.

Let X, X_1, X_2 be integrable random variables on a probability space $(\Omega, \mathcal{F}, \mathbb{P})$ and let $\mathcal{G}, \mathcal{G}_1, \mathcal{G}_2$ be sub-σ-algebras of $\mathcal{F}$. Prove that:

(ii) (*Linearity*). $\mathbb{E}(a_1X_1 + a_2X_2 \mid \mathcal{G}) = a_1\mathbb{E}(X_1 \mid \mathcal{G}) + a_2\mathbb{E}(X_2 \mid \mathcal{G})$ a.s.
(iii) (*Monotonicity*). If $X_1 \leq X_2$ a.s., then $\mathbb{E}(X_1 \mid \mathcal{G}) \leq \mathbb{E}(X_2 \mid \mathcal{G})$ a.s.
(iv) (*The tower property*). If $\mathcal{G}_1 \subset \mathcal{G}_2$, then $\mathbb{E}(X \mid \mathcal{G}_1) = \mathbb{E}(\mathbb{E}(X \mid \mathcal{G}_2) \mid \mathcal{G}_1)$ a.s.
(v) For every bounded, $\mathcal{G}$-measurable random variable Z on Ω there holds: $\mathbb{E}(ZX \mid \mathcal{G}) = Z\mathbb{E}(X \mid \mathcal{G})$ a.s. Consequently: $\mathbb{E}[ZX] = \mathbb{E}[Z\mathbb{E}(X \mid \mathcal{G})]$.
(vi) (*Jensen's inequality*). If $\phi : \mathbb{R} \to \mathbb{R}$ is a convex function and $\phi \circ X$ is integrable, then:

$$\phi \circ \mathbb{E}(X \mid \mathcal{G}) \leq \mathbb{E}(\phi \circ X \mid \mathcal{G}) \text{ a.s.}$$

The following simple observation is a direct consequence of Theorem A.11:

Lemma A.17 *Let X be a integrable random variable on $(\Omega, \mathcal{F}, \mathbb{P})$ that is the product of two probability spaces $(\Omega_1, \mathcal{F}_1, \mathbb{P}_1)$ and $(\Omega_2, \mathcal{F}_2, \mathbb{P}_2)$. Define $\mathcal{G} = \{A \times \Omega_2\}_{A \in \mathcal{F}_1}$ which is a sub-σ-algebra of $\mathcal{F}$ (viewed as a copy of $\mathcal{F}_1$ in $\mathcal{F}$).*

(i) For $\mathbb{P}_1$-a.e. $\omega_1 \in \Omega_1$, we have:

$$\mathbb{E}(X \mid \mathcal{G})(\omega_1) = \int_{\Omega_2} X(\omega_1, \omega_2)\, d\mathbb{P}_2(\omega_2) \quad a.s.$$

(ii) When $X = \mathbb{1}_{X_2 > X_1}$ is given by a random variable X_1 on $(\Omega_1, \mathcal{F}_1, \mathbb{P}_1)$ and a random variable X_2 on $(\Omega_2, \mathcal{F}_2, \mathbb{P}_2)$, then for $\mathbb{P}_1$-a.e. $\omega_1 \in \Omega_1$ there holds:

$$\mathbb{E}(X \mid \mathcal{G})(\omega_1) = \int_{\Omega_2} \mathbb{1}_{X_2 > X_1(\omega_1)}\, d\mathbb{P}_2 = \mathbb{P}_2\big(X_2 > X_1(\omega_1)\big) \quad a.s.$$

We may think of the conditional expectation as a projection. Indeed, for a $\mathcal{G}$-measurable X we have $\mathbb{E}(X \mid \mathcal{G}) = X$, and one can prove that $\mathbb{E}(\cdot \mid \mathcal{G})$ is actually an orthogonal projection from $L^2(\Omega, \mathcal{F}, \mathbb{P})$ onto $L^2(\Omega, \mathcal{G}, \mathbb{P}_{|\mathcal{G}})$.

The following conditional convergence theorems are the counterparts of the Fatou, Lebesgue and the monotone convergence Theorems A.6, A.7:

Exercise A.18 Let $\{X_i\}_{i=1}^{\infty}$ be a sequence of integrable random variables on a probability space $(\Omega, \mathcal{F}, \mathbb{P})$. Let $\mathcal{G}$ be a sub-σ-algebra of $\mathcal{F}$. Prove that:

(i) If each X_i is a.s. nonnegative, then: $\mathbb{E}\big(\liminf_{i\to\infty} X_i \mid \mathcal{G}\big) \leq \liminf_{i\to\infty} \mathbb{E}\big(X_i \mid \mathcal{G}\big)$.

Assume that $\{X_i\}_{i=1}^{\infty}$ converge pointwise a.s. to a random variable X and that one of the following properties holds:

(ii) For each $i \in \mathbb{N}$ we have: $X_i \geq 0$ and $X_i \leq X_{i+1}$ a.s.
(iii) There exists $Z \in L^1(\Omega, \mathcal{F}, \mathbb{P})$ such that $|X_i| \leq Z$ a.s. for every $i \in \mathbb{N}$.

Then one can pass to the limit under conditional expectation:

$$\lim_{i\to\infty} \mathbb{E}\big(X_i \mid \mathcal{G}\big) = \mathbb{E}\big(X \mid \mathcal{G}\big) \quad \text{a.s.}$$

A.5 Independence

Definition A.19 Let $(\Omega_1, \mathcal{F}_1)$ and $(\Omega_2, \mathcal{F}_2)$ be two measurable spaces. The random variables $X_1 : \Omega \to \Omega_1$ and $X_2 : \Omega \to \Omega_2$ on the probability space $(\Omega, \mathcal{F}, \mathbb{P})$ are *independent*, if:

$$\mathbb{P}(\{X_1 \in A_1\} \cap \{X_2 \in A_2\}) = \mathbb{P}(X_1 \in A_1) \cdot \mathbb{P}(X_2 \in A_2) \quad \text{for all } A_1 \in \mathcal{F}_1, \;\; A_2 \in \mathcal{F}_2.$$

The following observation, which we leave as an exercise, expresses independence in terms of equality of the induced product measures.

Exercise A.20 Let X_1, X_2 be two random variables on the probability space $(\Omega, \mathcal{F}, \mathbb{P})$, with values in measurable spaces $(\Omega_1, \mathcal{F}_1)$ and $(\Omega_2, \mathcal{F}_2)$, respectively. For $i = 1, 2$ and every $A_i \in \mathcal{F}_i$, define:

$$\mathbb{P}_i(A_i) = \mathbb{P}(X_i \in A_i).$$

(i) Both $(\Omega_i, \mathcal{F}_i, \mathbb{P}_i)$, where $i = 1, 2$, are probability spaces.
(ii) Let $(\Omega_1 \times \Omega_2, \bar{\mathcal{F}})$ be the product of $(\Omega_1, \mathcal{F}_1)$ and $(\Omega_2, \mathcal{F}_2)$. Then $X_1 \times X_2 : (\Omega, \mathcal{F}) \to (\Omega_1 \times \Omega_2, \bar{\mathcal{F}})$ is a random variable. Further, $(\Omega_1 \times \Omega_2, \bar{\mathcal{F}}, \bar{\mathbb{P}})$ is a probability space, where for all $A \in \bar{\mathcal{F}}$ we define: $\bar{\mathbb{P}}(A) = \mathbb{P}((X_1 \times X_2) \in A)$.
(iii) Let $\bar{\bar{\mathbb{P}}}$ be the product measure of $\mathbb{P}_1$ and $\mathbb{P}_2$ on $(\Omega_1 \times \Omega_2, \bar{\mathcal{F}})$. The random variables X_1 and X_2 are independent if and only if $\bar{\bar{\mathbb{P}}} = \bar{\mathbb{P}}$.

The next observation is a direct consequence of the Fubini–Tonelli theorem:

Lemma A.21 *Let X_1, X_2 be two independent random variables with values in Ω_1, Ω_2, as in Definition A.19. Define the product probability space $(\Omega_1 \times \Omega_2, \bar{\mathcal{F}}, \bar{\mathbb{P}})$ as in Exercise A.20 and assume that $Z : \Omega_1 \times \Omega_2 \to \bar{\mathbb{R}}$ is a $\bar{\mathcal{F}}$-measurable random variable, that is either $\bar{\mathbb{P}}$-integrable or nonnegative.*

(i) The integral in the right hand side below is well defined and we have:

$$\int_\Omega Z(X_1(\omega), X_2(\omega))\, d\mathbb{P}(\omega) = \int_\Omega \int_\Omega Z(X_1(\omega_1), X_2(\omega_2))\, d\mathbb{P}(\omega_2)\, d\mathbb{P}(\omega_1).$$

(ii) Consider the collection of sets: $\tilde{\mathcal{F}}_1 = \big\{\{X_1 \in A_1\};\ A_1 \in \mathcal{F}_1\big\}$, *which is a sub-*$\sigma$*-algebra of* $\mathcal{F}$ *in* Ω. *Then, for* $\mathbb{P}$*-a.e.* $\omega_1 \in \Omega$:

$$\mathbb{E}\Big(Z \circ (X_1 \times X_2) \mid \tilde{\mathcal{F}}_1\Big)(\omega_1) = \int_\Omega Z(X_1(\omega_1), X_2(\omega_2))\, d\mathbb{P}(\omega_2).$$

For more than two random variables, independence is understood in a similar manner as in Definition A.19, namely:

Definition A.22 The given $n \geq 2$ random variables $\{X_i : \Omega \to \Omega_i\}_{i=1}^n$, defined on the probability space $(\Omega, \mathcal{F}, \mathbb{P})$ and valued in the respective measurable spaces $(\Omega, \mathcal{F}_i)$ are called *independent*, provided that:

$$\mathbb{P}\Big(\bigcap_{i=1}^n \{X_i \in A_i\}\Big) = \prod_{i=1}^n \mathbb{P}(X_i \in A_i) \quad \text{for all } A_i \in \mathcal{F}_i,\ \ i = 1 \dots n.$$

Clearly, if $\{X_i\}_{i=1}^n$ are independent, then they are also *pairwise independent*, that is every pair (X_i, X_j) is independent for $i \neq j$. The converse is not true:

Exercise A.23 Give an example of three pairwise independent random variables on $B_1(0) \subset \mathbb{R}^2$, for which condition in Definition A.22 does not hold.

Similarly as in Exercise A.20, it is easy to observe that $\{X_i\}_{i=1}^n$ are independent if and only if the probability measure defined on $\prod_{i=1}^n(\Omega_i, \mathcal{F}_i)$ as the push-forward of $\mathbb{P}$ through the multi-dimensional random variable $(X_1, \dots, X_n)$, coincides with the product measure of the individual push-forwards $\mathbb{P}(X_i \in A_i)$ on each $(\Omega_i, \mathcal{F}_i)$. We further have the following useful result:

Exercise A.24 Assume that $\{X_i : \Omega \to \Omega_i\}_{i=1}^n$ are independent random variables as in Definition A.22. Given are measurable functions:

$$f_s : \prod_{i=k_s+1}^{k_{s+1}} \Omega_i \to \Omega_s^0 \qquad \text{for } s = 1 \dots m,$$

valued in some measurable spaces $(\Omega_s^0, \mathcal{F}_s^0)_{s=1}^m$, and defined on the indicated products of the consecutive spaces $(\Omega_i, \mathcal{F}_i)_{i=1}^n$, where $0 = k_1 < k_2 \dots < k_{m+1} \leq n$. Prove that the random variables $\{f_s \circ (X_{k_s+1}, \dots, X_{k_{s+1}})\}_{s=1}^m$ are independent.

A.6 Martingales and Stopping Times

One of the central concepts in probability is the concept of a martingale.

Definition A.25 Let $(\Omega, \mathcal{F}, \mathbb{P})$ be a probability space.

(i) A *filtration* $\{\mathcal{F}_i\}_{i=0}^{\infty}$ of the σ-algebra $\mathcal{F}$ is an increasing sequence of sub-σ-algebras of $\mathcal{F}$, namely:

$$\mathcal{F}_0 \subset \mathcal{F}_1 \subset \ldots \mathcal{F}_i \subset \mathcal{F}_{i+1} \subset \ldots \subset \mathcal{F}.$$

(ii) A *martingale* relative to a filtration $\{\mathcal{F}_i\}_{i=0}^{\infty}$ (we also say *adapted to a filtration*) is a sequence of integrable random variables $\{X_i\}_{i=0}^{\infty}$, such that each $X_i : \Omega \to \bar{\mathbb{R}}$ is $\mathcal{F}_i$-measurable and there holds:

$$X_i = \mathbb{E}(X_{i+1} \mid \mathcal{F}_i) \text{ a.s.} \qquad \text{for all } i \geq 0. \tag{A.10}$$

(iii) A *submartingale/supermartingale* $\{X_i\}_{i=0}^{\infty}$ relative to a filtration $\{\mathcal{F}_i\}_{i=0}^{\infty}$ is defined as above, with the equality (A.10) replaced by an inequality: $X_i \leq \mathbb{E}(X_{i+1} \mid \mathcal{F}_i)$ for submartingale, and $X_i \geq \mathbb{E}(X_{i+1} \mid \mathcal{F}_i)$ for supermartingale, in each case valid a.s. with respect to $\mathbb{P}$.

It is easy to observe that expectation is constant along a martingale, namely:

$$\mathbb{E}[X_i] = \mathbb{E}[X_0] \qquad \text{for all } i \geq 0, \tag{A.11}$$

whereas it is increasing/decreasing along a sub/supermartingale. Also, Exercise A.16 (iv) yields then in view of (A.10):

$$X_i = \mathbb{E}(X_j \mid \mathcal{F}_i) \text{ a.s.} \qquad \text{for all } 0 \leq i \leq j, \tag{A.12}$$

with the equality replaced by "$\leq$" for a submartingale and by "$\geq$" for a supermartingale. We further have:

Exercise A.26

(i) If $\{\mathcal{F}_i\}_{i=0}^{\infty}$ is a filtration of $\mathcal{F}$ and $X \in L^1(\Omega, \mathcal{F}, \mathbb{P})$, then $X_i \doteq \mathbb{E}(X \mid \mathcal{F}_i)$ defines a martingale $\{X_i\}_{i=0}^{\infty}$ adapted to the given filtration.
(ii) If $\{X_i\}_{i=0}^{\infty}$ is a martingale adapted to a filtration $\{\mathcal{F}_i\}_{i=0}^{\infty}$ and $\phi : \mathbb{R} \to \mathbb{R}$ is convex/concave, then $\{\phi \circ X_i\}_{i=0}^{\infty}$ is a sub/supermartingale, provided that $\phi \circ X_i$ is integrable for all i.

Many important concepts involving sequences of random variables are based on how the future time state depends on the past. This necessitates the notion of "time" itself as a random variable, and leads to the notion of a stopping time:

Definition A.27 A random variable $\tau : \Omega \to \{0, 1, \ldots, +\infty\}$ is called a *stopping time* relative to a filtration $\{\mathcal{F}_i\}_{i=0}^{\infty}$ of $\mathcal{F}$ if the two conditions hold:

(i) $\{\tau \leq i\} \doteq \{\omega \in \Omega;\ \tau(\omega) \leq i\} \in \mathcal{F}_i$ for all $i \geq 0$,
(ii) $\mathbb{P}(\tau = +\infty) = 0$.

Roughly speaking, the defining property of a stopping time is that the decision to stop (and to possibly read the value of a specific random variable X_i) at a "time" i, is based only on information available up to this time. Clearly, a constant random variable taking values in $\mathbb{N} \cup \{0\}$ is a stopping time:

Example A.28

(i) Any two stopping times τ_1, τ_2 generate a stopping time:

$$\tau_1 \wedge \tau_2 \doteq \min\{\tau_1, \tau_2\}.$$

(ii) Let $\{\mathcal{F}_i\}_{i=0}^{\infty}$ be a filtration of $\mathcal{F}$ in the probability space $(\Omega, \mathcal{F}, \mathbb{P})$ and let $\{X_i\}_{i=0}^{\infty}$ be a sequence of random variables such that each $X_i : \Omega \to \bar{\mathbb{R}}$ is $\mathcal{F}_i$-measurable. Given $r \in \mathbb{R}$, define:

$$\tau_r(\omega) = \min\Big\{n \geq 0;\ \sum_{i=0}^{n} X_i(\omega) \geq r\Big\}.$$

If $\mathbb{P}(\tau_r = +\infty) = 0$, then the random variable τ_r is a stopping time.
(iii) If τ, satisfying $\tau \geq 1$ a.s. is a stopping time, then $\tau - 1$ might fail to be a stopping time, whereas $\tau + 1$ is always a stopping time.

We further have the following important properties:

Exercise A.29 Let τ be a stopping time relative to a filtration $\{\mathcal{F}_i\}_{i=0}^{\infty}$ of $\mathcal{F}$.

(i) Define:

$$\mathcal{F}_\tau = \Big\{A \in \mathcal{F};\ A \cap \{\tau \leq i\} \in \mathcal{F}_i \ \text{ for all } i\Big\}.$$

Then $\mathcal{F}_\tau$ is a sub-σ-algebra of $\mathcal{F}$ and τ is $\mathcal{F}_\tau$-measurable. Moreover, if τ_1, τ_2 are two stopping times satisfying $\tau_1 \leq \tau_2$, then $\mathcal{F}_{\tau_1} \subset \mathcal{F}_{\tau_2}$. Intuitively, $\mathcal{F}_\tau$ represents the information available at the random time τ.
(ii) Let $\{X_i\}_{i=0}^{\infty}$ be a sequence of random variables where each X_i is $\mathcal{F}_i$-measurable. Then the random variable X_τ given by:

$$X_\tau(\omega) = X_{\tau(\omega)}(\omega),$$

is defined $\mathbb{P}$-a.s. in Ω and it is $\mathcal{F}_\tau$-measurable. If each X_i is integrable and τ is bounded, then X_τ is integrable. Intuitively, X_τ represents the state of the process $\{X_i\}_{i=0}^{\infty}$ at a random time τ.

Lemma A.30 *Let $\{X_i\}_{i=0}^{\infty}$ be a martingale/sub/supermartingale and τ be a stopping time relative to a filtration $\{\mathcal{F}_i\}_{i=0}^{\infty}$. Then $\{X_{\tau\wedge i}\}_{i=0}^{\infty}$ is also a martingale/sub/supermartingale, respectively, relative to $\{\mathcal{F}_i\}_{i=0}^{\infty}$.*

Proof It is enough to prove the result in the submartingale case. According to Exercise A.29, each $X_{\tau\wedge i}$ is $\mathcal{F}_{\tau\wedge i}$- and thus $\mathcal{F}_i$-measurable, and $\mathbb{P}$-integrable.

To prove the desired inequality between the random variables $\mathbb{E}(X_{\tau\wedge(i+1)} \mid \mathcal{F}_i)$ and $X_{\tau\wedge i}$, take any $A \in \mathcal{F}_i$ and observe that:

$$\begin{aligned}\int_A \mathbb{E}(X_{\tau\wedge(i+1)} \mid \mathcal{F}_i)\, \mathrm{d}\mathbb{P} &= \int_A X_{\tau\wedge(i+1)}\, \mathrm{d}\mathbb{P} \\ &= \int_{A\cap\{\tau\le i\}} X_{\tau\wedge(i+1)}\, \mathrm{d}\mathbb{P} + \int_{A\cap\{\tau> i\}} X_{\tau\wedge(i+1)}\, \mathrm{d}\mathbb{P} \\ &= \int_{A\cap\{\tau\le i\}} X_{\tau\wedge i}\, \mathrm{d}\mathbb{P} + \int_{A\cap\{\tau> i\}} X_{i+1}\, \mathrm{d}\mathbb{P}.\end{aligned}$$

Since $A \cap \{\tau > i\} \in \mathcal{F}_i$, the last integral in the right hand side above is bounded from below by: $\int_{A\cap\{\tau>i\}} X_i\, \mathrm{d}\mathbb{P} = \int_{A\cap\{\tau>i\}} X_{\tau\wedge i}\, \mathrm{d}\mathbb{P}$. Consequently, we get:

$$\int_A \mathbb{E}(X_{\tau\wedge(i+1)} \mid \mathcal{F}_i)\, \mathrm{d}\mathbb{P} \ge \int_A X_{\tau\wedge i}\, \mathrm{d}\mathbb{P},$$

which achieves the result. □

The next result is the celebrated *Doob's optional stopping theorem*. It says that the expectation of a martingale at a stopping time is equal to the expectation at the initial time (under suitable assumptions), very much as in the case of a constant stopping time $\tau = i$ in (A.11). In other words, the possibility of stopping at an opportune moment gives no advantage, as long as one cannot foresee the future.

Theorem A.31 *Let $\{X_i\}_{i=0}^{\infty}$ be a martingale and τ a stopping time, relative to a filtration $\{\mathcal{F}_i\}_{i=0}^{\infty}$ on the probability space $(\Omega, \mathcal{F}, \mathbb{P})$. Assume that one of the following properties holds:*

(i) The stopping time τ is bounded.
(ii) There exists an integrable random variable $Z \in L^1(\Omega, \mathcal{F}, \mathbb{P})$ such that $|X_i| \le Z$ a.s. for every $i \ge 0$.

Then $X_\tau \in L^1(\Omega, \mathcal{F}, \mathbb{P})$ and:

$$\mathbb{E}[X_\tau] = \mathbb{E}[X_0]. \tag{A.13}$$

Proof By Lemma A.30, the sequence $\{X_{\tau\wedge i}\}_{i=0}^{\infty}$ is a martingale relative to $\{\mathcal{F}_i\}_{i=0}^{\infty}$. Thus: $\mathbb{E}[X_{\tau\wedge i}] = \mathbb{E}[X_{\tau\wedge 0}] = \mathbb{E}[X_0]$. On the other hand, it is straightforward that:

$$\lim_{i\to\infty} \mathbb{E}[X_{\tau\wedge i}] = \mathbb{E}[X_\tau], \tag{A.14}$$

since in case (i) we have $X_{\tau\wedge i} = X_\tau$ for sufficiently large i, while in case (ii) one uses the Lebesgue dominated convergence theorem to pass to the limit. The proof of (A.13) is done. □

We now observe that the same result as in Theorem A.31 is valid when conditions (i) and (ii), that were only used to justify convergence in (A.14), are replaced by the uniform integrability of $\{X_{\tau\wedge i}\}_{i=0}^\infty$. Recall that a sequence of integrable random variables $\{X_i\}_{i=0}^\infty$ is *uniformly integrable*, if it is bounded in $L^1(\Omega, \mathcal{F}, \mathbb{P})$ and *equiintegrable*, i.e.

$$\forall \epsilon > 0 \quad \exists \delta > 0 \quad \forall A \in \mathcal{F} \quad \forall i \geq 0 \qquad \mathbb{P}(A) < \delta \Rightarrow \int_A |X_i| \, \mathrm{d}\mathbb{P} < \epsilon.$$

An equivalent definition of uniform integrability, that is more adequate in probability theory, is given by:

Exercise A.32 A sequence of random variables $\{X_i\}_{i=0}^\infty$ on a probability space $(\Omega, \mathcal{F}, \mathbb{P})$ is uniformly integrable if and only if:

$$\forall \epsilon > 0 \quad \exists M \quad \forall i \geq 0 \qquad \int_{\{|X_i|>M\}} |X_i| \, \mathrm{d}\mathbb{P} < \epsilon.$$

Note that assumptions (i) and (ii) in Theorem A.31 imply the uniform integrability, which is a more general condition. We also have:

Exercise A.33 Let $\{X_i\}_{i=0}^\infty$ be a sequence of integrable random variables, such that: $\sup_{i\geq 0} \|X_{i+1} - X_i\|_{L^\infty(\Omega)} < +\infty$. Let τ be an integrable stopping time. Then $\{X_{\tau\wedge i}\}_{i=0}^\infty$ is uniformly integrable.

The following is a more general version of Theorem A.31, extended to sub/supermartingales:

Theorem A.34 *Let $\{X_i\}_{i=0}^\infty$ be a sequence of random variables and τ a stopping time, relative to a filtration $\{\mathcal{F}_i\}_{i=0}^\infty$ on the probability space $(\Omega, \mathcal{F}, \mathbb{P})$. Assume that $\{X_{\tau\wedge i}\}_{i=0}^\infty$ is uniformly integrable. Then $X_\tau \in L^1(\Omega, \mathcal{F}, \mathbb{P})$ and:*

(i) If $\{X_i\}_{i=0}^\infty$ is a submartingale, then $\mathbb{E}[X_\tau] \geq \mathbb{E}[X_0]$.
(ii) If $\{X_i\}_{i=0}^\infty$ is a supermartingale, then $\mathbb{E}[X_\tau] \leq \mathbb{E}[X_0]$.
(iii) If $\{X_i\}_{i=0}^\infty$ is a martingale, then $\mathbb{E}[X_\tau] = \mathbb{E}[X_0]$.

Proof Since the sequence $\{X_{\tau\wedge i}\}_{i=0}^\infty$ converges pointwise $\mathbb{P}$-a.s. to X_τ and it is bounded in $L^1(\Omega, \mathcal{F}, \mathbb{P})$, Fatou's lemma implies the integrability of X_τ.

By Lemma A.30 we obtain that $\mathbb{E}[X_{\tau\wedge i}] \geq \mathbb{E}[X_0]$ for all $i \geq 0$ for the case of submartingale, with the inequality "$\geq$" replaced by "$\leq$" and "$=$" for supermartingale and martingale cases, respectively. To conclude the proof, it suffices to check (A.14). Fix $\epsilon > 0$ and consider a decreasing sequence of subsets of Ω:

$$A_i = \bigcup_{j=i}^\infty \{|X_{\tau\wedge j} - X_\tau| > \epsilon\}.$$

Since $\mathbb{P}(\bigcap_{i=0}^{\infty} A_i) = 0$ in view of the pointwise convergence, we obtain:

$$\lim_{i\to\infty} \mathbb{P}(A_i) = 0. \tag{A.15}$$

Consequently, for sufficiently large i:

$$\int_\Omega |X_{\tau\wedge i} - X_\tau| \,\mathrm{d}\mathbb{P} \leq \int_{\Omega\setminus A_i} |X_{\tau\wedge i} - X_\tau| \,\mathrm{d}\mathbb{P} + \int_{A_i} |X_{\tau\wedge i}| + |X_\tau| \,\mathrm{d}\mathbb{P} < 3\epsilon,$$

where we used the equiintegrability assumptions and (A.15) to bound the second integral term above. Thus (A.14) follows and the proof is complete. □

By integrating on $A \in \mathcal{F}_\tau$ rather than on Ω, we similarly obtain:

Exercise A.35 Let $\{X_i\}_{i=0}^{\infty}$ be a martingale and let τ_1, τ_2 be two stopping times, relative to a filtration $\{\mathcal{F}_i\}_{i=0}^{\infty}$. Assume that $\tau_1 \leq \tau_2$ and that the sequence $\{X_{\tau_2\wedge i}\}_{i=0}^{\infty}$ is uniformly integrable. Then $\{X_{\tau_1\wedge i}\}_{i=0}^{\infty}$ is uniformly integrable as well and:

$$\mathbb{E}(X_{\tau_2} \mid \mathcal{F}_{\tau_1}) = X_{\tau_1} \quad \text{a.s.}$$

We remark that by the Dunford–Pettis theorem, the uniform integrability of a sequence of random variables $\{X_i\}_{i=0}^{\infty}$ signifies, precisely, its relative weak sequential compactness in L^1. Since the weak limit and the pointwise limit must coincide, it follows that, given a stopping time τ relative to the same filtration as $\{X_i\}_{i=0}^{\infty}$, the new sequence $\{X_{\tau\wedge i}\}_{i=0}^{\infty}$ weakly converges to X_τ provided that it is uniformly integrable (in fact, it also converges strongly). This suffices to pass to the limit (A.14) in the proof of Doob's optional stopping and deduce the result in Theorem A.34.

A.7 Convergence of Martingales

Let $\{X_i\}_{i=0}^{\infty}$ be a sequence of random variables on a probability space $(\Omega, \mathcal{F}, \mathbb{P})$. For any two numbers $a < b$, we define the nonnegative random variable $N_{a,b}$ which counts the number of *upcrossings* of the interval $[a, b]$, by $\{X_i\}_{i=0}^{\infty}$, namely:

$$N_{a,b}(\omega) \doteq \sup \Big\{ n;\ \text{exist integers } 0 \leq s_1 < t_1 < \ldots s_n < t_n \text{ such that}$$
$$X_{s_i}(\omega) \leq a \ \text{ and } \ X_{t_i}(\omega) \geq b \text{ for all } i = 1\ldots n \Big\} \quad \text{for all } \omega \in \Omega.$$

It is easy to observe that $N_{a,b}$ is indeed $\mathcal{F}$-measurable.

The following *Dubins upcrossing inequality* controls the number of upcrossings of a nonnegative supermartingale:

Lemma A.36 *Let* $\{X_i\}_{i=0}^\infty$ *be a nonnegative supermartingale relative to a filtration* $\{\mathcal{F}_i\}_{i=0}^\infty$ *on the probability space* $(\Omega, \mathcal{F}, \mathbb{P})$. *Then:*

$$\mathbb{P}\big(N_{a,b} \geq n\big) \leq \Big(\frac{a}{b}\Big)^n \qquad \textit{for all } 0 \leq a < b \textit{ and all } n \geq 0.$$

Proof Define the auxiliary random variables:

$$\begin{aligned} \tau_0 \doteq 0, \quad \sigma_0 &\doteq \inf\{i \geq 0;\ X_i \leq a\}, \quad \tau_1 \doteq \inf\{i \geq \sigma_0;\ X_i \geq b\}, \\ \sigma_n &\doteq \inf\{i \geq \tau_n;\ X_i \leq a\}, \quad \tau_{n+1} \doteq \inf\{i \geq \sigma_n;\ X_i \geq b\} \quad \text{for all } n \in \mathbb{N}, \end{aligned}$$

with convention that $\inf = +\infty$ if the infimized set is empty. There holds:

$$\{N_{a,b} \geq n\} = \{\tau_n < +\infty\} \qquad \text{for all } n \in \mathbb{N}. \tag{A.16}$$

Note that each σ_n and τ_n satisfies (i) in Definition A.27 of stopping time. In particular, although the event $\{\tau_n = +\infty\}$ may have positive probability, Lemma A.30 yields that the sequence $\{X_{\tau_n \wedge i}\}_{i=0}^\infty$ is a (nonnegative) supermartingale. Hence:

$$\mathbb{E}(X_{\tau_n \wedge i} \mid \mathcal{F}_j) \leq X_{\tau_n \wedge j} \quad \text{a.s.} \qquad \text{for all } 0 \leq j \leq i. \tag{A.17}$$

We now estimate:

$$\begin{aligned} b \cdot \mathbb{P}(\tau_n \leq i) &\leq \int_{\{\tau_n \leq i\}} X_{\tau_n \wedge i}\ \mathrm{d}\mathbb{P} = \sum_{j=0}^{i} \int_{\{\tau_n \leq i\} \cap \{\sigma_{n-1} = j\}} X_{\tau_n \wedge i}\ \mathrm{d}\mathbb{P} \\ &\leq \sum_{j=0}^{i} \int_{\{\sigma_{n-1} = j\}} X_{\tau_n \wedge i}\ \mathrm{d}\mathbb{P}. \end{aligned}$$

To continue, observe that since $\{\sigma_{n-1} = j\} \in \mathcal{F}_j$, we may use (A.17) in:

$$\begin{aligned} \sum_{j=0}^{i} \int_{\{\sigma_{n-1} = j\}} X_{\tau_n \wedge i}\ \mathrm{d}\mathbb{P} &= \sum_{j=0}^{i} \int_{\{\sigma_{n-1} = j\}} \mathbb{E}(X_{\tau_n \wedge i} \mid \mathcal{F}_j)\ \mathrm{d}\mathbb{P} \\ &\leq \sum_{j=0}^{i} \int_{\{\sigma_{n-1} = j\}} X_{\tau_n \wedge j}\ \mathrm{d}\mathbb{P} \\ &= \int_{\{\sigma_{n-1} \leq i\}} X_{\sigma_{n-1}}\ \mathrm{d}\mathbb{P} \leq a \cdot \mathbb{P}(\sigma_{n-1} \leq i) \leq a \cdot \mathbb{P}(\tau_{n-1} \leq i). \end{aligned}$$

Consequently:

$$\mathbb{P}(\tau_n \le i) \le \frac{a}{b} \cdot \mathbb{P}(\tau_{n-1} \le i),$$

which after passing to the limit with $i \to \infty$ in both the increasing sequences of probabilities $\{\mathbb{P}(\tau_n \le i)\}_{i=0}^{\infty}$ and $\{\mathbb{P}(\tau_{n-1} \le i)\}_{i=0}^{\infty}$, we obtain:

$$\mathbb{P}(\tau_n < +\infty) \le \frac{a}{b} \cdot \mathbb{P}(\tau_{n-1} < +\infty) \qquad \text{for all } n \in \mathbb{N}.$$

It follows that by (A.16) that:

$$\mathbb{P}\big(N_{a,b} \ge n\big) \le \Big(\frac{a}{b}\Big)^n \cdot \mathbb{P}(\tau_0 < +\infty) = \Big(\frac{a}{b}\Big)^n,$$

as claimed. □

Corollary A.37 *Let $\{X_i\}_{i=0}^{\infty}$ be a supermartingale, bounded from below a.s., by some constant $c \in \mathbb{R}$. Then, there exists a random variable X such that:*

$$\lim_{i\to\infty} X_i = X \ \ a.s. \tag{A.18}$$

Proof Without loss of generality, we may assume that $c = 0$ and that all random variables X_i are all nonnegative. By Lemma A.36 it follows that:

$$\mathbb{P}(N_{a,b} = +\infty) = 0 \qquad \text{for all } 0 \le a < b. \tag{A.19}$$

On the other hand, there holds:

$$\big\{\liminf_{i\to\infty} X_i < \limsup_{i\to\infty} X_i\big\} \subset \bigcup_{0\le a<b,\ a,b\in\mathbb{Q}} \{N_{a,b} = +\infty\}$$

where, by (A.19), the set in the right hand side above has probability 0. Consequently: $\liminf_{i\to\infty} X_i = \limsup_{i\to\infty} X_i$ a.s., which proves the claim. □

We finally directly deduce:

Corollary A.38 *Let $\{X_i\}_{i=0}^{\infty}$ be a bounded martingale relative to a filtration $\{\mathcal{F}_i\}_{i=0}^{\infty}$ on the probability space $(\Omega, \mathcal{F}, \mathbb{P})$. Then, there exists an integrable random variable X satisfying (A.18). We also have: $\lim_{i\to\infty} \int_\Omega |X_i - X|\, d\mathbb{P} = 0$ and the martingale is* closed *by X, namely:*

$$\mathbb{E}(X \mid \mathcal{F}_i) = X_i \ \ a.s. \qquad \textit{for all } \ i \ge 0.$$

Proof The first claim follows from Corollary A.37, the second by the dominated convergence theorem, and the third by Exercise A.18 (iii):

$$0 = \lim_{j\to\infty} \mathbb{E}(X - X_j \mid \mathcal{F}_i) = \mathbb{E}(X - X_i \mid \mathcal{F}_i) \quad \text{a.s.}$$

where we used the tower property: $\mathbb{E}(X_j \mid \mathcal{F}_i) = X_i$ a.s. for all $j \geq i$. □

Appendix B
Background in Brownian Motion

In this chapter we recall definitions and basic results on Brownian motion: the Lévy construction, uniqueness of the Wiener measure, the Markov properties and Brownian motion harmonic extensions.

In our presentation we strive for completeness and clarity from the point of view of a reader familiar with Analysis and to a lesser extent with Probability.

In this chapter we recall definitions and preliminary facts on Brownian motion: the Lèvy construction, uniqueness of the Wiener measure, the Markov properties and Brownian motion harmonic extensions. These notions can be found in many modern graduate textbooks, for example in: Mörters and Peres (2010); Durrett (2010); Bishop and Peres (2017); Dudley (2004); Karatzas and Shreve (1991); in our presentation we strive for completeness and clarity from the point of view of a reader familiar with Analysis and to a lesser extent with Probability. The material below is used only in the "star" sections: 2.7*, 3.6*, 3.7* and 4.5*, and thus may be skipped at first reading.

B.1 Definition and Construction of Brownian Motion

Definition B.1 A family $\{\mathcal{B}_t^N\}_{t\geq 0}$ of vector-valued random variables $\mathcal{B}_t^N : \Omega \to \mathbb{R}^N$ on a probability space $(\Omega, \mathcal{F}, \mathbb{P})$ is a *N-dimensional Brownian motion*, provided that it has the following properties:

(i) For all $0 \leq s < t$ the random variable $\frac{1}{\sqrt{t-s}}(\mathcal{B}_t^N - \mathcal{B}_s^N)$ is $N(0, 1)$-normally distributed, i.e.

$$\mathbb{P}\big(\mathcal{B}_t^N - \mathcal{B}_s^N \in A\big) = \int_A \frac{1}{2\pi(t-s)^{N/2}} e^{-\frac{|x|^2}{2(t-s)}}\,\mathrm{d}x \quad \text{for all Borel } A \subset \mathbb{R}^N.$$

M. Lewicka, *A Course on Tug-of-War Games with Random Noise*, Universitext, https://doi.org/10.1007/978-3-030-46209-3

(ii) The process $\{\mathcal{B}_t^N\}_{t\geq 0}$ has independent increments, i.e. for any $n \geq 2$ disjoint intervals $\{(s_i, t_i)\}_{i=1}^n$, then n random variables $\{\mathcal{B}_{t_i}^N - \mathcal{B}_{s_i}^N\}_{i=1}^N$ are independent.
(iii) For every $\omega \in \Omega$, the function $[0, \infty) \ni t \mapsto \mathcal{B}_t^N(\omega) \in \mathbb{R}^N$ is continuous.

When $\mathcal{B}_0^N = 0$, we call $\{\mathcal{B}_t^N\}_{t\geq 0}$ a *standard N-dimensional Brownian motion.*

Let us first recall some related basic probability facts that are needed only in the present context.

Definition B.2 Let $X : \Omega \to \bar{\mathbb{R}}$ be a random variable on some probability space $(\Omega, \mathcal{F}, \mathbb{P})$. We say that X has *normal distribution* with *mean* $\mu \in \mathbb{R}$ and *variance* $\sigma^2 > 0$, provided that the push-forward of measure $\mathbb{P}$ by X is absolutely continuous with respect to the Lebesgue measure and its Radon–Nikodym derivative equals $\frac{1}{\sqrt{2\pi\sigma^2}} e^{-\frac{(x-\mu)^2}{2\sigma^2}}$. More precisely:

$$\mathbb{P}\big(X \in A\big) = \int_A \frac{1}{\sqrt{2\pi\sigma^2}} e^{-\frac{(x-\mu)^2}{2\sigma^2}}\, \mathrm{d}x \qquad \text{for all Borel } A \subset \mathbb{R}.$$

We then write: $X \sim N(\mu, \sigma^2)$.

Exercise B.3 In the above context, prove the following statements:

(i) If $X \sim N(\mu, \sigma^2)$ and $\gamma \neq 0$, then $\gamma X \sim N(\gamma\mu, \gamma^2\sigma^2)$.
(ii) If $X_1 \sim N(\mu_1, \sigma_1^2)$ and $X_2 \sim N(\mu_2, \sigma_2^2)$ are two independent random variables, then $X_1 + X_2 \sim N(\mu_1 + \mu_2, \sigma_1^2 + \sigma_2^2)$.
(iii) If both independent random variables X_1, X_2 have standard normal distribution $N(0, 1)$, then $\frac{1}{\sqrt{2}}(X_1 + X_2)$ and $\frac{1}{\sqrt{2}}(X_1 - X_2)$ are also independent with distributions $N(0, 1)$.

We start by constructing the $N = 1$ dimensional standard Brownian motion $\{B_t\}_{t\in[0,1]}$ on a particular probability space as in Definition B.1, that up to sets of measure 0 is defined as follows. Namely, we set $(\Omega, \mathcal{F}, \mathbb{P})$ to be the countable product of $\mathbb{R}$ equipped with the Borel σ-algebra and probability measure $\frac{1}{\sqrt{2\pi}} e^{-\frac{x^2}{2}}\, \mathrm{d}x$. We index the coordinates of: $\omega = \{\omega_q\}_{q=\frac{a}{2^k}} \in \Omega$ by ordered binary rationals $q = \frac{a}{2^k}$ where $a > 0$ is odd, $k \geq 0$ and $a < 2^k$, that results in having:

$$\omega = \big(\omega_1, \omega_{\frac{1}{2}}, \omega_{\frac{1}{4}}, \omega_{\frac{3}{4}}, \omega_{\frac{1}{8}}, \omega_{\frac{3}{8}}, \omega_{\frac{5}{8}}, \omega_{\frac{7}{8}}, \omega_{\frac{1}{16}}, \ldots\big).$$

Then we inductively define:

$$\begin{aligned} &B_0 \equiv 0, \qquad B_1(\omega) = \omega_1, \\ &B_{\frac{a}{2^{k+1}}}(\omega) = \frac{1}{2}\Big(B_{\frac{a-1}{2^{k+1}}}(\omega) + B_{\frac{a+1}{2^{k+1}}}(\omega) + \frac{1}{\sqrt{2^k}}\omega_{\frac{a}{2^{k+1}}}\Big) \qquad \text{for all } k \geq 0. \end{aligned} \tag{B.1}$$

Theorem B.4

(i) For every two binary rationals $0 \leq s < t \leq 1$, *the random variable* $\frac{1}{\sqrt{t-s}}(B_t - B_s)$ *has standard normal distribution. Equivalently, there holds:* $B_t - B_s \sim N(0, t-s)$.

(ii) If $\{(s_i, t_i)\}_{i=1}^n$ *are disjoint subintervals of* $[0, 1]$ *with binary endpoints, then* $\{B_{t_i} - B_{s_i}\}_{i=1}^n$ *are independent random variables.*

Proof

1. To show (i), we first observe that $B_1 - B_0 = \omega_1 \sim N(0, 1)$. Assume that the statement holds whenever $t - s = \frac{1}{2^k}$ with $k \geq 0$. Let now $s = \frac{a}{2^{k+1}}$, $t = \frac{a+1}{2^{k+1}} \in [0, 1]$ for some odd $a > 0$. From (B.1) we obtain:

$$\begin{aligned} B_t - B_s &= B_{\frac{a+1}{2^{k+1}}} - \frac{1}{2}\Big(B_{\frac{a-1}{2^{k+1}}} + B_{\frac{a+1}{2^{k+1}}} + \frac{1}{\sqrt{2^k}}\omega_{\frac{a}{2^{k+1}}}\Big) \\ &= \frac{1}{2}\Big(B_{\frac{a+1}{2^{k+1}}} - B_{\frac{a-1}{2^{k+1}}}\Big) - \frac{1}{2\sqrt{2^k}}\omega_{\frac{a}{2^{k+1}}}. \end{aligned} \tag{B.2}$$

Observe that by the inductive assumption and by Exercise B.3 (i) we have: $\frac{1}{2}\big(B_{\frac{a+1}{2^{k+1}}} - B_{\frac{a-1}{2^{k+1}}}\big) \sim N(0, \frac{1}{2^{k+2}})$ and $-\frac{1}{2\sqrt{2^k}}\omega_{\frac{a}{2^{k+1}}} \sim N(0, \frac{1}{2^{k+2}})$. Also, these two random variables are independent in view of Exercise A.24, so Exercise B.3 (ii) yields: $B_t - B_s \sim N(0, \frac{1}{2^{k+1}})$ as claimed.

Similarly, when $s = \frac{a-1}{2^{k+1}}$, $t = \frac{a}{2^{k+1}} \in [0, 1]$ for some odd $a > 0$, then:

$$\begin{aligned} B_t - B_s &= \frac{1}{2}\Big(B_{\frac{a-1}{2^{k+1}}} + B_{\frac{a+1}{2^{k+1}}} + \frac{1}{\sqrt{2^k}}\omega_{\frac{a}{2^{k+1}}}\Big) - B_{\frac{a-1}{2^{k+1}}} \\ &= \frac{1}{2}\Big(B_{\frac{a+1}{2^{k+1}}} - B_{\frac{a-1}{2^{k+1}}}\Big) + \frac{1}{2\sqrt{2^k}}\omega_{\frac{a}{2^{k+1}}} \sim N(0, \frac{1}{2^{k+1}}), \end{aligned} \tag{B.3}$$

and the statement in (i) is hence validated on all binary intervals of the two indicated types. The general statement will result by Exercise A.24, provided we show (ii).

2. We start by noting that $\sqrt{2}(B_{1/2} - B_0) = \frac{1}{\sqrt{2}}(\omega_1 + \omega_{1/2})$ and $\sqrt{2}(B_1 - B_{1/2}) = \frac{1}{\sqrt{2}}(\omega_1 - \omega_{1/2})$ are independent by Exercise B.3 (iii), so $B_{1/2} - B_0$ and $B_1 - B_{1/2}$ are independent. Now we again proceed by induction, with the induction step argument similar to that in the proof of Exercise B.3 (iii). Assume that the statement in (ii) is true for the family of intervals $\{(\frac{j}{2^k}, \frac{j+1}{2^k})\}_{j=0}^{2^k-1}$ at some $k \geq 1$. To show the same statement at $k+1$, consider the 2^{k+1} random variables $\{B_{\frac{j+1}{2^{k+1}}} - B_{\frac{j}{2^{k+1}}}\}_{j=0}^{2^{k+1}-1}$, which we view as components of the following $\mathbb{R}^{2^{k+1}}$-valued random variable:

$$X = \Big[\Big(B_{\frac{2i+2}{2^{k+1}}} - B_{\frac{2i+1}{2^{k+1}}}\Big), \Big(B_{\frac{2i+1}{2^{k+1}}} - B_{\frac{2i}{2^{k+1}}}\Big)\Big]_{i=0}^{2^k-1}.$$

By (B.2), (B.3) we observe that:

$$X = \frac{1}{\sqrt{2^k}}\Big[\frac{\sqrt{2^k}}{2}\Big(B_{\frac{2i+2}{2^{k+1}}} - B_{\frac{2i}{2^{k+1}}}\Big) - \frac{1}{2}\omega_{\frac{2i+1}{2^{k+1}}}, \frac{\sqrt{2^k}}{2}\Big(B_{\frac{2i+2}{2^{k+1}}} - B_{\frac{2i}{2^{k+1}}}\Big) + \frac{1}{2}\omega_{\frac{2i+1}{2^{k+1}}}\Big]_{i=0}^{2^k-1}$$
$$= \frac{1}{\sqrt{2^{k+1}}}RY,$$

where $Y = \Big[\sqrt{2^k}\Big(B_{\frac{2i+2}{2^{k+1}}} - B_{\frac{2i}{2^{k+1}}}\Big), \omega_{\frac{2i+1}{2^{k+1}}}\Big]_{i=0}^{2^k-1}$ is a $\mathbb{R}^{2^{k+1}}$-valued random variable, and where:

$$R = \mathrm{diag}\Bigg(\begin{bmatrix} \frac{1}{\sqrt{2}} & -\frac{1}{\sqrt{2}} \\ \frac{1}{\sqrt{2}} & \frac{1}{\sqrt{2}} \end{bmatrix}, \ldots, \begin{bmatrix} \frac{1}{\sqrt{2}} & -\frac{1}{\sqrt{2}} \\ \frac{1}{\sqrt{2}} & \frac{1}{\sqrt{2}} \end{bmatrix}\Bigg) \in \mathbb{R}^{2^{k+1}\times 2^{k+1}}$$

is a rotation of $\mathbb{R}^{2^{k+1}}$ equal to the composition of 2^k independent two-dimensional rotations, each by $\pi/4$ angle. The induction assumption implies that the components of Y are independent and $N(0,1)$-distributed. To deduce that components of X are independent, we consider Borel sets $\{A_i\}$ in:

$$\mathbb{P}\Big(X \in \prod_{j=0}^{2^{k+1}-1} A_i\Big) = \mathbb{P}\Big(Y \in \sqrt{2^{k+1}}R^{-1}\big(\prod_{j=0}^{2^{k+1}-1} A_i\big)\Big)$$
$$= \int_{\sqrt{2^{k+1}}R^{-1}(\prod A_i)} \frac{1}{\sqrt{2\pi}^{2^{k+1}}} e^{-\frac{|x|^2}{2}}\,\mathrm{d}x$$
$$= \int_{\prod(\sqrt{2^{k+1}}A_i)} \frac{1}{\sqrt{2\pi}^{2^{k+1}}} e^{-\frac{|x|^2}{2}}\,\mathrm{d}x = \prod_{j=0}^{2^{k+1}-1}\Big(\int_{\sqrt{2^{k+1}}A_i} \frac{1}{\sqrt{2\pi}} e^{-\frac{x^2}{2}}\,\mathrm{d}x\Big)$$
$$= \prod_{j=0}^{2^{k+1}-1} \mathbb{P}\big(\langle X, e_j\rangle \in A_i\big),$$

where again we invoked the rotational invariance of the density function and the fact that components of X are $N(0, \frac{1}{2^{k+1}})$-distributed in view of (i). This implies that for each $k \geq 1$, the random variables $\{B_{t_j} - B_{s_j}\}_{j=1}^{2^k}$ are independent, along the basic partition intervals $\{(s_j, t_j)\}_{j=1}^{2^k}$ of length $\frac{1}{2^k}$. We now conclude the statement (ii) in the general case as well, by Exercise A.24. This ends the proof of Theorem B.4. □

So far, we verified the validity of conditions (i) and (ii) in Definition B.1 for the binary rationals in $[0, 1]$. We presently complete the construction in (B.1) to have $\{B_t\}_{t\in[0,1]}$, via condition (iii).

Lemma B.5 *Define an increasing sequence of events* $\{A_{k_0}\}_{k_0=1}^{\infty}$ *in* Ω *by setting:*

$$A_{k_0} = \Big\{\omega \in \Omega;\ \big|B_{\frac{j+1}{2^k}}(\omega) - B_{\frac{j}{2^k}}(\omega)\big| \leq 2\sqrt{\frac{k}{2^k}} \quad \textit{for all } k \geq k_0,\ j = 0, \ldots, 2^k - 1\Big\}.$$

Then: $\mathbb{P}\Big(\bigcup_{k_0}^{\infty} A_{k_0}\Big) = 1.$

Proof Since $\sqrt{2^k}\big(B_{\frac{j+1}{2^k}} - B_{\frac{j}{2^k}}\big) \sim N(0,1)$, it follows that for every $k \geq 1$ and every $j = 0, \ldots, 2^k - 1$ there holds:

$$\begin{aligned}\mathbb{P}\Big(\big|B_{\frac{j+1}{2^k}}(\omega) - B_{\frac{j}{2^k}}(\omega)\big| > 2\sqrt{\frac{k}{2^k}}\Big) &= \int_{|x|>2\sqrt{k}} e^{-\frac{x^2}{2}}\,\mathrm{d}x = 2\int_{2\sqrt{k}}^{\infty} e^{-\frac{x^2}{2}}\,\mathrm{d}x \\ &\leq \frac{1}{2\sqrt{k}}\int_{2\sqrt{k}}^{\infty} xe^{-\frac{x^2}{2}}\,\mathrm{d}x = \frac{1}{\sqrt{k}}e^{-2k}.\end{aligned}$$

Consequently:

$$\mathbb{P}(\Omega \setminus A_{k_0}) \leq \sum_{k\geq k_0}\sum_{j=0}^{2^k-1} \mathbb{P}\Big(\big|B_{\frac{j+1}{2^k}}(\omega) - B_{\frac{j}{2^k}}(\omega)\big| > 2\sqrt{\frac{k}{2^k}}\Big) \leq \sum_{k\geq k_0} 2^k e^{-2k}.$$

The right hand side above converges to 0 as $k_0 \to \infty$, because the geometric series $\sum_k (2e^{-2})^k$ converges. Hence $\mathbb{P}\big(\bigcap_{k_0=1}^{\infty}(\Omega \setminus A_{k_0})\big) = 0$, implying the claim. □

Corollary B.6 *For* $\mathbb{P}$*-a.e.* $\omega \in \Omega$*, the path* $t \mapsto B_t(\omega)$ *may be uniquely extended to a Hölder continuous function on* $[0,1]$*, such that:*

(i) For every $0 \leq s < t \leq 1$ *we have:* $B_t - B_s \sim N(0, t-s)$.
(ii) If $\{(s_i, t_i)\}_{i=1}^n$ *are pairwise disjoint subintervals of* $[0,1]$*, then the random variables* $\{(B_{t_1} - B_{si}\}_{i=1}^n$ *are independent.*

Proof

1. For $k_0 \geq 1$ and $\omega \in A_{k_0}$. Let $0 \leq s < t \leq 1$ be two binary rationals such that $|t-s| < \frac{1}{2^{k_0}}$. Then, there exists $k_1 \geq k_0$ satisfying:

$$\frac{1}{2^{k_1+1}} \leq |t-s| < \frac{1}{2^{k_1}}, \tag{B.4}$$

and we may express the interval $[s,t]$ as the union of intervals of the type $[\frac{j-1}{2^k}, \frac{j}{2^k}]$ with only $k > k_1$ present and with each such generation k being represented by at most two intervals (possibly one or none). Consequently,

$|B_t(\omega) - B_s(\omega)|$ is bounded by the sum of increments along the indicated subdivision intervals. Recalling the defining property of A_{k_0}, we get:

$$|B_t(\omega) - B_s(\omega)| \leq 2 \sum_{k>k_1} 2\sqrt{\frac{k}{2^k}} \leq C\sqrt{\frac{k_1}{2^{k_1}}},$$

where $C > 0$ is a universal constant as in Exercise B.7.

Fix $\alpha > 0$. Clearly, $k_1 \leq C_\alpha (2^{k_1})^\alpha$ for any $k_1 \geq 1$, where the constant $C_\alpha > 0$ depending only on α. Consequently:

$$|B_t(\omega) - B_s(\omega)| \leq C_\alpha \Big(\frac{1}{2^{k_1}}\Big)^{\frac{1-\alpha}{2}} \leq C_\alpha |t-s|^{\frac{1-\alpha}{2}}$$

in virtue of (B.4). It follows that $t \mapsto B_t(\omega)$ is Hölder continuous (with any prescribed exponent $\beta < \frac{1}{2}$) and as such it may be uniquely extended to a β-Hölder path $[0, 1] \ni t \mapsto B_t(\omega)$, whose norm is bounded in function of β and k_0 only. By Lemma B.5, we see that this extension can be performed for $\mathbb{P}$-a.e. $\omega \in \Omega$, and that:

$$B_t = \lim_{n\to\infty} B_{t_n} \quad \text{a.s. in } \Omega, \text{ when binary rationals } t_n \to t. \tag{B.5}$$

2. It remains to check properties (i) and (ii). Given $0 \leq t < s \leq 1$, let $t_n \to t$ and $s_n \to s$ be some approximating binary rational sequences. Then:

$$Z_n = \frac{1}{\sqrt{t_n - s_n}}(B_{t_n} - B_{s_n}) \sim N(0, 1)$$

by Theorem B.4 (i). Since Z_n converge $\mathbb{P}$-a.s. to $Z = \frac{1}{\sqrt{t-s}}(B_t - B_s)$, we obtain the following convergence, valid for every $f \in C_c(\mathbb{R})$:

$$\int_{\mathbb{R}} f(x)\frac{1}{\sqrt{2\pi}}e^{-\frac{x^2}{2}}\,\mathrm{d}x = \int_\Omega f \circ Z_n \,\mathrm{d}\mathbb{P} \to \int_\Omega f \circ Z \,\mathrm{d}\mathbb{P}, \qquad \text{as } n \to \infty$$

and thus $Z \sim N(0, 1)$ as claimed in (i). Condition (ii) can be shown similarly, in view of Theorem B.4 (ii). □

Exercise B.7 Show that there exists $C \geq 1$ with the property that:

$$\forall k_1 \geq 1 \qquad \sum_{k>k_1} \sqrt{\frac{k}{2^k}} \leq C\sqrt{\frac{k_1}{2^{k_1}}}.$$

We finally state:

Theorem B.8 *There exists a standard N-dimensional Brownian motion, satisfying all properties in Definition B.1.*

Proof We first extend the one-dimensional process $\{B_t\}_{t\in[0,1]}$ defined in (B.1) and Corollary B.6 for $t \in [0,1]$, to the positive reals $t \geq 0$. Let $(\Omega_0, \mathcal{F}_0, \mathbb{P}_0)$ be the new probability space, where $\Omega_0 = \bigcup_{k_0\geq 1} A_{k_0}$ and $\mathcal{F}_0 = \{A \cap \Omega_0;\ A \in \mathcal{F}\}$ with $\mathbb{P}_0 = \mathbb{P}_{|\mathcal{F}_0}$. Define:

$$(\Omega_{\mathbb{N}}, \mathcal{F}_{\mathbb{N}}, \mathbb{P}_{\mathbb{N}}) = (\Omega_0, \mathcal{F}_0, \mathbb{P}_0)^{\mathbb{N}} = \big\{\omega = (\omega^{(1)}, \omega^{(2)}, \ldots);\ \omega^{(i)} \in \Omega_0 \quad \text{for all } i \geq 1\big\}$$

and let $\{B_t^{(i)}\}_{t\in[0,1]}$ be the one-dimensional Brownian motion in Corollary B.6 on the i-th coordinate space of $\Omega_{\mathbb{N}}$, namely: $B_t^{(i)}(\omega) = B_t(\omega^{(i)})$. Then we set:

$$\mathcal{B}_t = \Big(\sum_{j=1}^{i} B_1^j\Big) + B_{t-i}^{(i+1)} \qquad \text{for } t \in [i, i+1],\ i \geq 0.$$

Second, we define $\{\mathcal{B}_t^N\}_{t\geq 0}$ on the probability space $(\Omega_{\mathbb{N}}, \mathcal{F}_{\mathbb{N}}, \mathbb{P}_{\mathbb{N}})^N$ as the vector-valued process $\mathcal{B}_t^N = (\mathcal{B}_t^{(1)}, \ldots, \mathcal{B}_t^{(N)})$, which on each coordinate $i = 1 \ldots N$ coincides with the above one-dimensional construction: $\mathcal{B}_t^{(i)} = \mathcal{B}_t \circ \pi_i$ on the i-th coordinate probability space. Clearly, $\{\mathcal{B}_t^N\}_{t\geq 0}$ satisfies all the required properties. □

We right away deduce a simple useful fact:

Lemma B.9 *A Brownian motion exits any ball $B_r(0)$ almost surely, i.e.*

$$\mathbb{P}\big(\exists t \geq 0;\ |\mathcal{B}_t^N| \geq r\big) = 1 \qquad \textit{for all } r > 0.$$

Proof The claim follows, since for any fixed $r, T > 0$ we have:

$$\begin{aligned}\mathbb{P}\big(\exists t \geq 0;\ |\mathcal{B}_t^N| \geq r\big) \geq \mathbb{P}\big(|\mathcal{B}_T^N| \geq r\big) &= \int_{\mathbb{R}^N \setminus B_r(0)} \frac{1}{(2\pi T)^{N/2}} e^{-\frac{|x|^2}{2T}}\,\mathrm{d}x \\ &= \int_{\mathbb{R}^N \setminus B_{r/\sqrt{T}}(0)} \frac{1}{(2\pi)^{N/2}} e^{-\frac{|y|^2}{2}}\,\mathrm{d}y,\end{aligned}$$

the integral in the right hand side goes to 1 as $T \to \infty$. □

Exercise B.10 Show that the 1-dimensional Brownian motion hits points a.s.:

$$\forall \theta \in \mathbb{R} \qquad \mathbb{P}\big(\exists t \geq 0 \quad \mathcal{B}_t^1 = \theta\big) = 1.$$

Remark B.11 According to the observation in Lemma B.9, Brownian motion will a.s. enter the open set $\mathbb{R}^N \setminus \bar{B}_r(0)$. One may ask if the same is true for bounded sets, say for a ball $B_r(x_0) \not\ni 0$. The answer is affirmative in dimensions $N = 1, 2$, namely: $\mathbb{P}(\exists t >;\ \mathcal{B}_t^N \in B_r(x_0)) = 1$, whereas for $N \geq 3$ we have: $\mathbb{P}(\exists t >;\ \mathcal{B}_t^N \in B_r(x_0)) = \frac{r^{N-2}}{|x_0|^{N-2}} < 1$ when $|x_0| > r$.

This fact is related to the recurrence and transience properties. For $N = 1$, Brownian motion is *point-recurrent*: $\mathbb{P}(\exists t_n \to \infty;\ \mathcal{B}_{t_n}^1 = x) = 1$ for every $x \in \mathbb{R}$. For $N = 2$, it is *neighbourhood-recurrent*: $\mathbb{P}(\exists t_n \to \infty;\ \mathcal{B}_{t_n}^2 \in B_r(x_0)) = 1$ for every $B_r(x_0) \subset \mathbb{R}^2$, but it is not point-recurrent. For $N \geq 3$, Brownian motion is *transient*: $\mathbb{P}(\lim_{t\to\infty} |\mathcal{B}_t^N| = +\infty) = 1$.

B.2 The Wiener Measure and Uniqueness of Brownian Motion

We now prove the uniqueness of Brownian motion. In order to not be restricted by the choice of the probability space $(\Omega, \mathcal{F}, \mathbb{P})$ in Definition B.1, this is done by proving uniqueness of the associated Wiener measure.

Definition B.12 Let $\{\mathcal{B}_t^N\}_{t\geq 0}$ be a standard Brownian N-dimensional motion. The *Wiener measure* μ_W is the probability measure on the Banach space $E = C([0,1], \mathbb{R}^N)$ equipped with the σ-algebra of its Borel subsets, such that:

$$\mu_W(A) = \mathbb{P}\Big(\omega;\ \big([0,1] \ni t \mapsto \mathcal{B}_t^N(\omega) \in \mathbb{R}^N\big) \in A\Big)$$
$$\text{for all Borel } A \subset E. \tag{B.6}$$

Thus, μ_W is the push-forward of $\mathbb{P}$ by the indicated in (B.6) measurable map from Ω to E. Indeed, we observe that:

Exercise B.13

(i) The space $E = C([0,1], \mathbb{R}^N)$ is separable. The countable dense subset of E consists, for example, of all polynomials on $[0,1]$ with rational coefficients.
(ii) The σ-algebra of Borel subsets of E is generated by the closed balls.
(iii) The mapping $\Omega \ni \omega \mapsto \big([0,1] \ni t \mapsto \mathcal{B}_t^N(\omega) \in \mathbb{R}^N\big) \in E$ is measurable.

Automatically, for all $\psi \in C_b(E)$ we have:

$$\int_E \psi(f)\, \mathrm{d}\mu_W(f) = \int_\Omega \psi\big([0,1] \ni t \mapsto \mathcal{B}_t^N(\omega)\big)\, \mathrm{d}\mathbb{P}(\omega).$$

Denoting the right hand side above by $F(\psi)$, it follows that $F \in C_b(E)^*$, namely F is a linear, continuous and positive (i.e. $F(\psi) \geq 0$ for all $\psi \geq 0$) functional on the Banach space $C_b(E)$. When restricted to the unit ball, F is also continuous with respect to the topology of uniform convergence on compact sets. Hence, μ_W is naturally the measure obtained by invoking the Riesz representation theorem (on nonlocally compact space E, see Parthasarathy 1967).

Exercise B.14 Let $\mu_1 \neq \mu_2$ be two distinct probability measures on $C_b(E)$. Show that there exists $0 \leq t_1 < t_2 \ldots < t_n \leq 1$ and $\tilde{\psi} \in C_c(\mathbb{R}^{Nn})$ such that defining:

$$\psi(f) = \tilde{\psi}\big(f(t_1), \ldots, f(t_n)\big) \qquad \text{for all } f \in E, \tag{B.7}$$

the function $\psi \in C_b(E)$ satisfies: $\int_E \psi \, \mathrm{d}\mu_1 \neq \int_E \psi \, \mathrm{d}\mu_2$.

Theorem B.15 *The Wiener measure μ_W is independent of the process $\{\mathcal{B}_t^N\}_{t\in[0,1]}$, as long as $\mathcal{B}_0^N = 0$ and conditions (i)–(iii) (restricted to the interval $[0,1]$) in Definition B.1 hold.*

Proof Let ψ be as in (B.7) for some $0 \leq t_1 < \ldots < t_n \leq 1$ and $\tilde{\psi} \in C_c(\mathbb{R}^{Nn})$. Then:

$$\begin{aligned} F(\psi) &= \int_\Omega \psi\big(t \mapsto \mathcal{B}_t^N(\omega)\big) \, \mathrm{d}\mathbb{P}(\omega) = \int_\Omega \tilde{\psi}\big(\mathcal{B}_{t_1}^N(\omega), \ldots, \mathcal{B}_{t_n}^N(\omega)\big) \, \mathrm{d}\mathbb{P}(\omega) \\ &= \int_\Omega \tilde{\phi}\big(\mathcal{B}_{t_1}^N(\omega) - \mathcal{B}_0^N(\omega), \mathcal{B}_{t_2}^N(\omega) - \mathcal{B}_{t_1}^N(\omega), \ldots, \mathcal{B}_{t_n}^N(\omega) - \mathcal{B}_{t_{n-1}}^N(\omega)\big) \, \mathrm{d}\mathbb{P}(\omega), \end{aligned}$$

where $\tilde{\phi} \in C_c(\mathbb{R}^{Nn})$ is given by:

$$\tilde{\phi}\big(x^{(1)}, \ldots, x^{(n)}\big) = \tilde{\psi}\big(x^{(1)}, x^{(1)} + x^{(2)}, \ldots, x^{(1)} + \ldots + x^{(n)}\big).$$

As the increments $\{\mathcal{B}_{t_{i+1}}^N - B_{t_i}^N\}_{i=0}^{n-1}$ are independent and normally distributed:

$$F(\psi) = \int_{\mathbb{R}^{Nn}} \tilde{\phi}\big(x^{(1)}, \ldots, x^{(n)}\big) \prod_{i=0}^{n-1} \frac{1}{(2\pi(t_{i+1} - t_i))^{N/2}} e^{-\frac{|x^{(i)}|^2}{2(t_{i+1}-t_i)}} \, \mathrm{d}(x^{(1)}, \ldots, x^{(n)}).$$

The above quantity depends only on ψ, and it is independent of the particular choice of the process $\{\mathcal{B}_t^N\}_{t\in[0,1]}$. Thus μ_W is uniquely defined, by Exercise B.14. □

Exercise B.16 Show that for every selection of points $0 = t_0 < t_1 < t_2 \ldots < t_n \leq 1$ and Borel sets $A_i \subset \mathbb{R}^N$, $i = 1 \ldots n$ there holds:

$$\begin{aligned} &\mu_W\big(f \in E;\ f(t_i) \in A_i \quad \text{for all } i = 1 \ldots n\big) = \\ &\int_{\prod_{i=1}^n A_i} \prod_{i=1}^{n} \frac{1}{(2\pi(t_i - t_{i-1}))^{N/2}} e^{-\frac{|x_i - x_{i-1}|^2}{2(t_i - t_{i-1})}} \, \mathrm{d}(x_1, \ldots, x_n). \end{aligned}$$

The above formula is sometimes taken as the definition of the Wiener measure.

Exercise B.17

(i) Consider the space $E_0 = C([0,\infty),\mathbb{R}^N)$ with topology of uniform convergence on compact intervals $[0,T]$, for all $T>0$. As in Exercise B.13, show that the Borel σ-algebra $\mathcal{F}_0$ in E_0 is generated by sets of the type: $A_{g,T,\epsilon} = \{f \in E_0;\ \|f-g\|_{L^\infty([0,T])} \leq \epsilon\}$, where g are polynomials with rational coefficients and $T, \epsilon > 0$ are rational numbers.

(ii) Show that $\mu_W(A) = \mathbb{P}\big(\omega;\ (t \mapsto \mathcal{B}_t^N(\omega)) \in A\big)$ defines a probability measure on $(E_0, \mathcal{F}_0)$.

(iii) Prove that each standard N-dimensional Brownian motion $\{\mathcal{B}_t^N\}_{t\geq 0}$ induces the same Wiener measure μ_W as in (ii).

B.3 The Markov Properties

The following statement is the *Markov property of Brownian motion*:

Theorem B.18 *If $\{\mathcal{B}_t^N\}_{t\geq 0}$ is a standard Brownian motion as in Definition B.1, then for every $s>0$ the process $\{\mathcal{B}_{t+s}^N - \mathcal{B}_s^N\}_{t\geq 0}$ is also a standard Brownian motion, which is independent of $\{\mathcal{B}_t^N\}_{t\in[0,s]}$. More precisely, the latter property states that if $0 \leq s_1 < \ldots < s_k \leq s$ and $0 < t_1 < \ldots < t_n$ for some $k, n \geq 1$, then the following two vector-valued random variables:*

$$\big(\mathcal{B}_{s_1}^N, \ldots, \mathcal{B}_{s_k}^N\big) \quad \textit{and} \quad \big(\mathcal{B}_{t_1+s}^N - \mathcal{B}_s^N, \ldots, \mathcal{B}_{t_n+s}^N - \mathcal{B}_s^N\big)$$

are independent.

Proof The fact that $\{\bar{\mathcal{B}}_t = \mathcal{B}_{t+s}^N - \mathcal{B}_s^N\}_{t\geq 0}$ satisfies conditions (i)–(iii) of Definition B.1 is self-evident. To show that $\{\mathcal{B}_t^N\}_{t\in[0,s]}$ and $\{\bar{\mathcal{B}}_t\}_{t\geq 0}$ are independent, fix two Borel sets $A \subset \mathbb{R}^{Nk}$ and $C \subset \mathbb{R}^{Nn}$. Then:

$$\big\{\big(\mathcal{B}_{s_1}^N, \ldots, \mathcal{B}_{s_k}^N\big) \in A\big\} = \big\{\big(\mathcal{B}_{s_1}^N - \mathcal{B}_0^N, \mathcal{B}_{s_2}^N - \mathcal{B}_{s_1}^N, \ldots, \mathcal{B}_{s_k}^N - \mathcal{B}_{s_{k-1}}^N\big) \in M_k(A)\big\}$$

and:

$$\begin{aligned}&\big\{\big(\bar{\mathcal{B}}_{t_1}, \ldots, \bar{\mathcal{B}}_{t_n}\big) \in C\big\} \\ &\quad = \big\{\big(\mathcal{B}_{t_1+s}^N - \mathcal{B}_s^N, \mathcal{B}_{t_2+s}^N - \mathcal{B}_{t_1+s}^N, \ldots, \mathcal{B}_{t_n+s}^N - \mathcal{B}_{t_{n-1}+s}^N\big) \in M_n(C)\big\},\end{aligned}$$

where for each $i \geq 1$ we define the invertible matrix:

$$M_i = \begin{bmatrix} Id_N & & & & \\ -Id_N & Id_N & & & \\ & -Id_N & Id_N & & \\ & & & \ddots & \\ & & & -Id_N & Id_N \end{bmatrix} \in \mathbb{R}^{Ni\times Ni}.$$

Using the fact that the intervals $(0, s_1), (s_1, s_2), \ldots, (s_{k-1}, s_k), (s, t_1+s), (t_1+s, t_2+s), \ldots, (t_{n-1}+s, t_n+s)$ are disjoint, we invoke the independence of increments property of $\{\mathcal{B}_t^N\}_{t\geq 0}$ which in view of Exercise A.24 yields:

$$\begin{aligned}
&\mathbb{P}\Big(\big\{(\mathcal{B}_{s_1}^N, \ldots, \mathcal{B}_{s_k}^N) \in A\big\} \cap \big(\bar{\mathcal{B}}_{t_1}, \ldots, \bar{\mathcal{B}}_{t_n}\big) \in C\big\}\Big) \\
&= \mathbb{P}\Big(\big(\mathcal{B}_{s_1}^N - \mathcal{B}_0^N, \mathcal{B}_{s_2}^N - \mathcal{B}_{s_1}^N, \ldots, \mathcal{B}_{s_k}^N - \mathcal{B}_{s_{k-1}}^N\big) \in M_k(A)\Big) \\
&\quad \cdot \mathbb{P}\Big(\big(\mathcal{B}_{t_1+s}^N - \mathcal{B}_s^N, \mathcal{B}_{t_2+s}^N - \mathcal{B}_{t_1+s}^N, \ldots, \mathcal{B}_{t_n+s}^N - \mathcal{B}_{t_{n-1}+s}^N\big) \in M_n(C)\Big) \\
&= \mathbb{P}\Big(\big(\mathcal{B}_{s_1}^N, \ldots, \mathcal{B}_{s_k}^N\big) \in A\Big) \cdot \mathbb{P}\Big(\big(\bar{\mathcal{B}}_{t_1}, \ldots, \bar{\mathcal{B}}_{t_n}\big) \in C\Big),
\end{aligned}$$

as needed. □

Definition B.19 Let $\{\mathcal{B}_t^N\}_{t\geq 0}$ be a standard N-dimensional Brownian motion as in Definition B.1.

(i) For every $t \geq 0$ we define $\mathcal{F}_t \subset \mathcal{F}$ to be the smallest sub-σ algebra of $\mathcal{F}$ such that for each $s \in [0, t]$ the random variable $\mathcal{B}_s^N : \Omega \to \mathbb{R}^N$ is $\mathcal{F}_t$-measurable. Clearly, $\mathcal{F}_s \subset \mathcal{F}_t$ when $s \leq t$.
(ii) We say that $\tau : \Omega \to [0, \infty]$ is a *stopping time* for $\{\mathcal{B}_t^N\}_{t\geq 0}$, provided that $\{\tau \leq t\} \in \mathcal{F}_t$ for all $t \geq 0$ and $\mathbb{P}(\tau = +\infty) = 0$. We then further define:

$$\mathcal{F}_\tau = \big\{A \in \mathcal{F};\ A \cap \{\tau \leq t\} \in \mathcal{F}_t \quad \text{for all } t \geq 0\big\}.$$

Exercise B.20 In the above context, prove the following assertions:

(i) $\mathcal{F}_\tau$ is indeed a σ-algebra and τ is $\mathcal{F}_\tau$-measurable. For a constant stopping time $\tau \equiv s \geq 0$ we have: $\mathcal{F}_\tau = \mathcal{F}_s$.
(ii) If two stopping times satisfy $\tau_1 \leq \tau_2$, then $\mathcal{F}_{\tau_1} \subset \mathcal{F}_{\tau_2}$.
(iii) For every $k \geq 0$, the random variable $\tau_k = \frac{1}{2^k}\lceil 2^k \tau \rceil$ is another stopping time and $\mathcal{B}_{\tau_k}^N$ defined as $\mathcal{B}_{\tau_k}^N(\omega) = \mathcal{B}_{\tau_k(\omega)}^N(\omega)$ is a random variable. Since τ_k converge to τ pointwise a.s. in Ω, it follows immediately that $\mathcal{B}_\tau^N$ is a random variable, where $\mathcal{B}_\tau^N(\omega) = \mathcal{B}_{\tau(\omega)}^N(\omega)$.

Lemma B.21 *If τ is a stopping time for the Brownian motion $\{\mathcal{B}_t^N\}_{t\geq 0}$, then the random variable $\mathcal{B}_\tau^N$ is $\mathcal{F}_\tau$-measurable.*

Proof For every $k \geq 0$, define the random variable $\eta_k = \frac{1}{2^k}\lfloor 2^k \tau \rfloor$. We remark that since $\{\eta_k \leq t\} = \{\tau < \frac{1}{2^k}\lceil 2^k t \rceil\}$, then η_k is indeed measurable but it is not, in general, a stopping time. For Borel $A \subset \mathbb{R}^N$ and $t \geq 0$, consider the event:

$$\begin{aligned}
\{\mathcal{B}_{\eta_k}^N \in A\} \cap \{\tau \leq t\} &= \bigcup_{n\geq 0} \Big(\{\mathcal{B}_{\frac{n}{2^k}}^N \in A\} \cap \big\{\tau \in \big[\frac{n}{2^k}, \frac{n+1}{2^k}\big)\big\} \cap \{\tau \leq t\}\Big) \\
&= \bigcup_{0\leq n\leq 2^k t} \Big(\{\mathcal{B}_{\frac{n}{2^k}}^N \in A\} \cap \big\{\tau \in \big[\frac{n}{2^k}, \frac{n+1}{2^k}\big)\big\} \cap \{\tau \leq t\}\Big).
\end{aligned}$$

Now, for each $\frac{n}{2^k} \leq t$ we have: $\big\{\mathcal{B}^N_{\frac{n}{2^k}} \in A\big\} \in \mathcal{F}_{\frac{n}{2^k}} \subset \mathcal{F}_t$ and also: $\big\{\tau \in \big[\frac{n}{2^k}, \frac{n+1}{2^k}\big)\big\} \cap \{\tau \leq t\} \in \mathcal{F}_t$, which follows by observing that $\{\tau < s\} = \bigcup_{m\geq 0}\{\tau \leq s - \frac{1}{m}\} \in \mathcal{F}_s$. We conclude that the event in the left hand side of (62) belongs to $\mathcal{F}_t$, which readily implies that each random variable $\mathcal{B}^N_{\eta_k}$ is $\mathcal{F}_\tau$-measurable. Since η_k converge to τ as $k \to \infty$ and τ is a.s. finite, we see that $\mathcal{B}^N_{\eta_k}$ converge to $\mathcal{B}^N_\tau$ pointwise a.s. in Ω, proving the claim. □

Exercise B.22 Show that independence of $\{\mathcal{B}^N_t\}_{t\in[0,s]}$ and $\{\mathcal{B}^N_{t+s} - \mathcal{B}^N_s\}_{t\geq 0}$ stated in Theorem B.18, is equivalent to $\{\mathcal{B}^N_{t+s} - \mathcal{B}^N_s\}_{t\geq 0}$ being independent of $\mathcal{F}_s$, in the sense that for all $0 \leq t_1 < \ldots < t_n$, all $\{A_i\}_{i=1}^n$ Borel subsets of $\mathbb{R}^N$ and all $C \in \mathcal{F}_s$:

$$\mathbb{P}\Big(\bigcap_{i=1}^n \{\mathcal{B}^N_{t_i+s} - \mathcal{B}^N_s \in A_i\} \cap C\Big) = \mathbb{P}\Big(\bigcap_{i=1}^n \{\mathcal{B}^N_{t_i+s} - \mathcal{B}^N_s \in A_i\}\Big) \cdot \mathbb{P}(C).$$

Lemma B.23 *Let K be a closed subset of $\mathbb{R}^N$. Then:*

$$\tau_K(\omega) = \min\{t \geq 0;\ \mathcal{B}^N_t(\omega) \in K\}$$

is a stopping time, provided that $\mathbb{P}(\tau_K = +\infty) = 0$. In particular, if $0 \in \mathcal{D} \subset \mathbb{R}^N$ is open and bounded, then $\tau_{\partial\mathcal{D}}$ is a stopping time.

Proof Given $t \geq 0$, observe that:

$$\{\tau_K \leq t\} = \bigcap_{m\geq 1} \bigcup_{q\in[0,t]\cap\mathbb{Q}} \{\omega;\ \mathrm{dist}(\mathcal{B}^N_q(\omega), K) \leq \frac{1}{m}\} \in \mathcal{F}_t.$$

This implies that τ_K is a stopping time, in view of the assumed a.s. finiteness of τ_K. The second claim follows from Lemma B.9 and the continuity of the Brownian paths $t \mapsto \mathcal{B}^N_t(\omega)$. □

The following statement is an extension of Theorem B.18 to non-constant stopping times, known as the *strong Markov property of Brownian motion.*

Theorem B.24 *Let $\{\mathcal{B}^N_t\}_{t\geq 0}$ be a standard N-dimensional Brownian motion as in Definition B.1, and let τ be a stopping time. Then the process:*

$$\{\mathcal{B}^N_{t+\tau} - \mathcal{B}^N_\tau\}_{t\geq 0}$$

is also a standard Brownian motion (upon a possible modification of Ω by a set of $\mathbb{P}$-measure 0 in order to achieve (iii)). This process is independent of $\mathcal{F}_\tau$, namely: for all $0 \leq t_1 < \ldots < t_n$, all $\{A_i\}_{i=1}^n$ Borel subsets of $\mathbb{R}^N$ and all $C \in \mathcal{F}_\tau$:

$$\mathbb{P}\Big(\bigcap_{i=1}^n \{\mathcal{B}^N_{t_i+\tau} - \mathcal{B}^N_\tau \in A_i\} \cap C\Big) = \mathbb{P}\Big(\bigcap_{i=1}^n \{\mathcal{B}^N_{t_i+\tau} - \mathcal{B}^N_\tau \in A_i\}\Big) \cdot \mathbb{P}(C).$$

Proof

1. Since for $t \geq 0$, the random variable $t + \tau$ is also a stopping time, Exercise B.20 (iii) implies that all $\bar{\mathcal{B}}_t = \mathcal{B}^N_{t+\tau} - \mathcal{B}^N_\tau$ are measurable. It is also clear that for $\omega \in \{\tau < \infty\}$ the path $[0, \infty) \ni t \mapsto \bar{\mathcal{B}}_t(\omega) \in \mathbb{R}^N$ is continuous.

   Recall the definition of the approximating stopping times: $\big\{\tau_k = \frac{1}{2^k}\lceil 2^k\tau\rceil\big\}_{k\geq 0}$. For every $0 \leq s < t$ and a Borel set $A \subset \mathbb{R}^N$ we get:

$$\begin{aligned}\mathbb{P}\big(\mathcal{B}^N_{t+\tau_k} - \mathcal{B}^N_{s+\tau_k} \in A\big) &= \sum_{i\geq 0}\mathbb{P}\Big(\big\{\mathcal{B}^N_{t+\frac{i}{2^k}} - \mathcal{B}^N_{s+\frac{i}{2^k}} \in A\big\} \cap \big\{\tau_k = \frac{i}{2^k}\big\}\Big)\\ &= \sum_{i\geq 0}\mathbb{P}\Big(\mathcal{B}^N_{t+\frac{i}{2^k}} - \mathcal{B}^N_{s+\frac{i}{2^k}} \in A\Big)\cdot \mathbb{P}\big(\tau_k = \frac{i}{2^k}\big)\\ &= \int_A \frac{1}{(2\pi(t-s))^{N/2}}e^{-\frac{|x|^2}{2(t-s)}}\,\mathrm{d}x \cdot \sum_{i\geq 0}\mathbb{P}\big(\tau_k = \frac{i}{2^k}\big),\end{aligned}$$

   since $\big\{\mathcal{B}^N_{t+\frac{i}{2^k}} - \mathcal{B}^N_{s+\frac{i}{2^k}} \in A\big\}$ is independent of $\big\{\tau_k = \frac{i}{2^k}\big\} \in \mathcal{F}_{\frac{i}{2^k}} \subset \mathcal{F}_{s+\frac{i}{2^k}}$ in view of Exercise B.22. Above, we also used the property (i) in Definition B.1. Thus:

$$\mathbb{P}\big(\mathcal{B}^N_{t+\tau_k} - \mathcal{B}^N_{s+\tau_k} \in A\big) = \int_A \frac{1}{(2\pi(t-s))^{N/2}}e^{-\frac{|x|^2}{2(t-s)}}\,\mathrm{d}x,$$

   to the effect that: $\mathcal{B}^N_{t+\tau_k} - \mathcal{B}^N_{s+\tau_k} \sim N(0, t-s)$. As $\mathcal{B}^N_{t+\tau_k} - \mathcal{B}^N_{s+\tau_k}$ converges to $\bar{\mathcal{B}}^N_t - \bar{\mathcal{B}}_s$ as $k \to \infty$, pointwise $\mathbb{P}$-a.s.in Ω, it follows that $\bar{\mathcal{B}}^N_t - \bar{\mathcal{B}}_s \sim N(0, t-s)$ as well, proving property (i) in Definition B.1 for the process $\{\bar{\mathcal{B}}_t\}_{t\geq 0}$. Property (ii) is shown in a similar manner: first by calculations as above for each τ_k, in view of (i) and (ii) and Markov's property, and then by approximating to τ.
2. It remains to show the independence of $\{\bar{\mathcal{B}}_t\}_{t\geq 0}$ from $\mathcal{F}_\tau$. Firstly, for a fixed n-tuple $0 \leq t_1 < t_2 \ldots < t_n$, Borel subsets $\{A_j\}_{j=1}^n$ of $\mathbb{R}^N$, and $C \in \mathcal{F}_{\tau_k}$ with some $k \geq 0$, we observe that:

$$\begin{aligned}&\mathbb{P}\Big(\bigcap_{j=1}^n\{\mathcal{B}^N_{t_j+\tau_k} - \mathcal{B}^N_{\tau_k} \in A_j\} \cap C\Big)\\ &= \sum_{i\geq 0}\mathbb{P}\Big(\bigcap_{j=1}^n\{\mathcal{B}^N_{t_j+\frac{i}{2^k}} - \mathcal{B}^N_{\frac{i}{2^k}} \in A_j\} \cap \{\tau_k = \frac{i}{2^k}\} \cap C\Big)\\ &= \sum_{i\geq 0}\mathbb{P}\Big(\bigcap_{j=1}^n\{\mathcal{B}^N_{t_j+\frac{i}{2^k}} - \mathcal{B}^N_{\frac{i}{2^k}} \in A_j\}\Big)\cdot \mathbb{P}\Big(\{\tau_k = \frac{i}{2^k}\} \cap C\Big)\\ &= \mathbb{P}\Big(\bigcap_{j=1}^n\{\mathcal{B}^N_{t_j+\tau_k} - \mathcal{B}^N_{\tau_k} \in A_j\}\Big)\cdot \mathbb{P}(C),\end{aligned} \tag{B.8}$$

because $\left\{\tau_k = \frac{i}{2^k}\right\} \cap C \in \mathcal{F}_{\frac{i}{2^k}}$ and $\bigcap_{j=1}^n \left\{\mathcal{B}^N_{t_j+\frac{i}{2^k}} - \mathcal{B}^N_{\frac{i}{2^k}} \in A_j\right\}$ are independent of $\mathcal{F}_{\frac{i}{2^k}}$ by the Markov property.

We have also used the fact that $\mathbb{P}\big(\bigcap_{j=1}^n \left\{\mathcal{B}^N_{t_j+\frac{i}{2^k}} - \mathcal{B}^N_{\frac{i}{2^k}} \in A_j\right\}\big)$ does not depend on $i \geq 0$ and equals $\mathbb{P}\big(\bigcap_{j=1}^n \left\{\mathcal{B}^N_{t_j} \in A_j\right\}\big)$. To see this last assertion, we argue as in the proof of Theorem B.15. We write, for any $s \geq 0$:

$$\bigcap_{j=1}^n \{\mathcal{B}^N_{t_j+s} - \mathcal{B}^N_s \in A_j\}$$
$$= \Big\{\big(\mathcal{B}^N_{t_1+s} - \mathcal{B}^N_s, \mathcal{B}^N_{t_2+s} - \mathcal{B}^N_{t_1+s}, \ldots, \mathcal{B}^N_{t_n+s} - \mathcal{B}^N_{t_{n-1}+s}\big) \in M_n^{-1}\big(\prod_{j=1}^n A_j\big)\Big\},$$

where M_n is the following invertible matrix:

$$M_i = \begin{bmatrix} Id_N & & & \\ Id_N & Id_N & & \\ & & & \\ Id_N & Id_N & \ddots & \\ Id_N & Id_N & & Id_N \end{bmatrix} \in \mathbb{R}^{Nn\times Nn}.$$

Consequently:

$$\mathbb{P}\Big(\bigcap_{j=1}^n \{B^N_{t_j+s} - \mathcal{B}^N_s \in A_j\}\Big)$$
$$= \int_{M_n^{-1}\left(\prod_{j=1}^n A_j\right)} \prod_{i=0}^{n-1} \frac{1}{(2\pi(t_{i+1}-t_i))^{N/2}} e^{-\frac{|x^{(i)}|^2}{2(t_{i+1}-t_i)}} \,\mathrm{d}(x^{(1)}\ldots x^{(n)})$$

is constant in s and hence equal to $\mathbb{P}\big(\bigcap_{j=1}^n \left\{\mathcal{B}^N_{t_j} \in A_j\right\}\big)$, as claimed.

3. We now apply (B.8) to $C = \Omega$ to obtain that: $\mathbb{P}\big(\bigcap_{j=1}^n \left\{\mathcal{B}^N_{t_j+\tau_k} - \mathcal{B}^N_{\tau_k} \in A_j\right\}\big) = \mathbb{P}\big(\bigcap_{j=1}^n \left\{\mathcal{B}^N_{t_j} \in A_j\right\}\big)$. In conclusion:

$$\mathbb{P}\Big(\bigcap_{j=1}^n \{\mathcal{B}^N_{t_j+\tau_k} - \mathcal{B}^N_{\tau_k} \in A_j\} \cap C\Big)$$
$$= \mathbb{P}\Big(\bigcap_{j=1}^n \{\mathcal{B}^N_{t_j+\tau_k} - \mathcal{B}^N_{\tau_k} \in A_j\}\Big) \cdot \mathbb{P}(C), \tag{B.9}$$

establishing the desired independence for any stopping time τ_k. For the case of τ, we proceed by approximation, noting that $\mathcal{F}_\tau \subset \bigcap_{k\geq 0} \mathcal{F}_{\tau_k}$. Fix $C \in \mathcal{F}_\tau$; then by (B.9) we observe that:

$$\int_C \phi\big((\mathcal{B}^N_{t_j+\tau_k} - \mathcal{B}^N_{\tau_k})_{j=1}^n\big) \,\mathrm{d}\mathbb{P} = \int_\Omega \phi\big((\mathcal{B}^N_{t_j+\tau_k} - \mathcal{B}^N_{\tau_k})_{j=1}^n\big) \,\mathrm{d}\mathbb{P} \cdot \mathbb{P}(C),$$

for all $k \geq 0$ and all $\phi \in C_c(\mathbb{R}^{Nn})$. Since $\mathcal{B}^N_{t_j+\tau_k} - \mathcal{B}^N_{\tau_k}$ converges as $k \to \infty$ to $\mathcal{B}^N_{t_j+\tau} - \mathcal{B}^N_{\tau}$ pointwise $\mathbb{P}$-a.s., it follows that:

$$\int_C \phi\big((\mathcal{B}^N_{t_j+\tau} - \mathcal{B}^N_{\tau})^n_{j=1}\big)\, \mathrm{d}\mathbb{P} = \int_\Omega \phi\big((\mathcal{B}^N_{t_j+\tau} - \mathcal{B}^N_{\tau})^n_{j=1}\big)\, \mathrm{d}\mathbb{P} \cdot \mathbb{P}(C),$$

Exercise B.25 Deduce the following "reverse strong Markov property". Let $\{\mathcal{B}^N_t\}_{t\geq 0}$, $\{\bar{\mathcal{B}}^N_t\}_{t\geq 0}$ be two standard Brownian motions on some probability space $(\Omega, \mathcal{F}, \mathbb{P})$. Let τ be a stopping time as in Definition B.19 (ii). Define:

$$\bar{\bar{\mathcal{B}}}_t = \begin{cases} \mathcal{B}^N_t & \text{for } t \leq \tau \\ \bar{\mathcal{B}}^N_t + \mathcal{B}^N_\tau - \bar{\mathcal{B}}^N_\tau & \text{for } t > \tau. \end{cases}$$

Then $\{\bar{\bar{\mathcal{B}}}_t\}_{t\geq 0}$ is a N-dimensional Brownian motion on $(\Omega, \mathcal{F}, \mathbb{P})$.

B.4 Brownian Motion and Harmonic Extensions

Let $\mathcal{D} \subset \mathbb{R}^N$ be open, bounded and connected. Let $F : \partial\mathcal{D} \to \mathbb{R}$ be a continuous function, which without loss of generality we view as the restriction to $\partial\mathcal{D}$ of some $F \in C_c(\mathbb{R}^N)$. Given $x \in \mathcal{D}$, we consider the stopping time:

$$\tau_x = \min\{t \geq 0;\ \mathcal{B}^N_t \in \partial\mathcal{D} - x\}$$

as in Lemma B.9 and define:

$$u(x) = \int_\Omega F(x + \mathcal{B}^N_{\tau_x})\, \mathrm{d}\mathbb{P}. \tag{B.10}$$

The main observation is that u is harmonic in $\mathcal{D}$. Classically (see Mörters and Peres 2010; Doob 1984; Parthasarathy 1967; Durrett 2010; Kallenberg 2002), it is proved by taking $\bar{B}_r(x) \subset \mathcal{D}$ and writing:

$$\begin{aligned} u(x) &= \mathbb{E}\Big[F \circ (x + \mathcal{B}^N_{\tau_x})\Big] = \mathbb{E}\Big[\mathbb{E}\big(F \circ (x + \mathcal{B}^N_{\tau_x}) \mid \mathcal{F}_{\tau^r_x}\big)\Big] \\ &= \mathbb{E}\Big[u \circ (x + \mathcal{B}^N_{\tau^r_x})\Big] = \fint_{\partial B_r(x)} u(y)\, \mathrm{d}\sigma^{N-1}(y), \end{aligned}$$

where the first equality is just the definition in (B.10), the second follows by the tower property of conditional expectation with respect to the stopping time:

$$\tau^r_x = \min\{t \geq 0;\ \mathcal{B}^N_t \in \partial B_r(x)\}, \tag{B.11}$$

the third one results from the strong Markov property (Theorem B.24), and the fourth from the rotational invariance of the standard Brownian motion. Thus, u satisfies the spherical mean value property so it is harmonic. We now give details of proofs of the above statements.

Exercise B.26 Show that u in (B.10) is Borel, using the outline below.

(i) Consider a new probability space $(\tilde{\Omega}, \tilde{\mathcal{F}}, \tilde{\mathbb{P}})$ obtained by taking the product of $(\Omega, \mathcal{F}, \mathbb{P})$ and the set $\mathcal{D}$ equipped with its Borel σ-algebra and the normalized Lebesgue measure $\frac{1}{|\mathcal{D}|}\,\mathrm{d}x$. Check that the process $\{\tilde{\mathcal{B}}_t = x + \mathcal{B}_t^N(\omega)\}_{t\geq 0}$ is a N-dimensional Brownian motion on $(\tilde{\Omega}, \tilde{\mathcal{F}}, \tilde{\mathbb{P}})$.

(ii) Check that the associated filtration for $\{\tilde{\mathcal{B}}_t\}_{t\geq 0}$ is $\{\tilde{\mathcal{F}}_t\}_{t\geq 0}$ where each $\tilde{\mathcal{F}}_t$ is the product of $\mathcal{F}_t$ with the σ-algebra of Borel subsets of $\mathcal{D}$. Shown that $\tilde{\tau} = \min\{t \geq 0;\ \tilde{\mathcal{B}}_t \in \partial\mathcal{D}\}$ is a stopping time. Thus, $\mathcal{F} \circ \tilde{\mathcal{B}}_{\tilde{\tau}}$ is a bounded random variable on $(\tilde{\Omega}, \tilde{\mathcal{F}})$.

(iii) Use Fubini's theorem to conclude that: $x \mapsto \int_\Omega (F \circ \tilde{\mathcal{B}}_{\tilde{\tau}})(\omega, x)\,\mathrm{d}\mathbb{P}(x)$ is Borel.

Exercise B.27

(i) Consider the measurable space $(\mathbb{E}_0, \mathcal{F}_0)$ introduced in Exercise B.17. Then:

$$E \doteq \Big\{f \in E_0;\ f(0) = 0 \ \text{ and } \ |f(t)| > \operatorname{diam} \mathcal{D} \ \text{ for some } t > 0\Big\} \in \mathcal{F}_0.$$

(ii) Let $\psi : \mathcal{D} \times E \to \mathbb{R}^N$ be given by:

$$\psi(x, f) = x + f\big(\min\{t \geq 0;\ x + f(t) \in \partial\mathcal{D}\}\big).$$

Prove that ψ is measurable with respect to the product σ-algebra on $\mathcal{D} \times E$ of: the σ-algebra of Borel subsets of $\mathcal{D}$ and the restriction of $\mathcal{F}_0$ to E.

Theorem B.28 *For u as in (B.10) and any $\bar{B}_r(x) \subset \mathcal{D}$ there holds:*

$$u(x) = \int_\Omega u(x + \mathcal{B}^N_{\tau_x^r})\,\mathrm{d}\mathbb{P}.$$

Proof Given $\bar{B}_r(x) \subset \mathcal{D}$, let $\mathcal{F}_{\tau_x^r}$ be the sub-σ-algebra of $\mathcal{F}$ generated by the stopping time τ_x^r in (B.11). Define the $\mathcal{D}$-valued random variable on $(\Omega, \mathcal{F})$ by: $X_1 = x + \mathcal{B}^N_{\tau_x^r}$ and note that X_1 is $\mathcal{F}_{\tau_x^r}$-measurable. Further, define the E-valued random variable on $(\Omega, \mathcal{F})$ in:

$$X_2(\omega) = \Big([0, \infty) \ni t \mapsto \mathcal{B}^N_{\tau_x^r + t} - \mathcal{B}^N_{\tau_x^r} \in \mathbb{R}^N\Big).$$

As in Exercise B.27, it follows that X_2 is indeed measurable with respect to the Borel σ-algebra in E, whereas by Theorem B.24 we see that X_1 and X_2 are independent.

To show the latter, it is enough to observe that each preimage basis set:

$$X_2^{-1}\big(A_{g,T,\epsilon}\big) = \bigcap_{q\in[0,T]\cap\mathbb{Q}} \Big\{\mathcal{B}^N_{\tau^r_x+q} - \mathcal{B}^N_{\tau^r_x} \in \bar{B}_\epsilon(g(q))\Big\}$$

belongs to the σ-algebra generated by the Brownian motion $\{\mathcal{B}^N_{\tau^r_x+t} - \mathcal{B}^N_{\tau^r_x}\}_{t\geq 0}$ which is independent of $\mathcal{F}_{\tau^r_x}$.

Let now μ_1, μ_2 be the push-forwards of $\mathbb{P}$ via X_i for $i = 1, 2$, respectively:

$$\mu_1(A) = \mathbb{P}(X_1 \in A) \quad \text{for all Borel } A \subset \mathcal{D}$$
$$\mu_2(C) = \mathbb{P}(X_2 \in C) \quad \text{for all Borel } C \subset E.$$

Recall that with ψ given in Exercise B.27, we have:

$$x + \mathcal{B}^N_{\tau_x} = x + \mathcal{B}^N_{\tau^r_x} + \mathcal{B}^N_{\tau_x} - \mathcal{B}^N_{\tau^r_x} = \psi\circ(X_1, X_2).$$

Thus there holds:

$$\begin{aligned} u(x) &= \int_\Omega F(x+\mathcal{B}^N_{\tau_x})\,\mathrm{d}\mathbb{P} = \int_\Omega F\circ\psi\circ(X_1,X_2)\,\mathrm{d}\mathbb{P} \\ &= \int_{\mathcal{D}\times E} F\circ\psi\,\mathrm{d}(\mu_1\times\mu_2) = \int_{\mathcal{D}}\int_E \big(F\circ\psi\big)(z,f)\,\mathrm{d}\mu_2(f)\,\mathrm{d}\mu_1(z) \\ &= \int_{\mathcal{D}}\int_\Omega \big(F\circ\psi\big)(z, X_2(\omega))\,\mathrm{d}\mathbb{P}(\omega)\,\mathrm{d}\mu_1(z), \end{aligned}$$

where we used the independence of X_1 and X_2 in view of Exercise (A.20) and Fubini's theorem. Finally:

$$u(x) = \int_{\mathcal{D}} u(z)\,\mathrm{d}\mu_1(z) = \int_\Omega u\circ X_1\,\mathrm{d}\mathbb{P} = \int_\Omega u\big(x+\mathcal{B}^N_{\tau^r_x}\big)\,\mathrm{d}\mathbb{P},$$

which completes the proof. □

Corollary B.29 *Let u be as in (B.10). Then, for any $\bar{B}_r(x)\subset\mathcal{D}$ there holds:*

$$u(x) = \fint_{\partial B_r(x)} u(y)\,d\sigma^{N-1}(y).$$

Consequently, u is a harmonic function in $\mathcal{D}$.

Proof Let τ^r_x be the stopping time in (B.11) and define μ to be the push-forward of $\mathbb{P}$ on $\partial B_r(0)$ via the measurable mapping $\mathcal{B}^N_{\tau^r_x}$, so that:

$$\mu(A) = \mathbb{P}\big(\mathcal{B}^N_{\tau^r_x}\in A\big) \quad \text{for all Borel } A\subset\partial B_r(0).$$

We now argue that μ is rotationally invariant. Recall the measurable space $(E, \mathcal{F}_{0|E})$ defined in Exercise B.27, where the function $\psi : E \to \mathbb{R}^N$ given by:

$$\psi(f) = f\big(\min\{t \geq 0;\ f(t) \in \partial B_r(0)\}\big)$$

is measurable. For any Borel $A \subset \partial B_r(0)$, the composition $\mathbb{1}_A \circ \psi$ is thus a bounded random variable on the probability space $(E, \mathcal{F}_{0|E}, \mu_W)$ with the Wiener measure μ_W on $C([0,\infty), \mathbb{R}^N)$ given in Exercise B.17. Consequently:

$$\begin{aligned}\int_E \mathbb{1}_A \circ \psi \, \mathrm{d}\mu_W &= \int_\Omega (\mathbb{1}_A \circ \psi)(t \mapsto \mathcal{B}_t^N(\omega)) \, \mathrm{d}\mathbb{P}(\omega) \\ &= \int_\Omega \mathbb{1}_A\big(\mathcal{B}_{\tau_x^r}^N(\omega)\big) \, \mathrm{d}\mathbb{P}(\omega) = \mu(A).\end{aligned}$$

On the other hand, for any rotation $R \in SO(N)$, the process $\{R^T \circ \mathcal{B}_t^N\}_{t \geq 0}$ is also a standard Brownian motion, and hence it generates the same Wiener measure:

$$\begin{aligned}\int_E \mathbb{1}_A \circ \psi \, \mathrm{d}\mu_W &= \int_\Omega (\mathbb{1}_A \circ \psi)(t \mapsto R^T \mathcal{B}_t^N(\omega)) \, \mathrm{d}\mathbb{P}(\omega) \\ &= \int_\Omega \mathbb{1}_A \circ R^T \mathcal{B}_{\tau_x^r}^N \, \mathrm{d}\mathbb{P} = \mu(RA).\end{aligned}$$

It follows that $\mu(A) = \mu(RA)$, so μ is indeed a rotationally invariant probability measure on $\partial B_r(0)$ that must coincide with the normalized spherical measure σ^{N-1} by Exercise 2.11. Finally, Theorem B.28 yields:

$$u(x) = \int_{\partial B_r(0)} u(x+y) \, \mathrm{d}\mu(y) = \fint_{\partial B_r(0)} u(x+y) \, \mathrm{d}\sigma^{N-1}(y) = \fint_{\partial B_r(x)} u \mathrm{d}\sigma^{N-1}.$$

Recalling that u is Borel as derived in Exercise B.26, harmonicity of u results from Remark C.20. □

Appendix C
Background in PDEs

In this chapter we recall definitions and both background and auxiliary material on the chosen topics in PDEs: Lebesgue and Sobolev spaces, semicontinuous functions, harmonic and p-harmonic functions, regularity theory and relation of various notions of solutions to the p-Laplacian: weak, p-harmonic, viscosity.

In this chapter we recall definitions and preliminary facts on the chosen topics in PDEs: Sobolev spaces, regularity theory and various notions of solutions to the p-Laplacian: weak, p-harmonic, viscosity. The presented material may be found in graduate textbooks on Analysis and PDEs such as: Brezis (2011); Evans (2010); Adams and Fournier (2003), and further in classical monographs about the nonlinear potential theory: Lindqvist (2019); Heinonen et al. (2006).

C.1 Lebesgue L^p Spaces and Sobolev $W^{1,p}$ Spaces

Let $U \subset \mathbb{R}^N$ be an open (possibly unbounded) set. In this section we consider the measurable space $(U, \mathcal{L}_N(U), |\cdot|)$ where $\mathcal{L}_N(U)$ is the σ-algebra of Lebesgue-measurable subsets of U, and $|\cdot|$ is the N-dimensional Lebesgue measure. As in Definition A.5, the space $L^1(U) \doteq L^1(U, \mathcal{L}_N(U), |\cdot|)$ consists of (Lebesgue) measurable functions $f : U \to \bar{\mathbb{R}}$ such that:

$$\|f\|_{L^1(U)} \doteq \int_U |f| \, \mathrm{d}x < +\infty,$$

where the (Lebesgue) integral is defined as in Appendix A.2. We identify with a given f all functions g coinciding with f in $U \setminus A$, for some zero measure set $A \subset U$; we say then that g is a *representative* of f. We continue with the convention that when a certain property is satisfied outside of a set of zero measure, we say that it holds *almost everywhere* and write: for a.e. $x \in U$.

M. Lewicka, *A Course on Tug-of-War Games with Random Noise*, Universitext, https://doi.org/10.1007/978-3-030-46209-3

Clearly, the basic results in previous sections regarding the monotone convergence and the Lebesgue dominated convergence (Theorem A.7) as well as the iterated integrals (formula (A.3) and the Fubini–Tonelli theorem) still hold.

Definition C.1

(i) Let $p \in [1, \infty)$. We define the space:

$$L^p(U) \doteq \Big\{f : U \to \bar{\mathbb{R}} \text{ measurable; } \ |f|^p \in L^1(U)\Big\}$$
$$= \Big\{f : U \to \bar{\mathbb{R}} \text{ measurable; } \ \|f\|_{L^p(U)} \doteq \Big(\int_U |f|^p \,\mathrm{d}x\Big)^{1/p} < +\infty\Big\}.$$

(ii) For $p = +\infty$ we define the space:

$$L^\infty(U) \doteq \Big\{f : U \to \bar{\mathbb{R}} \text{ measurable; } \ \|f\|_{L^\infty(U)} \doteq \operatorname*{ess\,sup}_{x\in U} f(x) < +\infty\Big\},$$

where for a given set $A \subset U$ one has:

$$\operatorname*{ess\,sup}_{x\in A} f(x) \doteq \inf\Big\{\sup_{x\in A\setminus B} |f(x)|; \ \ B \subset A \text{ and } |B| = 0\Big\}.$$

Theorem C.2 *For every $p \in [1, \infty]$, the Lebesgue space $L^p(U)$ is a Banach space with norm $\|\cdot\|_{L^p(U)}$. When $p \in (1, \infty)$ then $L^p(U)$ is reflexive. When $p \in [1, \infty)$ then $L^p(U)$ is separable.*

We write: $f \in L^p_{loc}(U)$, if $f \in L^p(V)$ for every open bounded set V such that $\bar{V} \subset U$. When $f \in L^1_{loc}(U)$, we say that f is *locally integrable* (clearly, $f \in L^p_{loc}(U)$ for any p implies the local integrability of f). The *Lebesgue differentiation theorem* states that a locally integrable function coincides almost everywhere with its (existing a.e.) limit of averages on shrinking neighbourhoods:

Theorem C.3 *Let $f \in L^1_{loc}(U)$. Then:*

$$f(x) = \lim_{r\to 0} \fint_{B_r(x)} f(y)\,\mathrm{d}y \quad \textit{for a.e. } x \in U.$$

Recall that the linear space $C_c^\infty(U)$ consists of smooth *test functions* $\phi : U \to \mathbb{R}$ whose support: *supp* $\phi \doteq \overline{\{\phi = 0\}}$ is compact and contained in U. The following density result is valid for all $p \in [1, \infty)$:

$$L^p(U) = \text{closure}_{L^p(U)}\big(C_c^\infty(U)\big) \tag{C.1}$$

and we have the very useful *fundamental theorem of calculus of variations*:

Theorem C.4 *If* $f \in L^1_{loc}(U)$ *and*

$$\int_U f\phi \, \mathrm{d}x = 0 \quad \textit{for all } \phi \in C_c^\infty(U),$$

then $f = 0$ *a.e. in* U.

For every $p \in [1, \infty]$ define the *conjugate exponent* p' by requesting that: $\frac{1}{p} + \frac{1}{p'} = 1$. Then *Hölder's inequality* asserts:

$$\int_U fg \, \mathrm{d}x \leq \|f\|_{L^p(U)} \cdot \|g\|_{L^{p'}(U)} \qquad \text{for all } \; f \in L^p(U), \; g \in L^{p'}(U). \tag{C.2}$$

The main feature of conjugate exponents is that for $p \in [1, \infty)$ the space $L^{p'}(U)$ can be identified with the dual space $[L^p(U)]^*$ to $L^p(U)$, in the sense that all linear continuous functionals on $L^p(U)$ have the form $f \mapsto \int_U fg \, \mathrm{d}x$ for some $g \in L^{p'}(U)$. In this context, we say that a sequence $\{f_n \in L^p(U)\}_{n=1}^\infty$ *converges weakly* to the limit $f \in L^p(U)$ (we write: $f_n \rightharpoonup f$ in $L^p(U)$) provided that:

$$\lim_{n\to\infty} \int_U f_n g \, \mathrm{d}x = \int_U fg \, \mathrm{d}x \qquad \text{for all } g \in L^{p'}(U).$$

Clearly, strong convergence implies weak convergence by (C.2), but the converse is, in general, not true. The following compactness property is related to the reflexivity statement in Theorem C.2:

Theorem C.5 *Assume that* $p \in (1, \infty)$. *Then every bounded sequence* $\{f_n \in L^p(U)\}_{n=1}^\infty$ *has a subsequence that converges weakly to some limit* $f \in L^p(U)$.

One of the central notions in analysis and PDEs is the notion of Sobolev spaces. Intuitively, Sobolev space consists of functions that can be assigned a "weak gradient" through integration by parts (against gradients of test functions), and that have a prescribed summability exponent.

Definition C.6 Let $p \in [1, \infty]$. Define the space:

$$W^{1,p}(U) \doteq \Big\{ f \in L^p(U); \; \text{there exists } g \in L^p(U, \mathbb{R}^N) \text{ such that:} \\ \int_U f \nabla\phi \, \mathrm{d}x = -\int_U \phi g \, \mathrm{d}x \; \text{ for all } \phi \in C_c^\infty(U) \Big\}.$$

The function g is called the *distributional (weak) gradient* of f and we write: $\nabla f \doteq g$.

By Theorem C.4 it is easy to observe that the distributional gradient is defined uniquely (in the a.e. sense). When $f \in C^1(U) \cap L^p(U)$ and $\nabla f \in L^p(U, \mathbb{R}^N)$, then $f \in W^{1,p}(U)$ and the distributional gradient coincides with the weak one. Conversely, when $f \in C(U) \cap W^{1,p}(U)$ (i.e. f has a continuous representative) and the distributional gradient $\nabla f \in C(U)$ then $f \in C^1(U)$. As in Theorem C.2 we have:

Theorem C.7 *For every $p \in [1, \infty]$, the Sobolev space $W^{1,p}(U)$ is a Banach space with norm $\|f\|_{W^{1,p}(U)} \doteq \|f\|_{L^p(U)} + \|\nabla f\|_{L^p(U,\mathbb{R}^N)}$. When $p \in (1, \infty)$ then $W^{1,p}(U)$ is reflexive. When $p \in [1, \infty)$ then $W^{1,p}(U)$ is separable.*

We write: $f \in W^{1,p}_{loc}(U)$, if $f \in W^{1,p}(V)$ for every open set V such that $\bar{V} \subset U$. The following easy properties of Sobolev functions are collected as an exercise:

Exercise C.8

(i) Let $\{f_n \in W^{1,p}(U)\}_{n=1}^\infty$ be a sequence of Sobolev functions converging in $L^p(U)$ to some f and such that $\{\nabla f_n\}_{n=1}^\infty$ converges in $L^p(U, \mathbb{R}^N)$ to some g. Then $f \in W^{1,p}(U)$ and $g = \nabla f$.
(ii) If $\rho \in C^1(\bar{U})$ and $f \in W^{1,p}(U)$, then $\rho f \in W^{1,p}(U)$ and the distributional gradient equals: $\nabla(\rho f) = \rho \nabla f + f \nabla \rho$.
(iii) If $\rho \in L^1(\mathbb{R}^N)$ and $f \in W^{1,p}_{loc}(\mathbb{R}^N)$, then $\rho * f \in W^{1,p}_{loc}(\mathbb{R}^N)$ and $\nabla(\rho * f) = \rho * \nabla f$. Recall that the *convolution* of two functions ρ and f is given by the formula: $(\rho * f)(x) = \int_{\mathbb{R}^N} \rho(y) f(x - y)\, dy$.
(iv) Let $\rho \in C^1(\mathbb{R})$ be such that $\rho(0) = 0$ and that ρ' is bounded. Then for every $f \in W^{1,p}(U)$ we have: $\rho \circ f \in W^{1,p}(U)$ and $\nabla(\rho \circ f) = (\rho' \circ f)\nabla f$.
(v) Given a diffeomorphism $h : U_1 \to U_2$ of class C^1 between two open sets $U_1, U_2 \subset \mathbb{R}^N$ such that ∇h and ∇h^{-1} are bounded, and given $f \in W^{1,p}(U_2)$, we have $f \circ h \in W^{1,p}(U_1)$ and $\nabla(f \circ h) = (\nabla f \circ h)\nabla h$.

We also have two observations regarding *truncations* of Sobolev functions:

Exercise C.9 Denote the *positive* and *negative parts* of $f : U \to \bar{\mathbb{R}}$ by:

$$f^+(x) \doteq \max\{f(x), 0\}, \qquad f^-(x) \doteq (-f)^+(x).$$

(i) If $f \in W^{1,p}(U)$, then $f^+ \in W^{1,p}(U)$ and:

$$\nabla f^+ = \begin{cases} \nabla f & \text{a.e. in } \{f > 0\} \\ 0 & \text{a.e. in } \{f \leq 0\}. \end{cases}$$

(ii) If $\{f_n \in W^{1,p}(U)\}_{n=1}^\infty$ converges to f in $W^{1,p}(U)$, then the truncated sequence $\{f_n^+\}_{n=1}^\infty$ converges in $W^{1,p}(U)$ to f^+.

Sobolev functions can be approximated by smooth functions, as stated in the celebrated *Meyers–Serrin theorem* (known as the "$H = W$" theorem):

Theorem C.10 *Let $p \in [1, \infty)$. Then:*

$$W^{1,p}(U) = closure_{W^{1,p}(U)}\big(C^\infty(U) \cap W^{1,p}(U)\big).$$

When ∂U has Lipschitz regularity (in fact, a less restrictive "segment condition" in Lemma C.58 (i), suffices) and $p < \infty$, then every $W^{1,p}(U)$ function can be approximated by a sequence of $C_c^\infty(\mathbb{R}^N)$ test functions. Approximation by test functions supported in U is achieved only in the following important subspace of $W^{1,p}(U)$, defined for $p \in [1, \infty)$:

$$W_0^{1,p}(U) \doteq \mathrm{closure}_{W^{1,p}(U)}\big(C_c^\infty(U)\big).$$

Clearly, $W_0^{1,p}(\mathbb{R}^N)$ is a Banach space and $W_0^{1,p}(\mathbb{R}^N) = W^{1,p}(\mathbb{R}^N)$. Intuitively, functions in $W_0^{1,p}(U)$ are the Sobolev functions with zero boundary values. In this context, the *Poincaré inequality* is valid:

Theorem C.11 *If $p \in [1, \infty)$ and U has finite width (i.e. its projection on some line in $\mathbb{R}^N$ is bounded), then there exists a constant C depending only on p and U, with:*

$$\|f\|_{L^p(U)} \leq C\,\|\nabla f\|_{L^p(U,\mathbb{R}^N)} \qquad \text{for all } \ f \in W_0^{1,p}(U).$$

We define the *weak convergence* in $W^{1,p}(U)$ for $p \in [1, \infty)$ as the weak convergence of the given sequence and the corresponding sequence of distributional gradients. More precisely, $\{f_n \in W^{1,p}(U)\}_{n=1}^\infty$ converges weakly in $W^{1,p}(U)$ to f provided that $f_n \rightharpoonup f$ in $L^p(U)$ and $\nabla f_n \rightharpoonup \nabla f$ in $L^p(U, \mathbb{R}^N)$. Similarly to Theorem C.5, we now have:

Theorem C.12 *Assume that $p \in (1, \infty)$. Then every bounded sequence $\{f_n \in W^{1,p}(U)\}_{n=1}^\infty$ has a subsequence that converges weakly to some $f \in W^{1,p}(U)$.*

The next result gathers the *embedding theorems*. Namely, Sobolev $W^{1,p}$ functions are actually of higher regularity than merely L^p and the increase in regularity is more pronounced for lower dimensions N.

Theorem C.13

(i) (Sobolev embedding theorem). *Let $p \in [1, N)$ and define the Sobolev conjugate exponent p^* by: $\frac{1}{p^*} = \frac{1}{p} - \frac{1}{N}$. Then the following embedding is continuous:*

$$i : W^{1,p}(\mathbb{R}^N) \to L^{p^*}(\mathbb{R}^N).$$

(ii) (Sobolev embedding theorem, critical case). *For every* $q \in [N, \infty)$ *we have the continuous embedding:*

$$i : W^{1,N}(\mathbb{R}^N) \to L^q(\mathbb{R}^N).$$

(iii) (Morrey embedding theorem). *If* $p \in (N, \infty)$, *then we have the continuous embedding:*

$$i : W^{1,p}(\mathbb{R}^N) \to L^\infty(\mathbb{R}^N).$$

Moreover, there exists a constant $C > 0$ *depending only on* p *and* N, *such that for all* $f \in W^{1,p}(\mathbb{R}^N)$ *there holds:*

$$|f(x) - f(y)| \leq C|x - y|^{1-\frac{N}{p}} \|\nabla f\|_{L^p(\mathbb{R}^N,\mathbb{R}^N)} \quad \text{for a.e. } x, y \in \mathbb{R}^N. \tag{C.3}$$

In particular, f *has a Hölder continuous representative for which the above inequality holds for all* $x, y \in \mathbb{R}^N$.

The assertions of Theorem C.13 remain true when the domain $\mathbb{R}^N$ is replaced by U satisfying the *cone condition* i.e. such that each $x \in U$ is the vertex of some cone $C_x \subset U$ which is a rigid motion of one finite cone. The Hölder continuity bound in (C.3) additionally requires Lipschitz continuity of ∂U and the constant C depends then on p and U. When U is bounded and regular then all the subcritical Sobolev embeddings are compact, in view of the *Rellich–Kondrachov theorem* below:

Theorem C.14 *Let* U *be bounded and satisfy the cone condition. Then, the following embeddings are compact:*

$$i : W^{1,p}(U) \to L^q(U) \quad \text{for all } q \in [1, p^*), \text{ when } p \in [1, N),$$

$$i : W^{1,N}(U) \to L^q(U) \quad \text{for all } q \in [1, \infty),$$

$$i : W^{1,p}(U) \to C_{bdd}(U) \quad \text{when } p \in (N, \infty),$$

where by $C_{bdd}(U)$ *we denoted the space of continuous bounded functions on* U. *In particular,* $i : W^{1,p}(U) \to L^p(U)$ *is compact for all* p *and* N. *When* ∂U *is additionally Lipschitz continuous, then for* $p \in (N, \infty)$ *the embedding:*

$$i : W^{1,p}(U) \to C^{0,\alpha}(\bar{U})$$

is compact for every Hölder exponent $\alpha \in (0, 1 - \frac{N}{p})$.

C.2 Semicontinuous Functions

In this section we recall the definition and first properties of semicontinuous functions. These are functions whose values at points close to a given domain point remain approximately below or approximately above (as opposed to "approximately close", requested for continuous functions) the indicated value.

Definition C.15 A function $f : U \to (-\infty, +\infty]$ defined on an open subset U of $\mathbb{R}^N$ is called *lower-semicontinuous* if:

$$f(x) \leq \liminf_{y \to x} f(y) \qquad \text{for all } x \in \mathcal{D}, \tag{C.4}$$

where we denote:

$$\liminf_{y \to x} f(y) \doteq \lim_{r \to 0} \inf \big\{ f(y);\ y \in B_r(x) \cap (\mathcal{D} \setminus \{x\}) \big\}.$$

If $(-f)$ is lower-semicontinuous, then f is called *upper-semicontinuous*.

Clearly, $f : U \to \mathbb{R}$ is continuous if an only if it is both lower- and upper-semicontinuous. We summarize other basic properties below:

Exercise C.16

(i) A function f is lower-semicontinuous if and only if the set $\{f > \lambda\}$ is open for every $\lambda \in \mathbb{R}$. In particular, lower-semicontinuous functions are Borel-regular.
(ii) A lower-semicontinuous function $f : U \to (-\infty, +\infty]$ attains its infimum on any compact set $K \subset U$ (but it does not have to attain its supremum). In particular, every lower-semicontinuous function is locally bounded from below.
(iii) Supremum of an arbitrary family of lower-semicontinuous functions is lower-semicontinuous. Minimum of finitely many lower-semicontinuous functions is lower-semicontinuous.
(iv) For every lower-semicontinuous f on U there exists a nondecreasing sequence $\{f_n \in C^\infty(U)\}_{n=1}^\infty$ converging pointwise to f as $n \to \infty$.
(v) If $f : U \to (-\infty, +\infty]$ is locally bounded from below, then the function $g(x) = \liminf_{y \to x} f(y)$ is lower-semicontinuous in U.

For functions that are defined only up to zero measure sets, Definition C.15 and the construction in Exercise C.16 (v) has the following natural counterpart:

Exercise C.17

(i) Assume that a measurable function $f : U \to \bar{\mathbb{R}}$ is *locally essentially bounded from below*, i.e. for every compact set $K \subset U$ there exists a measure zero set A such that $\inf_{K\setminus A} f > -\infty$. Prove that the following function is lower-semicontinuous in U:

$$g(x) = \operatorname{ess} \liminf_{y\to x} f(y),$$

where $\operatorname{ess}\liminf_{y\to x} f(y) \doteq \lim_{r\to 0} \operatorname{ess}\inf_{B_r(x)} f$.

(ii) Prove that $f \in L^1_{loc}(U)$ has a continuous representative if and only if:

$$\operatorname{ess} \liminf_{y\to x} f(y) = \operatorname{ess} \limsup_{y\to x} f(y) \quad \text{for all } x \in U. \tag{C.5}$$

Finally, we recall the *Ascoli–Arzelá theorem*, valid for sequences of continuous functions on compacts:

Theorem C.18 *Let $K \subset \mathbb{R}^N$ be a compact set and assume that the sequence $\{f_n \in C(K)\}_{n=1}^\infty$ satisfies:*

(i) (Equiboundedness). *There exists $C \geq 0$ such that $\|f_n\|_{L^\infty(K)} \leq C$ for all $n \geq 1$.*
(ii) (Equicontinuity). *For every $\epsilon > 0$ there exists $\delta > 0$ such that if $x, y \in K$ satisfy $|x - y| < \delta$, then $|f_n(x) - f_n(y)| < \epsilon$ for all $n \geq 1$.*

Then $\{f_n\}_{n=1}^\infty$ has a subsequence, converging uniformly in K to some $f \in C(K)$.

C.3 Harmonic Functions

In this section we review the basic properties of one of the most important equations of mathematical physics, that is the *Laplace equation*. The same material and a further discussion can be found, among others, in the graduate textbooks by Evans (2010) or Gilbarg and Trudinger (2001).

Let $\mathcal{D} \subset \mathbb{R}^N$ be an open, bounded, connected set. We say that a function $u \in C^2(\mathcal{D})$ is *harmonic* in $\mathcal{D}$, provided that:

$$\Delta u = \sum_{i=1}^N \frac{\partial^2 u}{(\partial x_i)^2} = 0 \quad \text{in } \mathcal{D}. \tag{C.6}$$

Other notions of harmonicity and the derivation of (C.6) as the Euler–Lagrange equation of the energy $\mathcal{I}_2(u) = \int_\mathcal{D} |\nabla u|^2$ are discussed in the general case of the p-Laplacian, $p \in (1, \infty)$, in Sect. C.4.

Theorem C.19 *Let $u \in C(\mathcal{D})$. Then the following conditions are equivalent:*

(i) u satisfies the mean value property on spheres*:*

$$u(x) = \fint_{\partial B_r(x)} u(y)\, d\sigma^{N-1}(y) \qquad \text{for all } \bar{B}_r(x) \subset \mathcal{D}. \tag{C.7}$$

Here, σ^{N-1} is the spherical measure as in Example A.9.

(ii) u satisfies the mean value property on balls*:*

$$u(x) = \fint_{B_r(x)} u(y)\, dy \qquad \text{for all } \bar{B}_r(x) \subset \mathcal{D}. \tag{C.8}$$

(iii) $u \in C^2(\mathcal{D})$ and u is harmonic in $\mathcal{D}$.

Moreover, if the above conditions are satisfied, then $u \in C^\infty(\mathcal{D})$.

Proof

1. For a fixed $x \in \mathcal{D}$ and $r \in (0, \mathrm{dist}(x, \partial\mathcal{D}))$, denote:

$$\phi^x(r) \doteq \fint_{\partial B_r(x)} u(y)\, \mathrm{d}\sigma^{N-1}(y) = \fint_{\partial B_1(x)} u(x + r(y - x))\, \mathrm{d}\sigma^{N-1}(y),$$

so that integrating in "polar coordinates" as in (A.2), we obtain:

$$\int_{B_r(x)} u(y)\, \mathrm{d}y = \int_0^r |\partial B_s(x)| \cdot \phi_x(s)\, \mathrm{d}s \qquad \text{for all } \bar{B}_r(x) \subset \mathcal{D}.$$

Now, if u satisfies (C.7), then $\phi^x(r) = u(x)$ for all $r \in (0, \mathrm{dist}(x, \partial\mathcal{D}))$ and so $\int_{B_r(x)} u(y)\, \mathrm{d}y = u(x) \int_0^r |\partial B_s(x)|\, \mathrm{d}s = u(x) \cdot |B_r(x)|$, which gives (C.8). On the other hand, assuming (C.8) and noting that $\frac{\mathrm{d}}{\mathrm{d}r}|B_r(x)| = Nr^{N-1}|B_1(0)| = r^{N-1}|\partial B_1(0)| = |\partial B_r(x)|$, implies:

$$\begin{aligned} 0 &= \frac{\mathrm{d}}{\mathrm{d}r} \fint_{B_r(x)} u(y)\, \mathrm{d}y \\ &= \frac{1}{|B_r(x)|^2}\Big(|\partial B_r(x)| \cdot \phi^x(r) \cdot |B_r(x)| - \big(\frac{\mathrm{d}}{\mathrm{d}r}|B_r(x)|\big) \int_{B_r(x)} u(y)\, \mathrm{d}y\Big) \\ &= \frac{|\partial B_r(x)|}{|B_r(x)|}\Big(\phi^x(r) - \fint_{B_r(x)} u(y)\, \mathrm{d}y\Big) \qquad \text{for all } r \in (0, \mathrm{dist}(x, \partial\mathcal{D})). \end{aligned}$$

It follows then that $\phi_r(x) = \fint_{B_r(x)} u(y)\, \mathrm{d}y = u(x)$, which is (C.8). We have thus proved the equivalence of conditions (i) and (ii).

2. To show that (iii) implies (i), use the divergence theorem:

$$\begin{aligned}\frac{\mathrm{d}}{\mathrm{d}r}\phi^x(r) &= \fint_{\partial B_1(x)} \big\langle \nabla u(x + r(y-x)), y - x\big\rangle \,\mathrm{d}\sigma^{N-1}(y) \\ &= \fint_{\partial B_r(x)} \big\langle \nabla u(y), \frac{y-x}{r}\big\rangle \,\mathrm{d}\sigma^{N-1}(y) = \fint_{\partial B_r(x)} \frac{\partial u}{\partial n}\,\mathrm{d}\sigma^{N-1}(y) \\ &= \frac{1}{|B_r(x)|}\int_{B_r(x)} \Delta u(y)\,\mathrm{d}y = 0,\end{aligned} \tag{C.9}$$

where we denote the outward unit normal vector by n. Hence, the function $r \mapsto \phi^x(r)$ is constant, coinciding with its limit value $\lim_{r\to 0}\fint_{\partial B_r(x)} u(y)\,\mathrm{d}\sigma^{N-1}(y)$ that equals $u(x)$ in view of continuity of u. We thus get (C.7).

3. To prove that (i) implies $u \in C^\infty(\mathcal{D})$ and (iii), let $\{\phi_\epsilon\}_{\epsilon>0}$ be a family of smooth, radially symmetric mollifiers. Namely: $\phi_\epsilon(x) \doteq \frac{1}{\epsilon^N}\phi(\frac{x}{\epsilon})$ for some function $\phi \in C_c^\infty(B_1(0))$ that is nonnegative, radial and satisfies $\int_{B_1(0)}\phi(x)\,\mathrm{d}x = 1$. For every $B_\epsilon(x) \subset \mathcal{D}$ we then have:

$$\begin{aligned}(u * \phi_\epsilon)(x) &= \int_{B_\epsilon(0)} \phi_\epsilon(y)u(x-y)\,\mathrm{d}y = \int_0^\epsilon \int_{\partial B_r(0)} \phi_\epsilon(y)u(x-y)\,\mathrm{d}\sigma^{N-1}(y)\,\mathrm{d}r \\ &= \int_0^\epsilon (\phi_\epsilon)_{|\partial B_r(0)} \cdot |\partial B_r(0)| \cdot u(x)\,\mathrm{d}r = u(x)\int_{B_\epsilon(0)} \phi_\epsilon(y)\,\mathrm{d}y = u(x).\end{aligned}$$

Consequently, $u \in C^\infty(\mathcal{D})$. By the identities in (C.9) it follows that: $0 = \frac{\mathrm{d}}{\mathrm{d}r}\phi^x(r) = \frac{1}{|\partial B_r(x)|}\int_{B_r(x)}\Delta u(y)\,\mathrm{d}y$, for all $B_r(x) \subset \mathcal{D}$. Thus $\Delta u = 0$ in $\mathcal{D}$.

□

We thus see that every harmonic function is smooth. In fact, it is also analytic i.e. represented, locally in its domain, by a convergent power series.

Another remarkable result is that for a continuous and bounded function $u : \mathcal{D} \to \mathbb{R}$, the *strong converse mean value property* holds. That is, if for every $x \in \mathcal{D}$ there exists some radius $r(x) \in (0, \mathrm{dist}(x, \partial\mathcal{D}))$ such that $u(x) = \fint_{B_{r(x)}(x)} u(y)\,\mathrm{d}y$, then u must be harmonic in $\mathcal{D}$. Under the additional assumption that $u \in C(\bar{\mathcal{D}})$, this result is due to Volterra (1909) and Kellogg (1934). In the general case, it has been proved in Hansen and Nadirashvili (1993, 1994), where it is also shown that the same result is in general false when taking spherical (instead of ball) mean values.

Remark C.20 Equivalence of (i)–(iii) in Theorem C.19 remains valid under the assumption that $u : \mathcal{D} \to \mathbb{R}$ is bounded and Borel, where in (i) it is sufficient for the mean value property on spheres to hold for every $x \in \mathcal{D}$ and almost every $r \in (0, \mathrm{dist}(x, \partial\mathcal{D}))$. Indeed, for bounded Borel u, the integral in the right hand side of (C.7) is well defined for a.e. indicated r, in view of Fubini's theorem, and the weakened condition (i) implies (ii) by the same proof. The other implications remain verbatim the same.

Exercise C.21 Let $\mathcal{D} = B_r(0) \subset \mathbb{R}^N$ and let $F \in \mathcal{C}(\partial\mathcal{D})$. Then the function:

$$u(x) = (r^2 - |x|^2)r^{N-2} \fint_{\partial B_r(0)} \frac{F(y)}{|x-y|^N}\, \mathrm{d}\sigma^{N-1}(y)$$

is harmonic in $\mathcal{D}$ and it satisfies $\lim_{x\to x_0,\, x\in\mathcal{D}} u(x) = F(x_0)$ for all $x_0 \in \partial\mathcal{D}$. The function $(x, y) \mapsto \frac{(r^2-|x|^2)r^{N-2}}{|x-y|^N}$ is called the *Poisson kernel*.

Exercise C.22 Using the outline below, work out an alternative proof of the implication (ii)⇒(iii) of Theorem C.19.

(i) Let $u \in \mathcal{C}(\mathcal{D})$ satisfy (C.8). For a given $\bar{B}_r(x) \subset \mathcal{D}$, let $v \in \mathcal{C}(\bar{B}_r(x)) \cap \mathcal{C}^\infty(B_r(x))$ be given by the formula in Exercise C.21, to satisfy: $\Delta v = 0$ in $B_r(x)$ and $v = u$ on $\partial B_r(x)$. Applying the assumed mean value property of u together with the same property of the harmonic v, deduce contradiction from: $\max_{B_r(x)}(v-u) = v(x_0) - u(x_0) > 0$ at some $x_0 \in B_r(x)$.
(ii) Conclude that $v = u$ in $\bar{B}_r(x)$ and further, that u must be harmonic in $\mathcal{D}$.

The same outline above may be followed to prove the *maximum principle* for the Laplace equation:

Theorem C.23 *If $u \in \mathcal{C}(\bar{\mathcal{D}}) \cap \mathcal{C}^\infty(\mathcal{D})$ is harmonic in $\mathcal{D}$, then $\max_{\bar{\mathcal{D}}} u = \max_{\partial\mathcal{D}} u$. Moreover, if $\max_{\bar{\mathcal{D}}} u = u(x)$ at some $x \in \mathcal{D}$, then u is constant in $\mathcal{D}$.*

Proof Define $M \doteq \max_{\bar{\mathcal{D}}} u$ and assume that $M = u(x)$ for some $x \in \mathcal{D}$. Theorem C.19 then yields: $M \le \fint_{B_r(x)} u(y)\, \mathrm{d}y \le M$ for every $r \le \mathrm{dist}(x, \partial\mathcal{D})$. Consequently, $u = M$ in $B_r(x)$ and thus the set $u^{-1}(M)$ is both open and closed in $\mathcal{D}$. Therefore: $u = M$ in $\mathcal{D}$, proving both claims of the theorem. □

Corollary C.24 *For every $F \in \mathcal{C}(\partial\mathcal{D})$, the boundary value problem:*

$$\Delta u = 0 \quad \text{in } \mathcal{D}, \qquad\qquad u = F \quad \text{on } \partial\mathcal{D},$$

has at most one solution $u \in \mathcal{C}(\bar{\mathcal{D}}) \cap \mathcal{C}^\infty(\mathcal{D})$.

Our final statement resulting from the mean value property is the *Harnack inequality* for harmonic functions:

Theorem C.25 *Let V be an open, bounded, connected set such that $V \subset \bar{V} \subset \mathcal{D}$. There exists a constant $C > 0$ depending only on V and $\mathcal{D}$, such that:*

$$\sup_V u \le C \inf_V u,$$

for all nonnegative functions u that are harmonic in $\mathcal{D}$.

Proof Let $r = \frac{1}{2}\mathrm{dist}(V, \partial\mathcal{D})$. For any $x, y \in V$ with $|x - y| \leq r$ we have:

$$u(x) = \fint_{B_{2r}(x)} u(y)\,\mathrm{d}y \geq \frac{|B_r(x)|}{|B_{2r}(x)|} \fint_{B_r(x)} u(y)\,\mathrm{d}y = \frac{1}{2^N} u(y).$$

By switching the roles of x and y it further follows that:

$$\frac{1}{2^N} u(y) \leq u(x) \leq 2^N u(y).$$

By compactness, the set $\bar{V}$ may be covered by a finite number $n \geq 1$ of open balls with radius r. Thus we have:

$$\frac{1}{2^{nN}} u(y) \leq u(x) \leq 2^{nN} u(y) \qquad \text{for all } x, y \in V,$$

proving the stated inequality with $C = 2^{nN}$. □

C.4 The *p*-Laplacian and Its Variational Formulation

In this and the following sections we present the preliminaries of the classical theory of the p-Laplace equation (we restrict out attention to the bounded domains $\mathcal{D} \subset \mathbb{R}^N$), whose probabilistic (game-theoretical) treatment is the content of this book. We refer to the monographs by Heinonen et al. (2006) and Lindqvist (2019) for a thorough discussion; here we only introduce the basic definitions, prove facts needed in the future study and state main properties.

Let $\mathcal{D} \subset \mathbb{R}^N$ be an open, bounded, connected set and let $p \in (1, \infty)$. Consider the following Dirichlet integral:

$$\mathcal{I}_p(u) = \int_{\mathcal{D}} |\nabla u(x)|^p\,\mathrm{d}x \qquad \text{for all } u \in W^{1,p}(\mathcal{D}).$$

Minimizing the energy $\mathcal{I}_p$ among all functions u subject to some given boundary data, the condition for the vanishing of the first variation of $\mathcal{I}_p$ (see Lemma C.28 below) takes the form:

$$\int_{\mathcal{D}} \Big\langle |\nabla u|^{p-2}\nabla u, \nabla\eta \Big\rangle\,\mathrm{d}x = 0 \qquad \text{for all } \eta \in C_c^\infty(\mathcal{D}).$$

Assuming sufficient regularity of u, the classical divergence theorem yields:

$$\int_{\mathcal{D}} \eta\,\mathrm{div}\left(|\nabla u|^{p-2}\nabla u\right)\mathrm{d}x = 0 \qquad \text{for all } \eta \in C_c^\infty(\mathcal{D}),$$

by the fundamental theorem of calculus of variations (Theorem C.4), becoming:

$$\Delta_p u \doteq \operatorname{div}\Big(|\nabla u|^{p-2}\nabla u\Big) = 0 \quad \text{in } \mathcal{D}. \tag{C.10}$$

Definition C.26 The second order differential operator Δ_p is called the *p-Laplacian* and the partial differential equation (C.10) is called the *p-harmonic equation*.

An example of a p-harmonic function in the punctured space is provided by the following radially symmetric construction:

Exercise C.27 For a fixed $x_0 \in \mathbb{R}^N$, prove that the smooth radial function $u : \mathbb{R}^N \setminus \{x_0\} \to \mathbb{R}$ given by:

$$u(x) = \begin{cases} |x - x_0|^{\frac{p-N}{p-1}} & \text{if } p \neq N \\ \log|x - x_0| & \text{if } p = N \end{cases}$$

satisfies: $\Delta_p u = 0$ and $\nabla u \neq 0$ in $\mathbb{R}^N \setminus \{x_0\}$.

We now observe:

Lemma C.28 *Let $w \in W^{1,p}(\mathcal{D})$ for some $p \in (1, \infty)$. Then the problem:*

$$\textit{minimize } \Big\{\mathcal{I}_p(u);\ u - w \in W_0^{1,p}(\mathcal{D})\Big\} \tag{C.11}$$

has a unique solution $u \in W^{1,p}(\mathcal{D})$. Equivalently, u solves (C.11) if and only if $u - w \in W_0^{1,p}(\mathcal{D})$ and:

$$\int_{\mathcal{D}} \Big\langle |\nabla u|^{p-2}\nabla u, \nabla \eta\Big\rangle \,\mathrm{d}x = 0 \qquad \textit{for all } \eta \in C_c^\infty(\mathcal{D}), \tag{C.12}$$

where we set $|\nabla u(x)|^{p-2}\nabla u(x) = |0|^{p-2}0 \doteq 0$, whenever $\nabla u(x) = 0$ and $p < 2$.

Proof

1. We will frequently use the estimate:

$$|b|^p \geq |a|^p + p\big\langle |a|^{p-2}a, b - a\big\rangle \qquad \text{for all } a, b \in \mathbb{R}^N, \tag{C.13}$$

following from convexity of the function $x \mapsto |x|^p$ and from $\nabla|x|^p = p|x|^{p-2}x$.
To prove existence of a solution to (C.11), define:

$$I_{min} \doteq \inf_{u - w \in W_0^{1,p}(\mathcal{D})} \mathcal{I}_p(u) \in [0, \mathcal{I}_p(w)].$$

Consider a sequence $\{u_n\}_{n=1}^{\infty}$ satisfying $u_n - w \in W_0^{1,p}(\mathcal{D})$ and $\lim_{n\to\infty} \mathcal{I}_p(u_n) = I_{min}$. It is easy to note that $\{u_n\}_{n=1}^{\infty}$ is bounded in $W^{1,p}(\mathcal{D})$, because:

$$\begin{aligned}\|u_n\|_{L^p(\mathcal{D})} + \|\nabla u_n\|_{L^p(\mathcal{D})} &\leq \|u_n - w\|_{L^p(\mathcal{D})} + \|w\|_{L^p(\mathcal{D})} + \|\nabla u_n\|_{L^p(\mathcal{D})} \\ &\leq C\|\nabla u_n - \nabla w\|_{L^p(\mathcal{D})} + \|w\|_{L^p(\mathcal{D})} + \|\nabla u_n\|_{L^p(\mathcal{D})} \\ &\leq C\big(\|w\|_{W^{1,p}(\mathcal{D})} + \|\nabla u_n\|_{L^p(\mathcal{D})}\big) \leq C\big(\|w\|_{W^{1,p}(\mathcal{D})} + \mathcal{I}_p(u_n)^{1/p}\big) \leq C,\end{aligned}$$

where we used the Poincaré inequality in Theorem C.11 to estimate $\|u_n - w\|_{L^p(\mathcal{D})}$. Consequently, by Theorem C.12 $\{u_n\}_{n=1}^{\infty}$ has a subsequence (that we do not relabel) converging weakly in $W^{1,p}(\mathcal{D})$ to some limit u that must satisfy: $u_n - w \rightharpoonup u - w \in W_0^{1,p}(\mathcal{D})$. Moreover, by (C.13) we have:

$$\mathcal{I}_p(u_n) \geq \mathcal{I}_p(u) + p\int_{\mathcal{D}} \Big\langle |\nabla u|^{p-2}\nabla u, \nabla u_n - \nabla u\Big\rangle \,\mathrm{d}x. \tag{C.14}$$

Since $|\nabla u|^{p-2}\nabla u \in L^{p'}(\mathcal{D})$ with $p' = \frac{p}{p-1}$ and $\nabla(u_n - u)$ converges weakly to 0 in $L^p(\mathcal{D})$, it follows that the second term in the right hand side of (C.14) converges to 0. Therefore we obtain:

$$\mathcal{I}_p(u) \leq \lim_{n\to\infty} \mathcal{I}_p(u_n) = I_{min}.$$

This proves that u is a minimizer in (C.11).

2. For uniqueness, we use strict convexity of the function $x \mapsto |x|^p$. If u_1, u_2 both solve (C.11), then $u = \frac{1}{2}(u_1 + u_2)$ satisfies $u - w \in W_0^{1,2}(\mathcal{D})$ and:

$$\begin{aligned}I_{min} &\leq \mathcal{I}_p(u) = \int_{\{u_1=u_2\}} |u|^p \,\mathrm{d}x + \int_{\{u_1\neq u_2\}} |u|^p \,\mathrm{d}x \\ &\leq \frac{1}{2}\Big(\int_{\{u_1=u_2\}} |u_1|^p \,\mathrm{d}x + \int_{\{u_1=u_2\}} |u_2|^p \,\mathrm{d}x\Big) + \frac{1}{2}\Big(\int_{\{u_1\neq u_2\}} |u_1|^p \,\mathrm{d}x + \int_{\{u_1\neq u_2\}} |u_2|^p \,\mathrm{d}x\Big) \\ &= \frac{1}{2}\big(\mathcal{I}_p(u_1) + \mathcal{I}_p(u_2)\big) = I_{min},\end{aligned}$$

with the strict inequality on the set $\{u_1 \neq u_2\}$. Consequently, there must be: $|u_1 \neq u_2| = 0$ and so the two functions u_1, u_2 coincide almost everywhere in $\mathcal{D}$.

3. To show that the minimizer u in (C.11) satisfies (C.12), consider functions $u_\epsilon = u + \epsilon\eta$, for a given $\eta \in C_c^\infty(\mathcal{D})$ and each $c \in \mathbb{R}$. Clearly, $u_\epsilon - w \in W_0^{1,p}(\mathcal{D})$ and thus by (C.13) we get:

$$0 \geq \mathcal{I}_p(u) - \mathcal{I}_p(u_\epsilon) \geq -p\int_{\mathcal{D}} \Big\langle |\nabla u_\epsilon|^{p-2}\nabla u_\epsilon, \epsilon\nabla\eta\Big\rangle \,\mathrm{d}x.$$

Consequently:

$$(\operatorname{sgn}\epsilon)\cdot\int_{\mathcal{D}}\Big\langle|\nabla u_\epsilon|^{p-2}\nabla u_\epsilon,\nabla\eta\Big\rangle\,\mathrm{d}x\geq 0.$$

Since $|\nabla u_\epsilon|^{p-2}\nabla u_\epsilon$ converges in $L^{p'}(\mathcal{D})$ to $|\nabla u|^{p-2}\nabla u$ as $\epsilon\to 0$, we obtain $\int_{\mathcal{D}}\langle|\nabla u|^{p-2}\nabla u,\nabla\eta\rangle\,\mathrm{d}x=0$.

On the other hand, if for some $u\in W^{1,p}(\mathcal{D})$ there holds (C.12) with all $\eta\in C_c^\infty(\mathcal{D})$, then the same is true with all $\eta\in W_0^{1,p}(\mathcal{D})$. Let $v\in W^{1,p}(\mathcal{D})$ satisfy $v-w\in W_0^{1,p}(\mathcal{D})$. Then $v-u\in W_0^{1,p}(\mathcal{D})$ and thus:

$$0=\int_{\mathcal{D}}\Big\langle|\nabla u|^{p-2}\nabla u,\nabla v-\nabla u\Big\rangle\,\mathrm{d}x\leq\frac{1}{p}\big(\mathcal{I}_p(v)-\mathcal{I}_p(u)\big),$$

where we used (C.13). This proves $\mathcal{I}_p(u)\leq\mathcal{I}_p(v)$, establishing (C.11). □

C.5 Weak Solutions to the p-Laplacian

When it comes to the notion of a solution to the p-Laplace equation (C.10), several definitions are being used. In this and the next sections, we will introduce the following ones:

(i) *weak solutions*, defined via test functions under the integral;
(ii) *p-harmonic (potential-theoretic) solutions*, via comparison principle;
(iii) *viscosity solutions*, via test functions evaluated at points of contact.

Definition C.29 Let $\mathcal{D}\subset\mathbb{R}^N$ be an open, bounded, connected set and let $p\in(1,\infty)$. We say that $u\in W_{loc}^{1,p}(\mathcal{D})$ is a *weak solution* to $\Delta_p u=0$ in $\mathcal{D}$ if (C.12) holds. We say that u is a *weak supersolution* to $\Delta_p u=0$ in $\mathcal{D}$ if:

$$\int_{\mathcal{D}}\Big\langle|\nabla u|^{p-2}\nabla u,\nabla\eta\Big\rangle\,\mathrm{d}x\geq 0\qquad\text{for all }\ \eta\in C_c^\infty(\mathcal{D})\ \text{ with }\ \eta\geq 0.$$

When $(-u)$ is a weak supersolution, then we call u a *weak subsolution* to $\Delta_p u=0$ in $\mathcal{D}$. If no ambiguity arises, we will often simply write "weak solution" or "weak supersolution" instead of "weak solution to $\Delta_p u=0$ in $\mathcal{D}$" or "weak supersolution to $\Delta_p u=0$ in $\mathcal{D}$", etc.

Exercise C.30 Show that u is a weak solution if and only if it is both a weak supersolution and a weak subsolution.

It turns out that weak solutions to $\Delta_p u=0$ in $\mathcal{D}$ are automatically Hölder regular $C_{loc}^{1,\alpha}(\mathcal{D})$ with the Hölder exponent α depending only on N and p. The proof of this fact is rather complicated; it has been completed in Uralceva (1968); Uhlenbeck

(1977); Evans (1982) for $p \geq 2$ and in Lewis (1983) for the degenerate case of $p \in (1, 2)$, whereas the complete arguments covering all exponents $p \in (1, \infty)$ in a unified manner (in fact, as a special case of a more general class of quasilinear elliptic PDEs) can be found in DiBenedetto (1983) and Tolksdorf (1984). They are based on a sequence of fundamental properties of the weak super/ subsolutions, which are of independent interest and importance. We present some of them below.

Firstly, note that Hölder continuity of u follows easily in case $p > N$ by Morrey's embedding in Theorem C.13. For $p \in (1, N]$ the reasoning is more involved and it is achieved as follows.

Theorem C.31 *Every weak solution u to $\Delta_p u = 0$ in $\mathcal{D}$ is locally essentially bounded, i.e. $u \in L^\infty_{loc}(\mathcal{D})$.*

The next result is the celebrated *Harnack's inequality* that, as we shall see, implies Hölder continuity of u for any $p \in (1, \infty)$.

Theorem C.32 *Let u be a weak solution that is nonnegative in some ball $B_{2r}(x) \subset \mathcal{D}$. Then we have:*

$$\operatorname*{ess\,sup}_{B_r(x)} u \leq C \operatorname*{ess\,inf}_{B_r(x)} u, \tag{C.15}$$

where the constant $C \geq 1$ depends on N, p, but it is independent of $u, r, x \in \mathcal{D}$.

The proofs of Theorem C.31 and C.32 rely on the application of the Moser iteration technique in Moser (1961); Serrin (1963, 1964), see also the classical text by Ladyzhenskaya and Ural'tseva (1968). The following standard corollary is in order:

Corollary C.33 *Let u be a weak solution. Then:*

$$\lim_{r\to 0} \fint_{B_r(x)} u = \operatorname*{ess\,lim\,sup}_{y\to x} u(y) = \operatorname*{ess\,lim\,sup}_{y\to x} u(y) \quad \textit{for all } x \in \mathcal{D}.$$

In particular, u has a continuous representative.

Proof Fix $x \in \mathcal{D}$. For any $r < \frac{1}{3}\mathrm{dist}(x, \partial\mathcal{D})$, Theorem C.31 implies that $u \in L^\infty(B_{2r}(x))$. Applying Harnack's inequality (C.15) to the function $u - \operatorname{ess\,inf}_{B_{2r}(x)} u$, that is a weak solution, nonnegative in $B_{2r}(x)$, we get:

$$\begin{aligned} 0 &\leq \Big(\fint_{B_r(x)} u\Big) - \operatorname*{ess\,inf}_{B_r(x)} u \leq \operatorname*{ess\,sup}_{B_r(x)} u - \operatorname*{ess\,inf}_{B_{2r}(x)} u \\ &\leq C\big(\operatorname*{ess\,inf}_{B_r(x)} u - \operatorname*{ess\,inf}_{B_{2r}(x)} u\big). \end{aligned} \tag{C.16}$$

Observe that since the constant C is independent of r, the right hand side in (C.16) converges to 0 as $r \to 0$, because the function $r \mapsto \operatorname{ess\,inf}_{B_r(x)} u$ is bounded and nonincreasing. Consequently:

$$\operatorname{ess}\liminf_{y\to x} u(y) = \lim_{r\to 0} \fint_{B_r(x)} u.$$

By the same argument applied to the function $(-u)$ we get: $\operatorname{ess}\limsup_{y\to x} u(y) = \lim_{r\to 0} \fint_{B_r(x)} u$. Continuity of the representative $v(x) \doteq \lim_{r\to 0} \fint_{B_r(x)} u$, follows now by Exercise C.17 (ii). □

From now on, we will identify every weak solution u to $\Delta_p u = 0$ on $\mathcal{D}$ with its continuous representative. We now show that Harnack's inequality in fact yields Hölder's continuity:

Corollary C.34 *Let u be a (continuous) weak solution to $\Delta_p u = 0$ in $\mathcal{D}$. Then, for every compact set $K \subset \mathcal{D}$ we have:*

$$|u(x) - u(y)| \leq C\,|x-y|^\alpha \quad \textit{for all } x, y \in K, \tag{C.17}$$

where α depends only on N and p, while C depends on N, p, K and u.

Proof

1. Given a compact set K, let $\epsilon > 0$ be such that $K + B_{3\epsilon}(0) \subset \mathcal{D}$. Fix a radius $r \in (0, \epsilon]$ and let $x \in K$. We now apply Theorem C.32 to the following two nonnegative, continuous weak solutions: $u - (\inf_{B_{2r}(x)} u)$ and $(\sup_{B_{2r}(x)} u) - u$ on $B_{2r}(x)$. It follows that:

$$\sup_{B_r(x)} u - \inf_{B_{2r}(x)} u \leq C\big(\inf_{B_r(x)} u - \inf_{B_{2r}(x)} u\big),$$

$$\sup_{B_{2r}(x)} u - \inf_{B_r(x)} u \leq C\big(\sup_{B_{2r}(x)} u - \inf_{B_r(x)} u\big).$$

Adding the two inequalities and denoting by $\omega(x,r) \doteq \sup_{B_r(x)} u - \inf_{B_r(x)} u$ the oscillation of u on $B_r(x)$, we obtain: $\omega(x,r) + \omega(x,2r) \leq C\big(\omega(x,2r) + \omega(x,r)\big)$, which yields:

$$\omega(x,r) \leq \lambda\omega(x,2r) \tag{C.18}$$

with a constant $\lambda = \frac{C-1}{C+1} \in [0,1)$.
2. Let now $x, y \in K$. If $|x-y| \geq \epsilon$, then obviously:

$$|u(x) - u(y)| \leq 2\|u\|_{L^\infty(K)} \leq \frac{2\|u\|_{L^\infty(K)}}{\epsilon^\alpha}|x-y|^\alpha.$$

On the other hand, if $|x-y| \in (0, \epsilon)$, then denoting by $k \geq 0$ the integer such that $\frac{\epsilon}{2^{k-1}} \leq |x-y| < \frac{\epsilon}{2^k}$ and iterating $(k+1)$ times the bound (C.18), we get:

$$|u(x) - u(y)| \leq \omega(x, \frac{\epsilon}{2^k}) \leq \lambda^{k+1}\omega(x, 2\epsilon) \leq \left(\frac{1}{2^{k+1}}\right)^{\alpha} 2\|u\|_{L^\infty(K+B_{2r}(0))}$$
$$\leq \frac{2\|u\|_{L^\infty(K+B_{2r}(0))}}{\epsilon^\alpha}|x-y|^\alpha,$$

if only α satisfies $\lambda \leq \left(\frac{1}{2}\right)^\alpha$. Existence of such exponent α that depends only on N and p is guaranteed from $\lambda < 1$.

□

Note that, from the proof above the dependence of the constant C in (C.17) on u is only through the norm $\|u\|_{L^\infty(V)}$, where V is an open set such that $K \subset V \subset \bar{V} \subset \mathcal{D}$.

The next corollary to Harnack's inequality (C.15) is known as the *strong maximum principle*:

Corollary C.35 *Let u be a (continuous) weak solution to $\Delta_p u = 0$ in $\mathcal{D}$. If u attains its maximum (or minimum) in $\mathcal{D}$, then u must be constant.*

Proof Assume that $u(x_0) = \max_{\mathcal{D}} u$ for some $x_0 \in \mathcal{D}$ and apply Theorem C.32 to the nonnegative weak solution $u(x_0) - u$ on the ball $B_{2r}(x_0) \subset \mathcal{D}$. It follows that $\sup_{B_r(x_0)} \big(u(x_0) - u\big) \leq 0$, so there must be $u \equiv u(x_0)$ in $B_r(x_0)$. Consequently, the set $D_{max} = \{x \in \mathcal{D};\ u(x) = u(x_0)\}$ is open and nonempty. Since D_{max} is also closed in $\mathcal{D}$, we deduce that $D_{max} = \mathcal{D}$. The case of u attaining its minimum in $\mathcal{D}$ follows in the same manner. □

We remark that although our presentation is carried out in bounded domains $\mathcal{D}$, most of the properties remain true also in the unbounded case. In particular, the following *Liouville theorem* is a direct consequence of Corollary C.35 valid on $\mathbb{R}^N$: if a weak solution $u \in C(\mathbb{R}^N)$ to $\Delta_p u = 0$ is bounded, then it must be constant. In view of Harnack's inequality, it is in fact enough to have u bounded from below in order to conclude its constancy.

From the above, we deduce the first version of the *comparison principle*:

Corollary C.36 *Let u_1 and u_2 be two (continuous) weak solutions to $\Delta_p u = 0$ in $\mathcal{D}$. Assume that:*

$$\limsup_{y\to x} u_1(y) \leq \liminf_{y\to x} u_2(y) \quad \textit{for all } x \in \partial\mathcal{D},$$

where both limits are never simultaneously $-\infty$ or $+\infty$. Then $u_1 \leq u_2$ in $\mathcal{D}$.

Proof Note that $u_1 - u_2$ is a weak solution and that for all $x \in \partial\mathcal{D}$ we have:

$$\limsup_{y\to x}(u_1 - u_2)(y) \leq \limsup_{y\to x} u_1(y) - \liminf_{y\to x} u_2(y) \leq 0.$$

Consequently, if $(u_1-u_2)(x_0) > 0$ at some $x_0 \in \mathcal{D}$, then it easily follows that u_1-u_2 must attain its (positive) maximum in $\mathcal{D}$. By Corollary C.35 we get: $u_1 - u_2 \equiv C > 0$, contradicting the assumption. □

It is possible to prove the comparison principle in the a.e. sense and valid for weak super- and subsolutions, without referring to Harnack's inequality:

Exercise C.37

(i) Prove the following identity:

$$\big\langle |b|^{p-2}b - |a|^{p-2}a, b - a\big\rangle = \frac{1}{2}\big(|b|^{p-2} + |a|^{p-2}\big)|b - a|^2$$
$$+ \frac{1}{2}\big(|b|^{p-2} - |a|^{p-2}\big)\big(|b|^2 - |a|^2\big) \quad \text{for all } a, b \in \mathbb{R}^N.$$

Deduce that the expression in the left hand side above is strictly positive for all $a \neq b$.

(ii) Prove the following comparison principle, using only Definition C.29. Let u_1 be a weak subsolution and u_2 a weak supersolution to $\Delta_p u = 0$ in $\mathcal{D}$, satisfying $(u_1 - u_2)^+ \in W_0^{1,p}(\mathcal{D})$. Then $u_1 \leq u_2$ a.e. in $\mathcal{D}$.

Finally, we state without proof the mentioned ultimate regularity result, as a special case of the results in DiBenedetto (1983); Tolksdorf (1984).

Theorem C.38 *Let u be the weak (continuous) solution of $\Delta_p u = 0$ in $\mathcal{D}$. Then $\nabla u \in C^{0,\alpha}_{loc}(\mathcal{D})$, namely on every compact set $K \subset \mathcal{D}$, the inequality:*

$$|\nabla u(x) - \nabla u(y)| \leq C|x - y|^\alpha \quad \textit{for all } x, y \in K$$

is valid with exponent α depending only on N and p, and the constant C depending on N, p, $dist(K, \partial\mathcal{D})$ and $\|u\|_{L^\infty(V)}$, where V is an open set such that $K \subset V \subset \bar{V} \subset \mathcal{D}$. Further, u is analytic (i.e. it is represented by its Taylor polynomial) in the open set where $\nabla u \neq 0$.

We remark that the analyticity is due to Lewis (1980) and that the $C^{1,\alpha}_{loc}$ regularity is, in general, the best possible (see Lewis 1980; Tolksdorf 1984).

C.6 Potential Theory and p-Harmonic Functions

Motivated by the comparison principle in Corollary C.36, we introduce:

Definition C.39 A lower-semicontinuous function $u : \mathcal{D} \to (-\infty, +\infty]$ that is not equivalently equal to $+\infty$ in $\mathcal{D}$, is called *p-superharmonic* provided that on each open connected domain U satisfying $\bar{U} \subset \mathcal{D}$, the comparison with weak solutions holds. Namely: if $v \in C(\bar{U}) \cap W^{1,p}_{loc}(U)$ is a weak solution to $\Delta_p v = 0$ in U and $v \leq u$ on ∂U, then $v \leq u$ in U.

If the function $(-u)$ is p-superharmonic in $\mathcal{D}$, then we say that u is *p-subharmonic* in $\mathcal{D}$. A continuous $u : \mathcal{D} \to \mathbb{R}$ is called *p-harmonic* when it is both p-superharmonic and p-subharmonic.

It turns out that p-harmonic functions are exactly the continuous representatives of weak solutions. We start comparing the notions in Definitions C.29 and C.39 by stating the counterpart of Theorem C.31 and Corollary C.33:

Theorem C.40 *Every weak supersolution u to $\Delta_p u = 0$ in $\mathcal{D}$ is locally essentially bounded from below and $u(x) = \operatorname{ess\,lim\,inf}_{y\to x} u(y)$ for a.e. $x \in \mathcal{D}$. Consequently, u has a lower-semicontinuous representative which satisfies:*

$$u(x) = \operatorname*{ess\,lim\,inf}_{y\to x} u(y) \qquad \textit{for all} \quad x \in \mathcal{D}. \tag{C.19}$$

From now on, we will identify every weak supersolution u with its lower-semicontinuous representative satisfying (C.19) (this limiting property is due to Heinonen and Kilpeläinen 1988b), similarly as we chose to identify weak solutions with their continuous representatives. It is also immediate that $u \not\equiv +\infty$ and moreover we have:

Theorem C.41 *Let u be a (lower-semicontinuous and satisfying (C.19)) weak supersolution to $\Delta_p u = 0$ in $\mathcal{D}$. Then u is p-superharmonic.*

Proof It suffices to show the comparison property. Let $v \in C(\bar{U})$ be a weak solution in some open connected set U with $\bar{U} \subset \mathcal{D}$. Assume that $v \leq u$ on ∂U. Since for every $\epsilon > 0$ and every $x \in \partial U$ we have: $v(y) < v(x) + \frac{\epsilon}{2}$ in some open neighbourhood of x in $\bar{U}$, and since: $u(y) > u(x) - \frac{\epsilon}{2}$ for all y sufficiently close to x in $\mathcal{D}$ in view of Exercise C.16 (i), it follows that $v < u + \epsilon$ on some set of the form: $B_{\delta(x)}(x) \cap \bar{U}$. Consequently, for every $\epsilon > 0$ we have:

$$v < u + \epsilon \quad \text{in} \quad \{x \in U;\ \operatorname{dist}(x, \partial U) < \delta\},$$

if only $\delta > 0$ is sufficiently small. This yields $(v - u - \epsilon)^+ \in W^{1,p}_0(U)$ and we may apply Exercise C.37 (ii) to v and $u + \epsilon$, concluding that $v \leq u + \epsilon$ a.e. in U. Passing to the limit with $\epsilon \to 0$ it follows that $v \leq u$ a.e. in U. Thus, $v \leq u$ in U by (C.19). □

It is useful to record the following properties of p-superharmonic functions:

Lemma C.42

(i) Let $u : \mathcal{D} \to \bar{\mathbb{R}}$ and assume that for every $x \in D$ there exists its open neighbourhood $V_x \subset \mathcal{D}$ such that $u_{|V_x}$ is p-superharmonic in V_x. Then u is p-superharmonic in $\mathcal{D}$ (thus p-superharmonicity is a local property).

(ii) Minimum of finitely many p-superharmonic functions is p-superharmonic.

The following version of comparison principle is quite straightforward but it relies on the existence of a harmonic function taking the prescribed continuous boundary values as in Theorem C.49.

Exercise C.43 Let u_1 be p-subharmonic and u_2 be p-superharmonic in $\mathcal{D}$. Assume:

$$\limsup_{y\to x} u_1(y) \leq \liminf_{y\to x} u_2(y) \qquad \text{for all} \quad x \in \partial\mathcal{D}, \tag{C.20}$$

and that both quantities above are never simultaneously $-\infty$ or $+\infty$. Using the following outline, prove that $u_1 \leq u_2$ in $\mathcal{D}$.

(i) Show that for every $\epsilon > 0$ there exists $\delta > 0$ such that $u_1 - u_2 < \epsilon$ in the open set $V_\delta \doteq \{x \in \mathcal{D};\ \mathrm{dist}(x, \partial\mathcal{D}) < \delta\}$.
(ii) Fix $\epsilon > 0$ and let $U \subset \mathcal{D}$ be any open, connected subset satisfying $\partial U \subset V_\delta$. Since u_2 is lower-semicontinuous, Exercise C.16 (iv) implies that u_2 is a pointwise limit of a nondecreasing sequence $\{\phi_n \in C^\infty(\mathcal{D})\}_{n=1}^\infty$ as $n \to \infty$. Clearly, $\phi_n \leq u_2$ on ∂U for all $n \geq 1$. Show that $u_1 \leq \phi_n + \epsilon$ on ∂U, for a sufficiently large n.
(iii) Let $v_n \in C(\bar{U}) \cap W^{1,p}(U)$ denote the weak solution to $\Delta_p u = 0$ in a p-regular set U, such that $v_n = \phi_n$ on ∂U. Existence of such solution follows from Theorem C.49. Prove that for n large enough there holds: $u_1 \leq v_n + \epsilon \leq u_2 + \epsilon$ in U. Conclude the result exhausting $\mathcal{D}$ with p-regular sets.

To prove a statement converse to Theorem C.41, we first observe:

Lemma C.44 *For every p-superharmonic function u in $\mathcal{D}$, the set where it attains finite values $\{u \neq +\infty\}$ is dense in $\mathcal{D}$.*

Proof We first show that $u \equiv +\infty$ on $B_r(x_0) \subset \mathcal{D}$ where $r < \frac{1}{4}\mathrm{dist}(x_0, \partial\mathcal{D})$, implies that $u \equiv +\infty$ on $B_{2r}(x_0)$. By Exercise C.27, the function:

$$v(x) = \begin{cases} |x - x_0|^{\frac{p-N}{p-1}} & \text{if } p \neq N \\ -\log|x - x_0| & \text{if } p = N \end{cases} \tag{C.21}$$

is a weak solution in the annulus $U = B_{3r}(x_0) \setminus \bar{B}_r(x_0)$ and $v \in C(\bar{U})$. If $u \equiv +\infty$ in $\bar{B}_{3r}(x_0)$, then there is nothing to prove. Otherwise, we apply the comparison property in Definition C.39 to the well-defined p-superharmonic function: $u - \inf_{\bar{B}_{3r}(x_0)} u$ and to each of the weak solutions in the sequence:

$$\Big\{ v_k \doteq k\big(v - v(x_0 + 3re_1)\big) \Big\}_{k=1}^\infty.$$

Observing that $v_k \leq u - \inf_{\bar{B}_{3r}(x_0)} u$ on ∂U, it follows that the same inequality holds in U for every $k \geq 1$. Since $\lim_{k\to\infty} v_k(x) = +\infty$ for all $x \in B_{2r}(x_0) \setminus \bar{B}_r(x_0)$, we obtain that $u \equiv +\infty$ on $B_{2r}(x_0)$, as claimed.

Consider now the set $D_\infty \doteq \{u = +\infty\}$. By the above reasoning, if $B_\delta(x) \subset D_\infty$ for some $r > 0$ and $x \in \mathcal{D}$, then $B_{\frac{1}{2}\mathrm{dist}(x,\partial\mathcal{D})}(x) \subset \mathcal{D}$. Since the case $D_\infty = \mathcal{D}$ is excluded by Definition C.39, the proof is done. □

Note that the fundamental solution v in (C.21) is p-superharmonic in $\mathcal{D} \doteq B_r(x_0)$, but it is not a weak supersolution, merely because it fails to belong to $W^{1,p}_{loc}(\mathcal{D})$. However, one has (see Heinonen and Kilpeläinen 1988b):

Theorem C.45 *If u is p-superharmonic and locally bounded in $\mathcal{D}$, then u is a weak supersolution to $\Delta_p u = 0$ in $\mathcal{D}$. In particular, $u \in W^{1,p}_{loc}(D)$.*

Consequently, we conclude that (continuous) weak solutions coincide with p-harmonic functions.

C.7 Boundary Continuity of Weak Solutions to $\Delta_p u = 0$

Definition C.46 Let $\mathcal{D} \subset \mathbb{R}^N$ be open, bounded, connected and let $p \in (1, \infty)$. A boundary point $x \in \partial\mathcal{D}$ is called *p-regular* if for every $v \in C(\bar{\mathcal{D}}) \cap W^{1,p}(\mathcal{D})$ the following holds. Let $u \in C(\mathcal{D}) \cap W^{1,p}(\mathcal{D})$ be the unique solution to the problem:

$$\Delta_p u = 0 \quad \text{in } \mathcal{D} \qquad \text{and} \qquad u - v \in W^{1,p}_0(\mathcal{D}); \tag{C.22}$$

then there must be $\lim_{y \to x} u(y) = v(x)$.

To find a geometric description of regular points, we introduce:

Definition C.47

(i) Let A be a subset of an open, bounded set $U \subset \mathbb{R}^N$ and let $p \in (1, \infty)$. The *p-capacity* of A in U is:

$$C_p(A, U) \doteq \inf\Big\{ \int_U |\nabla u|^p \, \mathrm{d}x; \;\; u \in W^{1,p}_0(U) \text{ and } \\ u = 1 \text{ a.e. in an open neighbourhood of } A \Big\}.$$

If the set of admissible functions u above is empty, then $C_p(A, U) = +\infty$.

(ii) Let $A \subset \mathbb{R}^N$. For every $x_0 \in \mathbb{R}^N$, we define the *Wiener function* $(0, 1) \ni r \mapsto \delta_p(r; x_0, A)$:

$$\delta_p(r; x_0, A) \doteq \frac{C_p\big(A \cap B_r(x_0), B_{2r}(x_0)\big)}{C_p\big(B_r(x_0), B_{2r}(x_0)\big)}.$$

The Wiener function is left-continuous, and we define the *Wiener integral*:

$$W_p(x_0, A) \doteq \int_0^1 \frac{\delta_r(r; x_0, A)^{\frac{1}{p-1}}}{r} \mathrm{d}r.$$

We say that the set A is *p-thick* at x_0 if $W_p(x_0, A) = +\infty$. If $W_p(x_0, A) < +\infty$, then we say that A is *p-thin* at x_0.

We remark that one also has:

$$C_p(A, U) = \inf_{V \text{ open}, \, A \subset V \subset U} \; \sup_{K \text{ compact}, \, K \subset V} \inf \left\{ \int_U |\nabla \phi|^p; \;\; \phi \in C_c^\infty(\mathcal{D}) \text{ and } \phi \geq 1 \text{ in } K \right\}$$

and that $C_p(\cdot, \cdot)$ is a *Choquet capacity*, i.e. it obeys the following conditions:

(i) (Monotonicity). If $A_1 \subset A_2 \subset U$, then $C_p(A_1, U) \leq C_p(A_2, U)$.
(ii) (Continuity along increasing sequences). If $\{A_n \subset U\}_{n=1}^\infty$ is an increasing sequence, then $\lim_{n\to\infty} C_p(A_n, U) = C_p(\bigcup_{n=1}^\infty A_n, U)$.
(iii) (Continuity along decreasing sequences of compacts). If $\{A_n \subset U\}_{n=1}^\infty$ is a decreasing sequence of compact sets, then $\lim_{n\to\infty} C_p(A_n, U) = C_p(\bigcap_{n=1}^\infty A_n, U)$.

One can prove that when A is a C^1 manifold embedded in U, then $C_p(A, U) > 0$ if and only if: $\dim A \in (N - p, N]$. In particular, sets containing an open subset or a manifold of co-dimension 1 always have positive capacity, whereas points (nonempty sets) have positive capacity if an only if $p > N$.

We now state the following classical result:

Theorem C.48 *A boundary point $x_0 \in \partial\mathcal{D}$ is p-regular if and only if the set $\mathbb{R}^N \setminus \mathcal{D}$ is p-thick at x_0. We then have:*

$$\sup_{x \in B_r(x_0) \cap \mathcal{D}} |u(x) - v(x_0)| \leq \sup_{x, y \in B_r(x_0) \cap \partial\mathcal{D}} |v(x) - v(y)| + \Big(\sup_{x, y \in \partial\mathcal{D}} |v(x) - v(y)| \Big) \cdot \exp\Big(-C \int_r^1 \frac{\delta_p(\rho, x_0, \mathbb{R}^N \setminus \mathcal{D})^{\frac{1}{p-1}}}{\rho} \mathrm{d}\rho \Big)$$

for all $r \in (0, 1)$ and all $v \in C(\bar{\mathcal{D}}) \cap W^{1,p}(\mathcal{D})$ where $u \in C(\mathcal{D}) \cap W^{1,p}(\mathcal{D})$ is defined through (C.22). The constant C depends only on p and N.

Sufficiency of p-thickness for regularity was shown in Wiener (1924), see also Littman et al. (1963); Maz'ya (1976), while the converse was proved in Lindqvist and Martio (1985) for $p > N - 1$ and in Kilpeläinen and Malý (1994) for $p \in (1, N]$. We remark that when $p > N$ then all boundary points are regular since the corresponding Wiener integrals diverge in view of singletons having positive p-capacity. As a result, we get:

Theorem C.49 *If every $x \in \partial\mathcal{D}$ is p-regular (we say then that $\mathcal{D}$ satisfies the* Wiener p-condition*), then for all $v \in C(\bar{\mathcal{D}}) \cap W^{1,p}(\mathcal{D})$ there exists a unique solution $u \in C(\bar{\mathcal{D}}) \cap W^{1,p}(\mathcal{D})$ to:*

$$\Delta_p u = 0 \quad \text{in } \mathcal{D} \qquad \text{and} \qquad u = v \quad \text{on } \partial\mathcal{D}.$$

According to the *Kellogg property*, the set of boundary points that are not p-regular must have p-capacity 0 in its any open bounded superset. The following lemma from Heinonen et al. (2006) gathers examples of regular points:

Lemma C.50

(i) Assume that $A \subset \mathbb{R}^N$ satisfies the corkscrew condition *at $x_0 \in A$: there exists $c \in (0, 1]$ and $R > 0$ such that:*

$$B_{cr}(x_r) \subset A \cap B_r(x_0) \qquad \text{for some } x_r \in A \text{ and all } r \in (0, R).$$

Then $W_p(x_0, A) = +\infty$ for all $p \in (1, \infty)$.

(ii) If A contains an open cone with vertex at $x_0 \in A$ (i.e. A satisfies the cone condition *at x_0), then $W_p(x_0, A) = +\infty$ for all $p \in (1, \infty)$.*

It follows that all polyhedra, all balls and all sets with Lipschitz boundary satisfy the Wiener p-condition in Theorem C.49, i.e. each boundary point is p-regular. In particular, every open set $\mathcal{D} \subset \mathbb{R}^N$ can be exhausted by such regular open sets.

We now recall another fundamental notion related to boundary regularity.

Definition C.51 For every $F \in C(\partial\mathcal{D})$, the following function h_F is called the *Perron solution* to $\Delta_p u = 0$ in $\mathcal{D}$ with boundary values F:

$$\begin{aligned} h_F \doteq \inf \Big\{ v;\ & p\text{-superharmonic in } \mathcal{D}, \text{ bounded from below} \\ & \text{and such that } \liminf_{y \to x} v(y) \geq F(x) \text{ for all } x \in \partial\mathcal{D} \Big\} \\ = \sup \Big\{ v;\ & p\text{-subharmonic in } \mathcal{D}, \text{ bounded from above} \\ & \text{and such that } \limsup_{y \to x} v(y) \leq F(x) \text{ for all } x \in \partial\mathcal{D} \Big\}. \end{aligned}$$

The (well defined) above h_F is a (continuous) weak solution to $\Delta_p u = 0$ in $\mathcal{D}$.

The fact that the upper and the lower Perron solutions in Definition C.51 coincide (for any continuous function g given on the boundary of a bounded domain $\mathcal{D}$), is the famous *Wiener resolutivity theorem*, established for various cases in: Wiener (1925); Lindqvist and Martio (1985); Kilpeläinen (1989). An important observation is that if g extends to a continuous $W^{1,p}$ function in $\mathcal{D}$, then h_g automatically coincides with the variational solution:

Lemma C.52 *If $v \in C(\bar{\mathcal{D}}) \cap W^{1,p}(\mathcal{D})$, then $h_{v|\partial\mathcal{D}} - v \in W_0^{1,p}(\mathcal{D})$, so the corresponding Perron solution $h_{v|\partial\mathcal{D}} \in W^{1,p}(\mathcal{D}) \cap C(\mathcal{D})$ equals the unique variational solution u of (C.22).*

Corollary C.53 *A boundary point $x \in \partial\mathcal{D}$ is p-regular if and only if:*

$$\lim_{y \to x} h_F(y) = F(x) \qquad \textit{for every} \quad F \in C(\partial\mathcal{D}).$$

Proof Regularity as in the statement of the Corollary implies p-regularity in Definition C.46 in view of Lemma C.52. Conversely, let $x \in \partial\mathcal{D}$ be p-regular. Given $F \in C(\partial\mathcal{D})$, one proceeds by approximating it uniformly with functions $\{v_n \in C^\infty(\mathbb{R}^N)\}_{n=1}^\infty$, so that $h_{(v_n)|\partial\mathcal{D}} - v_n \in W_0^{1,p}(\mathcal{D})$. The p-regularity of x yields:

$$\lim_{y \to x} h_{(v_n)|\partial\mathcal{D}}(y) = v_n(x). \tag{C.23}$$

Observe that the sequence $\{h_{(v_n)|\partial\mathcal{D}}\}_{n=1}^\infty$ converges uniformly to h_F in $\mathcal{D}$, because of the bound: $\|h_{(v_n)|\partial\mathcal{D}} - h_F\|_{L^\infty(\mathcal{D})} \leq \|v_n - F\|_{L^\infty(\partial\mathcal{D})}$ that is a direct consequence of Definition C.51. Thus, the same boundary convergence as in (C.23) is valid for h_F, as claimed. □

In view of the comparison principle in Corollary C.36 we obtain:

Corollary C.54 *If $\mathcal{D}$ satisfies the Wiener p-condition, then for all $F \in C(\bar{\mathcal{D}})$ there exists a unique $u \in C(\bar{\mathcal{D}}) \cap W_{loc}^{1,p}(\mathcal{D})$ which is a weak solution to $\Delta_p u = 0$ in $\mathcal{D}$ and which satisfies: $u_{|\partial\mathcal{D}} = F$.*

To compare Theorem C.49 and Corollary C.54, we point out that in general, Perron's solution u fails to be $W^{1,p}(\mathcal{D})$ regular, even when $\mathcal{D}$ is p-regular (unless the given boundary function extends continuously to the boundary of $\mathcal{D}$ and belongs to $W^{1,p}(\mathcal{D})$). In this situation, u cannot be obtained from the minimization of the $\mathcal{I}_p$ energy, as shown in the following *Hadamard's example*:

Exercise C.55 Let $\mathcal{D} = B_1(0) \subset \mathbb{R}^2$ and define $u : \mathcal{D} \to \mathbb{R}$:

$$u(r, \theta) = \sum_{n=1}^{\infty} \frac{r^{n!} \sin(n!\theta)}{n^2},$$

in polar coordinates. Show that $u \in C(\bar{\mathcal{D}})$ and that it is a weak solution to $\Delta_2 u = 0$ in $\mathcal{D}$. Further, show that $u \in W_{loc}^{1,2}(\mathcal{D}) \setminus W^{1,2}(\mathcal{D})$.

The following equivalent description of boundary regularity, extending the classical results valid for $p = 2$, is due to Kilpeläinen and Malý (1994):

Theorem C.56 *A boundary point $x \in \partial\mathcal{D}$ is p-regular if and only if one of the following equivalent conditions holds:*

(i) There exists a weak solution $u \in C(\bar{\mathcal{D}}) \cap W^{1,p}(\mathcal{D})$ to $\Delta_p u = 0$ on $\mathcal{D}$, such that $u(x) = 0$ and $u > 0$ on $\bar{\mathcal{D}} \setminus \{x\}$.

(ii) There exists a barrier function *relative to $\mathcal{D}$, namely a p-superharmonic function $u : \mathcal{D} \to (-\infty, +\infty]$ such that:* $\liminf_{z\to y} u(z) > 0$ *for all $y \in \partial\mathcal{D} \setminus \{x\}$, while:* $\lim_{z\to x} u(z) = 0$.

It turns out that the method of barriers is a useful device in linear and nonlinear well-posedness theories. For example, we have:

Exercise C.57 Let $\mathcal{D} \subset \mathbb{R}^2$ be open, bounded, connected and assume that $x \in \partial\mathcal{D}$ satisfies the *line condition*, i.e. there exists a simple continuous arc $S \subset \mathbb{R}^2 \setminus \mathcal{D}$ such that $x \in S$. Let $B_r(x)$ have a radius $r < 1$ so small that S intersects $\partial B_r(x)$. Let $\tilde{B}$ be $B_r(x)$ less the part of S from x to the first hit of $\partial B_r(x)$. Then the function $z \mapsto \log(z - x)$ has a well-defined single-valued analytic branch in $\tilde{B}$. Show that:

$$v(z) \doteq -\mathrm{Re}\,\frac{1}{\log(z-x)}$$

is 2-harmonic in $\tilde{B}$ and that it satisfies: $\liminf_{z\to y} v(z) > 0$ for all $y \in (\partial\mathcal{D} \setminus \{x\}) \cap \tilde{B}$, while: $\lim_{z\to x} v(z) = 0$. Conclude the 2-regularity of x by modifying the function v to construct a 2-barrier at x.

In fact, in dimension $N = 2$ the line condition implies p-regularity for any $p \in (1, +\infty)$. Even more generally, we have:

Lemma C.58 *Let $\mathcal{D} \subset \mathbb{R}^2$ be open, bounded and connected. Let $x \in \partial\mathcal{D}$ have the property that the single point set $\{x\}$ is not a connected component of $\mathbb{R}^2 \setminus \mathcal{D}$. Then x is p-regular for any $p \in (1, \infty)$. In particular, if no connected component of $\mathbb{R}^2 \setminus \mathcal{D}$ is a single point, then $\mathcal{D}$ satisfies the Wiener 2-condition for all $p \in (1, \infty)$.*

The above result follows by a direct estimate of Wiener's integral:

$$W_p(x, \mathbb{R}^2 \setminus \mathcal{D}) \geq c \int_0^{r_0} \frac{1}{r}\,\mathrm{d}r = +\infty,$$

which is due to the bound on the Wiener function: $\delta_p(r; x, \mathbb{R}^2 \setminus \mathcal{D}) \geq c > 0$, valid for all $r > 0$ that are sufficiently small. Indeed, if $\delta_p(r_n; x, \mathbb{R}^2 \setminus \mathcal{D}) \to 0$ for some $r_n \to 0$ as $n \to \infty$, then a result in Heinonen and Kilpeläinen (1988a) implies existence of shrinking spheres $\partial B_{\rho_n}(x_0) \subset \mathcal{D}$ with $\rho_n \in (0, \frac{r_n}{2})$. This means that $\{x\}$ is a connected component of $\mathbb{R}^2 \setminus \mathcal{D}$, contradicting the assumption.

C.8 Viscosity Solutions to $\Delta_p u = 0$

We have yet another definition of solutions to (C.10):

Definition C.59 Let $u : \mathcal{D} \to (-\infty, +\infty]$ be lower-semicontinuous and not equivalently equal to $+\infty$ in $\mathcal{D}$. We say that u is a *viscosity p-supersolution* when for every $x_0 \in \mathcal{D}$ and every $\phi \in C^\infty(\mathcal{D})$ satisfying:

$$\phi(x) \leq u(x) \quad \text{for all } x \in \mathcal{D} \qquad \text{and} \qquad \phi(x_0) = u(x_0) \quad \text{with} \quad \nabla\phi(x_0) \neq 0$$

(we say that ϕ *touches u from below* at x_0), we have:

$$\Delta_p \phi(x_0) \leq 0.$$

If $(-u)$ is a viscosity p-supersolution, then we say that u is a *viscosity p-subsolution*. A continuous function $u : \mathcal{D} \to \mathbb{R}$ is called a *viscosity solution* to $\Delta_p u = 0$ in $\mathcal{D}$ if it is both a viscosity p-supersolution and a viscosity p-subsolution.

For an account of the modern theory of viscosity solutions, we refer to Crandall et al. (1992); Koike (2004). To motivate Definition C.59, recall that $u \in C^2(I)$ defined on an interval $I = (a, b)$ is convex if and only if $u'' \geq 0$ on I. For merely continuous u we have the equivalence:

Exercise C.60 A continuous function $u : I \to \mathbb{R}$ is convex if and only if $\phi''(x_0) \geq 0$ for every $x_0 \in I$ and every $\phi \in C^\infty(I)$ such that:

$$\phi(x_0) = u(x_0) \quad \text{and} \quad \phi > u \text{ in } I \setminus \{x_0\}. \tag{C.24}$$

It can be directly observed that:

Lemma C.61 *Every p-superharmonic function is a viscosity p-supersolution.*

Proof Let u be p-superharmonic in $\mathcal{D}$. Assume that for some test function $\phi \in C^\infty(\mathcal{D})$ that touches u from below at a given $x_0 \in \mathcal{D}$ where $\nabla\phi(x_0) \neq 0$, we have $\Delta_p\phi(x_0) > 0$. By possibly modifying ϕ through: $\phi - \epsilon|x - x_0|^4$ we may also, without loss of generality, assume that:

$$\phi(x) < u(x) \qquad \text{for all } x \in \mathcal{D} \setminus \{x_0\}. \tag{C.25}$$

Since $\Delta_p\phi \geq 0$ on some ball $B_r(x_0) \subset \mathcal{D}$, in virtue of Theorem C.41, ϕ is p-subharmonic in $B_r(x_0)$. By the lowersemicontinuity of u, the ordering in (C.25) implies:

$$m \doteq \min_{\partial B_r(x_0)} (u - \phi) > 0.$$

We now apply the comparison principle in Exercise C.43 to the p-superharmonic function u and the p-subharmonic function $\phi+m$, in view of $\phi+m \leq u$ on $\partial B_r(x_0)$. It follows that $\phi + m \leq u$ in $B_r(x_0)$, contradicting $\phi(x_0) = u(x_0)$. □

A more complicated argument in Juutinen et al. (2001) (whereas an alternative, simpler proof, based on approximation with the so-called infimal convolutions, is due to Julin and Juutinen 2012) yields:

Theorem C.62 *If u is a viscosity p-supersolution, then u is p-superharmonic.*

Concluding, Theorems C.62, C.45 and C.41 imply that locally bounded viscosity p-supersolutions coincide with weak supersolutions and also that the three notions of: viscosity solutions, p-harmonic functions and weak solutions to $\Delta_p u = 0$ in $\mathcal{D}$ are actually the same. In view of the comparison principle in Corollary C.36 and Corollary C.54 we obtain:

Corollary C.63 *Viscosity solutions $u \in C(\bar{\mathcal{D}})$ to:*

$$\Delta_p u = 0 \text{ in } \mathcal{D} \qquad \text{and} \qquad u = F \text{ on } \partial\mathcal{D},$$

with given boundary values $F \in C(\partial\mathcal{D})$ are unique and satisfy $u \in W^{1,p}_{loc}(\mathcal{D})$. If all boundary points $x \in \partial\mathcal{D}$ are p-regular, then such (unique) solution exists.

Appendix D
Solutions to Selected Exercises

This chapter contains solutions to selected problems in the book.

Exercise 2.7

(i) Given $x_1, x_2 \in \mathbb{R}^N \setminus A$, let $y_1, y_2 \in A$ be the respective minimizing points for the introduced definition of the extension. It then follows that:

$$F(x_1) - F(x_2) \leq \Big(F(y_2) + \frac{|x_1 - y_2|}{\mathrm{dist}(x_1, A)} - 1\Big) - \Big(F(y_2) + \frac{|x_2 - y_2|}{\mathrm{dist}(x_2, A)} - 1\Big)$$
$$= \frac{|x_1 - y_2|}{\mathrm{dist}(x_1, A)} - \frac{|x_2 - y_2|}{\mathrm{dist}(x_2, A)} \leq \frac{|x_1 - x_2|}{\mathrm{dist}(x_1, A)} + \frac{|x_1 - y_2| \cdot |x_1 - x_2|}{\mathrm{dist}(x_1, A) \cdot \mathrm{dist}(x_2, A)},$$

resulting in the continuity of F on $\mathbb{R}^N \setminus A$. To check continuity at $x_0 \in \partial A$, let $x \notin A$ and observe that for every $r > 0$ there holds:

$$F(x) - F(x_0) = \inf_A \Big(F(y) - F(x_0) + \frac{|x - y|}{\mathrm{dist}(x, A)} - 1\Big)$$
$$= \min\Big\{ \inf_{y \in A \cap B_r(x_0)} \big(F(y) - F(x_0)\big), \ \big(\inf_A F - \sup_A F\big) + \frac{r - |x - x_0|}{|x - x_0|} - 1\Big\}.$$

Since the second term in the above minimization converges to ∞ as $x \to x_0$ regardless of $r > 0$, we see that taking r small achieves $F(x) - F(x_0) \geq -\epsilon$ whenever $|x - x_0|$ is sufficiently small. For the opposite bound, set $y \in A$ so that $|x - y| = \mathrm{dist}(x, A)$. Then:

$$F(x) - F(x_0) \leq F(y) - F(x_0) \to 0 \quad \text{as } x \to x_0,$$

since $|y - x_0| \leq 2|x - x_0|$.

M. Lewicka, *A Course on Tug-of-War Games with Random Noise*, Universitext,
https://doi.org/10.1007/978-3-030-46209-3

Exercise 2.11

(i) We first prove that each function $f_n : \partial B_1(0) \to \mathbb{R}$ defined by:

$$f_n(x) = \frac{\mu(U \cap B(x, \frac{1}{n}))}{\mu(B(x, \frac{1}{n}))}$$

is Borel-regular. Indeed, given $r \in (0, 1)$ and a sequence $x_n \to x_0$ in $\partial B_1(0)$, it is easily seen that $\mathbb{1}_{B(x_0,r)} \le \liminf_{n\to\infty} \mathbb{1}_{B(x_n,r)}$, so by Fatou's lemma:

$$\int_U \mathbb{1}_{B(x_0,r)}\, \mathrm{d}\mu \le \liminf_{n\to\infty} \int_U \mathbb{1}_{B(x_n,r)}\, \mathrm{d}\mu$$

proving that the function $x \mapsto \int_U \mathbb{1}_{B(x,r)}\, \mathrm{d}\mu = \mu\big(U \cap B(x,r)\big)$ is lower-semicontinuous (see Definition C.15) and hence Borel. Recalling that $x \mapsto \mu(B(x,r))$ is constant by the rotational invariance, f_n must be Borel as well.

Observe that $\lim_{n\to\infty} f_n = 1$ in U. Applying Fatou's lemma again, we get:

$$\begin{aligned}\sigma^{N-1}(U) &\le \liminf_{n\to\infty} \int_U f_n(x)\, \mathrm{d}\sigma^{N-1}(x)\\ &= \liminf_{n\to\infty} \frac{1}{\mu(B(x, \frac{1}{n}))} \int_U \int_U \mathbb{1}_{B(x,\frac{1}{n})}(y)\, \mathrm{d}\mu(y)\, \mathrm{d}\sigma^{N-1}(x),\end{aligned}$$

whereas Fubini's theorem yields:

$$\begin{aligned}\int_U \int_U \mathbb{1}_{B(x,\frac{1}{n})}(y)\, \mathrm{d}\mu(y)\, \mathrm{d}\sigma^{N-1}(x) &= \int_U \int_U \mathbb{1}_{B(x,\frac{1}{n})}(y)\, \mathrm{d}\sigma^{N-1}(x)\, \mathrm{d}\mu(y)\\ &= \int_U \sigma^{N-1}\big(U \cap B(y, \frac{1}{n})\big)\, \mathrm{d}\mu(y) \le \sigma^{N-1}(B(x, \frac{1}{n})) \cdot \mu(U),\end{aligned}$$

achieving (2.11).

(ii) A symmetric argument as above gives: $\mu(U) \le \Big(\liminf_{n\to\infty} \frac{\mu(B(x,\frac{1}{n}))}{\sigma^{N-1}(B(x,\frac{1}{n}))}\Big) \cdot \sigma^{N-1}(U)$, for all open sets U. In particular, for $U = \partial B_1(0)$ we obtain:

$$\begin{aligned}1 \le \liminf_{n\to\infty} \frac{\mu(B(x, \frac{1}{n}))}{\sigma^{N-1}(B(x, \frac{1}{n}))} &= \Big(\limsup_{n\to\infty} \frac{\sigma^{N-1}(B(x, \frac{1}{n}))}{\mu(B(x, \frac{1}{n}))}\Big)^{-1}\\ &\le \Big(\liminf_{n\to\infty} \frac{\sigma^{N-1}(B(x, \frac{1}{n}))}{\mu(B(x, \frac{1}{n}))}\Big)^{-1} \le 1.\end{aligned}$$

Thus there exists the limit: $\lim_{n\to\infty} \frac{\sigma^{N-1}(B(x,\frac{1}{n}))}{\mu(B(x,\frac{1}{n}))} = 1$. By (2.11) this results in: $\sigma^{N-1}(U) \leq \mu(U)$ and, as before, in: $\mu(U) \leq \sigma^{N-1}(U)$ for all open sets $U \subset \partial B_1(0)$, proving that $\mu = \sigma^{N-1}$.

Exercise 2.16
For fixed $\eta, \delta > 0$ and each boundary point $y_0 \in \partial\mathcal{D}$, choose $\hat{\delta}(y_0) \in (0, \frac{\delta}{2})$ and $\hat{\epsilon}(y_0) \in (0,1)$ so that the bound in Definition 2.12 (a) holds for the parameters η and $\frac{\delta}{2}$. By compactness of $\partial\mathcal{D}$ we may choose its finite covering:

$$\partial\mathcal{D} \subset \bigcup_{i=1}^{n} B_{\hat{\delta}(y_{0,i})}(y_{0,i}),$$

corresponding to the boundary points $\{y_{0,i}\}_{i=1}^n$. Let $\hat{\delta} \in (0,\delta)$ be such that for every $y_0 \in \partial\mathcal{D}$ there holds: $B_{\hat{\delta}}(y_0) \subset B_{\hat{\delta}(y_{0,i})}(y_{0,i})$ for some $i = 1 \ldots n$. Let $\hat{\epsilon} = \min_{i=1\ldots n} \hat{\epsilon}(y_{0,i})$. It now follows that:

$$\mathbb{P}\big(X^{x_0} \in B_\delta(y_0)\big) \geq \mathbb{P}\big(X^{x_0} \in B_{\frac{\delta}{2}}(y_{0,i})\big) \geq 1 - \eta,$$

for all $\epsilon \in (0, \hat{\epsilon})$ and all $x_0 \in B_{\hat{\delta}}(y_0) \cap \mathcal{D}$, as needed.

Exercise 3.3
We compute:

$$\begin{aligned}
(r-q)|\nabla u|^{2-p}\Delta_p u &= (r-q)\Delta u + (r-q)(p-2)\Delta_\infty u \\
&= (r-q)\Delta u + \big((r-p)(q-2) + (p-q)(r-2)\big)\Delta_\infty u \\
&= (r-p)\big(\Delta u + (q-2)\Delta_\infty u\big) + (p-q)\big(\Delta u + (r-2)\Delta_\infty u\big) \\
&= (r-p)|\nabla u|^{2-q}\Delta_q u + (p-q)|\nabla u|^{2-r}\Delta_r u.
\end{aligned}$$

Exercise 3.7
(i) By symmetry and change of variable it follows that:

$$\fint_{B_\epsilon(0)} |y_1|^2 \,\mathrm{d}y = \frac{1}{N}\fint_{B_\epsilon(0)} |y|^2 \,\mathrm{d}y = \frac{\epsilon^2}{N}\fint_{B_1(0)} |y|^2 \,\mathrm{d}y.$$

For the last integral, we employ the hyperspherical coordinates $(r, \phi_1, \ldots, \phi_{N-1})$. Calling $S = (\sin^{N-2}\phi_1)(\sin^{N-3}\phi_2)\ldots(\sin\phi_{N-2})$, the volume element is given by: $r^{N-1}S\,\mathrm{d}r\,\mathrm{d}\phi_1\,\mathrm{d}\phi_1\dots\mathrm{d}\phi_{N-1}$. We thus get:

$$\fint_{B_1(0)}|y|^2\,\mathrm{d}y=\frac{\int_0^{2\pi}\int_0^{\pi}\dots\int_0^{\pi}\big(\int_0^1 r^{N+1}\mathrm{d}r\big)\,S\,\mathrm{d}\phi_1\,\mathrm{d}\phi_1\dots\mathrm{d}\phi_{N-1}}{\int_0^{2\pi}\int_0^{\pi}\dots\int_0^{\pi}\big(\int_0^1 r^{N-1}\mathrm{d}r\big)\,S\,\mathrm{d}\phi_1\,\mathrm{d}\phi_1\dots\mathrm{d}\phi_{N-1}}$$
$$=\frac{\int_0^1 r^{N+1}\mathrm{d}r}{\int_0^1 r^{N-1}\mathrm{d}r}=\frac{N}{N+2},$$

which implies that: $\fint_{B_\epsilon(0)}|y_1|^2\,\mathrm{d}y=\frac{\epsilon^2}{N+2}$.

Exercise 3.12

(i) By Definition C.15, the upper-semicontinuity of $x\mapsto\inf_{B_\epsilon(x)}v$ reads:

$$\inf_{B_\epsilon(x)}v\geq\lim_{r\to 0}\sup_{y\in B_r(x)\setminus\{x\}}\Big(\inf_{B_\epsilon(y)}v\Big),$$

and the function under the limit in the right hand side, is nonincreasing in r. Fix $\delta>0$ and let $y_\delta\in B_\epsilon(x)$ be such that $v(y_\delta)\leq\delta+\inf_{B_\epsilon(y)}v$. We observe that for every $r_0<\epsilon-|y_\delta-x|$ and every $y\in B_{r_0}(x)$, there holds: $y_\delta\in B_\epsilon(y)$. Consequently:

$$\lim_{r\to 0}\sup_{y\in B_r(x)\setminus\{x\}}\Big(\inf_{B_\epsilon(y)}v\Big)\leq\sup_{y\in B_{r_0}(x)}\Big(\inf_{B_\epsilon(y)}v\Big)\leq v(y_\delta)\leq\delta+\inf_{B_\epsilon(x)}v,$$

which ends the proof upon taking $\delta\to 0$.

Exercise 4.10

(i) Since $g_A'(t)=\frac{Ae^t}{A(e^t-1)+1}>0$ for $t\in(0,\infty)$, the function g_A is strictly increasing from $\lim_{t\to 0+}g_A(t)=0$ to $\lim_{t\to\infty}g_A(t)=\infty$. We observe that $(g_A(t)-t)'=g_A(t)-1=\frac{A-1}{A(e^t-1)+1}$ and that the function $t\mapsto\big(A(e^t-1)+1\big)^{-1}$ is strictly decreasing from 1 to 0 on $(0,\infty)$. Thus we directly obtain the claimed bounds of $g_A'(t)-1$. As a consequence $g_A(t)-t$ is strictly increasing/decreasing according to the sign of $(A-1)$. Since $\lim_{t\to 0+}(g_A(t)-t)=0$ and $\lim_{t\to\infty}(g_A(t)-t)=\lim_{t\to\infty}\log\big(A+\frac{1-A}{e^t}\big)=\log A$, we conclude the remaining bounds.

(ii) It suffices to check: $g_A\big(g_{1/A}(t)\big)=\log\big(A\exp(\log(\frac{1}{A}(e^t-1)+1))+1-A\big)=t$.

(iii) Applying (i) to $A=1-\epsilon<1$, we get: $v_\epsilon(x)-u(x)=g_{1-\epsilon}(u(x))-u(x)\in\big(\log(1-\epsilon),0\big)$, in view of the assumption $u(x)>0$. Further:

$$\nabla v_\epsilon(x)=g_{1-\epsilon}'(u(x))\nabla u(x),$$
$$\nabla^2 v_\epsilon(x)=g_{1-\epsilon}''(u(x))\nabla u(x)\otimes\nabla u(x)+g_{1-\epsilon}'(u(x))\nabla^2 u(x),$$
$$\Delta v_\epsilon(x)=g_{1-\epsilon}''(u(x))|\nabla u(x)|^2+g_{1-\epsilon}'(u(x))\Delta u(x),$$

so that, noting the fact that $\frac{\nabla v_\epsilon}{|\nabla v_\epsilon|} = \frac{\nabla u}{|\nabla u|}$, we directly arrive at the indicated formula for $\Delta_p v_\epsilon$ and at its simplified form in case of $\Delta_p u = 0$. The final estimates follow from $\epsilon < \frac{1}{2}$ in view of:

$$g'_{1-\epsilon}(t) = 1 - \frac{\epsilon}{(1-\epsilon)(e^t - 1) + 1} > 1 - \epsilon > \frac{1}{2},$$

$$g''_{1-\epsilon}(t) = \epsilon \frac{(1-\epsilon)e^t}{\big((1-\epsilon)e^t + \epsilon\big)^2} \geq \epsilon \frac{1}{(1-\epsilon)e^t} > \frac{\epsilon}{e^t}.$$

Exercise 5.10

(i) For a given $s \in \mathbb{N}$ and $A_1 \in \mathcal{F}_s$, consider the family $\mathcal{A}$ of sets $A_2 \in \mathcal{F}$ satisfying:

$$\tilde{\mathbb{P}}\big(\{Y_1 \in A_1\} \cap \{Y_2 \in A_2\}\big) = \tilde{\mathbb{P}}\big(Y_1 \in A_1\big) \cdot \big(Y_2 \in A_2\big).$$

It is easy to observe that $\mathcal{A}$ is a σ-algebra. Since by (5.22) $\mathcal{A}$ contains all subsets $A_2 \in \bigcup_{t\in\mathbb{N}} \mathcal{F}_t$ of Ω, we deduce that $\mathcal{A} = \mathcal{F}$. Similarly, fix now $A_2 \in \mathcal{F}$ and consider the family of sets $A_1 \in \mathcal{F}_{fin}$ for whom the equality displayed above holds true. Again, this family is closed with respect to countable unions and complements, and it contains all subsets $A_1 \in \bigcup_{s\in\mathbb{N}} \mathcal{F}_s \subset \mathcal{F}_{fin}$ of Ω_{fin}, so it equals $\mathcal{F}_{fin}$.

(ii) Since the subset $\{\exists n \leq \tau \quad X_n \notin B_{\delta_k}(y_0)\} \subset \Omega$ is $\mathcal{F}$-measurable, it suffices to prove that the function $g : \Omega_{fin} \times \Omega \to \Omega$ given below is measurable:

$$g\big(\{(w_n, a_n, b_n)\}_{n=1}^s, \{(w_n, a_n, b_n)\}_{n=s+1}^\infty\big) \doteq \{(w_n, a_n, b_n)\}_{n=1}^\infty.$$

To this end, we will show that $g^{-1}(A)$ belongs to the product σ-algebra of $\mathcal{F}_{fin}$ and $\mathcal{F}$, for the sets $A \in \mathcal{F}$ of the form: $A = \prod_{i=1}^\infty A_i$, where $A_i \in \mathcal{F}_1$ for all $i \in \mathbb{N}$ and all but finitely many indices i satisfy $A_i = \Omega_1$. Such sets generate the σ-algebra $\mathcal{F}$. Indeed, we have:

$$g^{-1}(A) = \bigcup_{n=1}^\infty \big(\prod_{i=1}^n A_i\big) \times \big(\prod_{i=n+1}^\infty A_i\big),$$

together with $\prod_{i=1}^n A_i \in \mathcal{F}_{fin}$ and $\prod_{i=n+1}^\infty A_i \in \mathcal{F}$.

Exercise 5.21

(i) The first bound in (5.36) is obvious. The second bound follows from: $k_2 > \rho\kappa_2 > \rho\kappa_1 + \rho(1+r) > k_1 + 1 + r$, in view of $\rho > 1$ and (5.39). For the third bound, observe that:

$$k_3 = \rho\kappa_3 - (\rho-1)r > \rho(\kappa_2 + 1) + \rho r - (\rho - 1)r = k_2 + 1 + r.$$

For the last bound, we have:

$$\frac{k_2(k_2+1)}{k_1}-1=\frac{\rho(\kappa_2+1)-1}{\kappa_1}(\kappa_2+1)-1=\rho\frac{\kappa_2(\kappa_2+1)}{\kappa_1}+\frac{(\rho-1)(\kappa_2+1)}{\kappa_1}-1$$
$$\geq \rho\kappa_3+\rho-1+(\rho-1)\frac{\kappa_2+1}{\kappa_1}>\rho\kappa_3>\rho\kappa_3-(\rho-1)r=k_3.$$

(ii) For $i = 1, 2$, the first inequality in (5.37) follows directly from (5.36). Otherwise, observe that for $m \geq 2$ we get:

$$k_{2m+1}-k_{2m}=\alpha^{m-1}(k_3-k_2)>k_3-k_2>1+r$$
$$k_{2m}-k_{2m-1}=\alpha^{m-2}(\alpha k_2-k_3+r)\geq\alpha k_2-k_3+r=\frac{k_3(k_2-k_1)-rk_2}{k_1}+r$$
$$>\frac{k_3(1+r)-rk_2}{k_1}+r=\frac{k_3+r(k_3-k_2)}{k_1}+r>1+r.$$

To prove the second bound in (5.37), check first that it holds for $i = 2$ by the last inequality in (5.36), whereas for $i = 3$ we have:

$$k_3(k_2+1)-(k_4+1)k_1=k_3+k_2r-k_1r-k_1>(k_2-k_1)(1+r)+1+r>0,$$

by (5.36). Let now $m \geq 2$ and we compute:

$$k_{2m}(k_2+1)=\alpha^{m-1}k_2(k_2+1)+\frac{\alpha^{m-1}-1}{\alpha-1}r+k_2+1$$
$$>\alpha^{m-1}(k_3+1)k_1+\frac{\alpha^{m-1}-1}{\alpha-1}rk_1=k_{2m-1}k_1+\alpha^{m-1}k_1\geq(k_{2m+1}+1)k_1$$
$$k_{2m-1}(k_2+1)=\alpha^{m-2}k_3(k_2+1)+\frac{\alpha^{m-2}-1}{\alpha-1}r(k_2+1)$$
$$\geq\alpha^{m-1}k_1k_2+\alpha^{m-2}k_1r+\alpha^{m-2}k_1+\frac{\alpha^{m-2}-1}{\alpha-1}r(k_2+1)$$
$$>k_1\Big(\alpha^{m-1}k_2+\frac{\alpha^{m-1}-1}{\alpha-1}r+\alpha^{m-2}\Big)=(k_{2m}+1)k_1.$$

Finally, the last expression in (5.37) is equivalent to the definition of the sequence $\{k_i\}_{i=1}^{\infty}$ via: $k_{i+2}=\alpha k_i+r$.

(iii) We proceed as in Exercise 5.10 (i) and (5.22). Let $C_1 \in \mathcal{F}_s$ and $C_2 \in \mathcal{F}_t$ for some $s, t \in \mathbb{N}$. Then:

$$\mathbb{P}(Y_1\in C_1)=\mathbb{P}_s\Big(C_1\cap\{j=s\}\cap\bigcap_{n<s}\{Z_n\notin A_{i+2}\}\Big),$$
$$\mathbb{P}(Y_2\in C_2)=\sum_{j\geq1}\mathbb{P}\Big(\big\{\{(w_n,a_n,b_n)\}_{n=j+1}^{\infty}\in C_2\big\}\cap\bigcap_{n<j}\{Z_n\notin A_{i+2}\}\Big)$$

$$= \mathbb{P}_t(C_2) \cdot \sum_{j\geq 1} \mathbb{P}_j\Big(\bigcap_{n<j}\{Z_n \notin A_{i+2}\}\Big) = \mathbb{P}_t(C_2)\cdot \mathbb{P}(\tilde{\Omega}),$$

$$\mathbb{P}\big(\{Y_1 \in C_1\} \cap \{Y_2 \in C_2\}\big) = P_s\Big(C_1 \cap \{j = s\} \cap \bigcap_{n<s}\{Z_n \notin A_{i+2}\}\Big) \cdot \mathbb{P}_t(C_2).$$

We thus have:

$$\tilde{\mathbb{P}}\big(\{Y_1 \in C_1\} \cap \{Y_2 \in C_2\}\big) = \tilde{\mathbb{P}}(Y_1 \in C_1) \cdot \tilde{P}(Y_2 \in C_2)$$

for all $C_1 \in \mathcal{F}_s$, $C_2 \in \mathcal{F}_t$ and all $s, t \in \mathbb{N}$. We conclude that the same identity must also be valid for all $C_1 \in \mathcal{F}_{fin}$ and $C_2 \in \mathcal{F}$. This proves independence of the random variables Y_1 and Y_2.

(iv) The support of the indicator function F is equal to:

$$\bigcup_{s\geq 1}\Big\{(\omega_1, \omega_2) \in \Omega_s \times \Omega;$$

$$\min\{n > s;\ Z_n(g(\omega_1, \omega_2)) \in B_{\frac{\epsilon}{2}}(y_0)\} < \min\{n > s;\ Z_n \in A_{i+1}\}\Big\},$$

where $g : \Omega_s \times \Omega \to \Omega$ is given by:

$$g\big(\{(w_n, a_n, b_n)\}_{n=1}^{s}, \{(w_n, a_n, b_n)\}_{n=s+1}^{\infty}\big) \doteq \{(w_n, a_n, b_n)\}_{n=1}^{\infty}.$$

For every $s \geq 1$, the set in the displayed union is measurable, in view of the measurability of g, proved as in Exercise 5.10 (ii).

Exercise 6.3

Apply the argument as in the proof of Theorem 6.1, we get:

$$\begin{aligned}
f_u(x; x_0, r) &= u(x) + \frac{1}{2}\Big\langle \nabla^2 u(x) : \gamma^2 |x - x_0|^2 \fint_{\partial B^{N-1}} \big(R(x)y\big)^{\otimes 2}\,\mathrm{d}y\Big\rangle + o(\epsilon^2)\\
&= u(x) + \frac{\gamma^2}{2(N-1)}\Big(|x - x_0|^2 \Delta u(x) - \big\langle \nabla^2 u(x) : (x - x_0)^{\otimes 2}\big\rangle\Big) + o(\epsilon^2)\\
&= u(x_0) + \langle \nabla u(x_0), x - x_0\rangle + \frac{\gamma^2}{2(N-1)}|x - x_0|^2 \Delta u(x_0)\\
&\quad + \Big(\frac{1}{2} - \frac{\gamma^2}{2(N-1)}\Big)\big\langle \nabla^2 u(x_0) : (x - x_0)^{\otimes 2}\big\rangle + o(\epsilon^2),
\end{aligned}$$

where we used: $\fint_{\partial B_1^d(0)} y^{\otimes 2}\,\mathrm{d}y = \frac{1}{d}\fint_{\partial B_1^d(0)} |y|^2\,\mathrm{d}y \cdot Id_d = \frac{1}{d} Id_d$, so that:

$$\fint_{\partial B^{N-1}} \big(R(x)y\big)^{\otimes 2}\,\mathrm{d}y = \frac{1}{N-1} R(x)\big(Id_N - e_N^{\otimes 2}\big)R(x)^T = \frac{1}{N-1}\Big(Id_N - \big(\frac{x - x_0}{|x - x_0|}\big)^{\otimes 2}\Big).$$

Calling $\bar{\bar{f}}_u$ the polynomial:

$$\bar{\bar{f}}_u(x; x_0, \epsilon) = u(x_0) + \langle \nabla u(x_0), x - x_0 \rangle + \Big\langle \frac{\gamma^2}{2(N-1)} \Delta u(x_0) Id_N + \Big(\frac{1}{2} - \frac{\gamma^2}{2(N-1)}\Big) \nabla^2 u(x_0) : (x - x_0)^{\otimes 2} \Big\rangle,$$

the claim in Step 2 of proof of Theorem 6.1 yields:

$$\frac{1}{2}\Big(\inf_{x \in B(x_0,\epsilon)} \bar{\bar{f}}_u(x; x_0, \epsilon) + \sup_{x \in B(x_0,\epsilon)} \bar{\bar{f}}_u(x; x_0, \epsilon) \Big) = u(x_0) + \frac{\gamma^2 \epsilon^2}{2(N-1)} \Big(\Delta u(x_0) + \Big(\frac{N-1}{\gamma^2} - 1\Big) \Delta_\infty u(x_0) \Big) + O(\epsilon^3).$$

Clearly, there holds $\frac{N-1}{\gamma^2} - 1 = p - 2$ for the anticipated scaling factor $\gamma = \sqrt{\frac{N-1}{p-1}}$.

Exercise 6.19
We only indicate the proof of (i). For every small $\hat{\delta} > 0$, define the open, bounded, connected set $\mathcal{D}^{\hat{\delta}}$ and the distance:

$$\mathcal{D}^{\hat{\delta}} = \big\{ q \in \mathcal{D};\ \mathrm{dist}(q, \mathbb{R}^N \setminus \mathcal{D}) > \hat{\delta} \big\} \quad \text{and} \quad d_\epsilon^{\hat{\delta}}(q) = \frac{1}{\epsilon} \min\{\epsilon, \mathrm{dist}(q, \mathbb{R}^N \setminus \mathcal{D}^{\hat{\delta}})\}.$$

Fix $\eta > 0$. In view of (6.27) and since without loss of generality the data F is constant outside of some large bounded superset of $\mathcal{D}$ in $\mathbb{R}^N$, there exists $\hat{\delta} > 0$ with:

$$|u_\epsilon(x + z) - u_\epsilon(x)| \leq \eta \qquad \text{for all } x \in \mathbb{R}^N \setminus \mathcal{D}^{\hat{\delta}}, \ |w| \leq \hat{\delta}, \ \epsilon \in (0, \hat{\epsilon}). \tag{D.1}$$

Fix $x_0, y_0 \in \bar{\mathcal{D}}$ with $|x_0 - y_0| \leq \frac{\hat{\delta}}{2}$ and let $\epsilon \in (0, \frac{\hat{\delta}}{2})$. Consider the following function $\tilde{u}_\epsilon \in C(\mathbb{R}^N)$:

$$\tilde{u}_\epsilon(x) = u_\epsilon(x - (x_0 - y_0)) + \eta.$$

Then, by (6.11) and recalling the definition of the averaging operator S_ϵ, we get:

$$\begin{aligned}(S_\epsilon \tilde{u}_\epsilon)(x) &= (S_\epsilon u_\epsilon)(x - (x_0 - y_0)) + \eta = u_\epsilon(x - (x_0 - y_0)) + \eta \\ &= \tilde{u}_\epsilon(x) \quad \text{for all } x \in \mathcal{D}^{\hat{\delta}}.\end{aligned} \tag{D.2}$$

because in $\mathcal{D}^{\hat{\delta}}$ there holds:

$$\mathrm{dist}(x - (x_0 - y_0), \mathbb{R}^N \setminus \mathcal{D}) \geq \mathrm{dist}(x, \mathbb{R}^N \setminus \mathcal{D}) - |x_0 - y_0| \geq \hat{\delta} - \frac{\hat{\delta}}{2} = \frac{\hat{\delta}}{2} > \epsilon.$$

It follows now from (D.2) that:

$$\tilde{u}_\epsilon = d_\epsilon^{\hat{\delta}}(S_\epsilon \tilde{u}_\epsilon) + \left(1 - d_\epsilon^{\hat{\delta}}\right)\tilde{u}_\epsilon \qquad \text{in } \mathbb{R}^N.$$

On the other hand, u_ϵ itself similarly solves the same problem above, subject to its own data u_ϵ on $\mathbb{R}^N \setminus \mathcal{D}^{\hat{\delta}}$. Since for every $x \in \mathbb{R}^N \setminus \mathcal{D}^{\hat{\delta}}$ we have: $\tilde{u}_\epsilon(x) - u_\epsilon(x) = u_\epsilon(x - (x_0 - y_0)) - u_\epsilon(x) + \eta \geq 0$ in view of (D.1), the monotonicity property in Theorem 6.5 yields:

$$u_\epsilon \leq \tilde{u}_\epsilon \qquad \text{in } \mathbb{R}^N.$$

Thus, in particular: $u_\epsilon(x_0) - u_\epsilon(y_0) \leq \eta$. Exchanging x_0 with y_0 we get the opposite inequality. In conclusion, $|u_\epsilon(x_0) - u_\epsilon(y_0)| \leq \eta$, which yields the claimed equicontinuity of $\{u_\epsilon\}_{\epsilon\to 0}$ in $\bar{\mathcal{D}}$.

Exercise A.16

(i) Since $\{Y_1 < Y_2\} \in \mathcal{G}$, the defining condition of $\mathbb{E}(X \mid \mathcal{G}) = Y_i$ yields:

$$\int_{\{Y_1<Y_2\}} Y_1 - Y_2 \,\mathrm{d}\mathbb{P} = \int_{\{Y_1<Y_2\}} X - X \,\mathrm{d}\mathbb{P} = 0.$$

Consequently $\mathbb{P}(Y_1 < Y_2) = 0$, while $\mathbb{P}(Y_1 > Y_2) = 0$ follows by exchanging Y_1 with Y_2. It follows that $\mathbb{P}(Y_1 \neq Y_2) = 0$, as claimed.

Exercise A.18

(ii) By Exercise A.16 (iii), it follows that $\mathbb{E}(X_i \mid \mathcal{G})$ is a.s. nonnegative and nondecreasing. Hence this sequence converges pointwise a.s. to a $\mathcal{G}$-measurable random variable Y. Since for every $A \in \mathcal{G}$ we have, in virtue of Theorem A.7:

$$\int_A Y \,\mathrm{d}\mathbb{P} = \lim_{i\to\infty} \int_A \mathbb{E}(X_i \mid \mathcal{G}) \,\mathrm{d}\mathbb{P} = \lim_{i\to\infty} \int_A X_i \,\mathrm{d}\mathbb{P} = \int_A X \,\mathrm{d}\mathbb{P},$$

we obtain that $Y = \mathbb{E}(X \mid \mathcal{G})$.

(i) The sequence $Z_i = \inf_{j\geq i} X_j$ is a.s. nondecreasing and nonnegative, so by the already established statement in (ii) and since $Z_i \leq X_i$ a.s. we get:

$$\mathbb{E}\big(\liminf_{i\to\infty} X_i \mid \mathcal{G}\big) = \mathbb{E}\big(\lim_{i\to\infty} Z_i \mid \mathcal{G}\big) = \lim_{i\to\infty} \mathbb{E}(Z_i \mid \mathcal{G}) \leq \liminf_{i\to\infty} \mathbb{E}(X_i \mid \mathcal{G}) \quad a.s.$$

(iii) Apply (i) to the a.s. nonnegative $Z + X_i,\ Z - X_i \in L^1(\Omega, \mathcal{F}, \mathbb{P})$, to get:

$$\mathbb{E}(Z \pm X \mid \mathcal{G}) \leq \liminf_{i\to\infty} \mathbb{E}(Z \pm X_i \mid \mathcal{G}) \quad a.s.$$

Consequently:

$$\mathbb{E}(X \mid \mathcal{G}) \leq \liminf_{i\to\infty} \mathbb{E}(X_i \mid \mathcal{G}) \leq \limsup_{i\to\infty} \mathbb{E}(X_i \mid \mathcal{G}) \leq \mathbb{E}(X \mid \mathcal{G}) \quad a.s.$$

which yields the claimed equality.

Exercise A.26

(i) The following equalities result from the definition of $\{X_i\}_{i=0}^\infty$ and the tower property of conditional expectation:

$$\mathbb{E}(X_{i+1} \mid \mathcal{F}_i) = \mathbb{E}\big(\mathbb{E}(X \mid \mathcal{F}_{i+1}) \mid \mathcal{F}_i\big) = \mathbb{E}(X \mid \mathcal{F}_i) = X_i \quad \text{a.s.}$$

(ii) For ϕ convex (hence automatically continuous) we have: $\mathbb{E}(\phi \circ X_{i+1} \mid \mathcal{F}_i) \geq \phi \circ \mathbb{E}(X_{i+1} \mid \mathcal{F}_i) = \phi \circ X_i$ a.s. by Exercise A.16 (vi), whereas for ϕ concave the opposite inequality holds.

Exercise A.29

(i) Clearly $\emptyset \cap \{\tau \leq i\} \in \mathcal{F}_i$ so $\emptyset \in \mathcal{F}_\tau$, and also if $A \in \mathcal{F}_\tau$, then:

$$(\Omega \setminus A) \cap \{\tau \leq i\} = \{\tau \leq i\} \setminus (A \cap \{\tau \leq i\}) \in \mathcal{F}_i.$$

Further, given $\{A_k \in \mathcal{F}_\tau\}_{k=1}^\infty$ it follows that $\{\tau \leq i\} \cap \bigcup_{k=1}^\infty A_k = \bigcup_{k=1}^\infty (A_k \cap \{\tau \leq i\}) \in \mathcal{F}_\tau$. Consequently, $\mathcal{F}_\tau$ is a σ-algebra.

To show that τ is $\mathcal{F}_\tau$-measurable we observe that $\{\tau \leq k\} \in \mathcal{F}_\tau$, because: $\{\tau \leq k\} \cap \{\tau \leq i\} = \{\tau \leq k \wedge i\} \in \mathcal{F}_{\tau \wedge i} \subset \mathcal{F}_i$.

Finally, for $\tau_1 \leq \tau_2$ and $A \in \mathcal{F}_{\tau_1}$ we get that $A \in \mathcal{F}_{\tau_2}$, because:

$$A \cap \{\tau_2 \leq i\} = (A \cap \{\tau_1 \leq i\}) \setminus \big(\{\tau_1 \leq i\} \setminus \{\tau_2 \leq i\}\big) \in \mathcal{F}_i.$$

(ii) We have: $\{X_\tau \leq r\} \cap \{\tau \leq i\} = \bigcup_{k=0}^i \big(\{X_k \leq r\} \cap \{\tau = k\}\big) \in \mathcal{F}_i$, because $\{X_k \leq r\} \cap \{\tau = k\} \in \mathcal{F}_k \subset \mathcal{F}_i$ for all $i \leq k$. Hence $\{X_\tau \leq r\} \in \mathcal{F}_\tau$ and X_τ is $\mathcal{F}_\tau$-measurable. Integrability follows by:

$$\int_\Omega |X_\tau| \, \mathrm{d}\mathbb{P} = \sum_{k=0}^{\max \tau} \int_{\{\tau = k\}} |X_k| \, \mathrm{d}\mathbb{P} \leq \sum_{k=0}^{\max \tau} \int_\Omega |X_k| \, \mathrm{d}\mathbb{P} < \infty.$$

Exercise A.32

If $\{X_i\}_{i=0}^\infty$ are uniformly integrable, with $\|X_i\|_{L^1} \leq C$ for all i. Fix $\epsilon > 0$ and let $\delta > 0$ be as in the equiintegrability condition. Then:

$$\mathbb{P}\Big(|X_i| > \frac{C}{\delta}\Big) = \frac{\delta}{C} \int_{\{|X_i| > C/\delta\}} \frac{C}{\delta} \, \mathrm{d}\mathbb{P} \leq \frac{\delta}{C} \int_{\{|X_i| > C/\delta\}} |X_i| \, \mathrm{d}\mathbb{P} \leq \delta,$$

so $\int_{\{|X_i|>M\}} |X_i| \, d\mathbb{P} < \epsilon$, as claimed. For the opposite direction, assume the validity of the condition in Exercise A.32. Let M be chosen for $\epsilon = 1$, then:

$$\int_\Omega |X_i| \, d\mathbb{P} = \int_{\{|X_i|>M\}} |X_i| \, d\mathbb{P} + \int_{\{|X_i|\leq M\}} |X_i| \, d\mathbb{P} \leq 1 + M,$$

proving equiboundedness. To show equiintegrability, fix $\epsilon > 0$. Then:

$$\int_A |X_i| \, d\mathbb{P} = \int_{A\cap\{|X_i|>M\}} |X_i| \, d\mathbb{P} + \int_{A\cap\{|X_i|\leq M\}} |X_i| \, d\mathbb{P} < \epsilon + M \cdot \mathbb{P}(A) < 2\epsilon,$$

if only $\mathbb{P}(A) < \frac{\epsilon}{M}$.

Exercise A.33

By assumption, for some $C > 0$ there holds: $|X_i| \leq Ci + |X_0|$ a.s. for all $i \geq 0$. Hence, for every $A \in \mathcal{F}$ we get:

$$\int_A |X_{\tau\wedge i}| \, d\mathbb{P} \leq \int_A C(\tau \wedge i) + |X_0| \, d\mathbb{P} \leq C \int_A \tau \, d\mathbb{P} + \int_A |X_0| \, d\mathbb{P}.$$

Taking $A = \Omega$ we see that the sequence $\{X_{\tau\wedge i}\}_{i=0}^\infty$ is bounded in $L^1(\Omega, \mathcal{F}, \mathbb{P})$. The equiintegrability likewise follows, from integrability of X_0 and τ.

Exercise B.3

(i) We have, by the change of variable formula: $\mathbb{P}\big(\gamma X \in A\big) = \mathbb{P}\big(X \in \frac{1}{\gamma}A\big) = \int_{\frac{1}{\gamma}A} \frac{1}{\sqrt{2\pi\sigma^2}} e^{-\frac{(x-\mu)^2}{2\sigma^2}} \, dx = \int_A \frac{1}{\sqrt{2\pi\sigma^2}|\gamma|} e^{-\frac{(y-\gamma\mu)^2}{2\sigma^2\gamma^2}} \, dy.$

(ii) Let $A \subset \mathbb{R}$ be a Borel set. Applying Lemma A.21 to the nonnegative random variable $Z(x, y) = \mathbb{1}_{\{x+y\in A\}}$ on $\mathbb{R}^2$, we obtain:

$$\begin{aligned}
\mathbb{P}\big(X_1 + X_2 \in A\big) &= \int_\Omega \mathbb{P}\big(X_2 \in A \setminus X_1(\omega_1)\big) \, d\mathbb{P}(\omega_1) \\
&= \int_\mathbb{R} \frac{1}{\sqrt{2\pi\sigma_1^2}} e^{-\frac{(x-\mu_1)^2}{2\sigma_1^2}} \cdot \frac{1}{\sqrt{2\pi\sigma_2^2}} \int_{A-x} e^{-\frac{(y-\mu_2)^2}{2\sigma_2^2}} \, dy \, dx \\
&= \int_A \frac{1}{\sqrt{2\pi\sigma_1^2} \cdot \sqrt{2\pi\sigma_2^2}} \int_R e^{-\frac{(x-\mu_1)^2}{2\sigma_1^2}} \cdot e^{-\frac{(z-x-\mu_2)^2}{2\sigma_2^2}} \, dx \, dz = \int_A f(z) \, dz,
\end{aligned}$$

where f is given as follows:

$$f(z) = \int_\mathbb{R} \frac{1}{\sqrt{2\pi\sigma_1^2} \cdot \sqrt{2\pi\sigma_2^2}} e^{-\frac{\sigma_2^2(x-\mu_1)^2+\sigma_1^2(z-x-\mu_2)^2}{2\sigma_1^2\sigma_2^2}} \, dx.$$

We now write $\sigma^2 = \sigma_1^2 + \sigma_2^2$ and compute the convolution:

$$f(z) = \int_{\mathbb{R}} \frac{1}{\sqrt{2\pi \frac{\sigma_1^2\sigma_2^2}{\sigma^2}}} \cdot \frac{1}{\sqrt{2\pi\sigma^2}} e^{-\frac{\left(x-\frac{\sigma_2^2\mu_1+\sigma_1^2(z-\mu_2)}{\sigma^2}\right)^2 - \left(\frac{\sigma_2^2\mu_1+\sigma_1^2(z-\mu_2)}{\sigma^2}\right)^2 + \frac{\sigma_1^2(z-\mu_2)^2+\sigma_2^2\mu_1^2}{\sigma^2}}{2\sigma_1^2\sigma_2^2/\sigma^2}} \, \mathrm{d}x$$

$$= \int_{\mathbb{R}} \frac{1}{\sqrt{2\pi \frac{\sigma_1^2\sigma_2^2}{\sigma^2}}} \cdot e^{-\frac{\left(x-\frac{\sigma_2^2\mu_1+\sigma_1^2(z-\mu_2)}{\sigma^2}\right)^2}{2\sigma_1^2\sigma_2^2/\sigma^2}} \, \mathrm{d}x \,\cdot$$

$$\cdot \frac{1}{\sqrt{2\pi\sigma^2}} \int_{\mathbb{R}} e^{-\frac{-\left(\sigma_2^2\mu_1+\sigma_1^2(z-\mu_2)\right)^2 + \sigma^2\sigma_1^2(z-\mu_2)^2+\sigma^2\sigma_2^2\mu_1^2}{2\sigma^2\sigma_1^2\sigma_2^2}} .$$

The integral above, after a change of variable, equals:

$$\int_{\mathbb{R}} \frac{1}{\sqrt{2\pi \frac{\sigma_1^2\sigma_2^2}{\sigma^2}}} e^{-\frac{x^2}{2(\sigma_1\sigma_2/\sigma)^2}} \, \mathrm{d}x = 1,$$

and thus we obtain the claimed property:

$$f(z) = \frac{1}{\sqrt{2\pi\sigma^2}} e^{-\frac{\sigma_1\sigma_2^2\mu_1^2+\sigma_1^2\sigma_2^2(z-\mu_2)^2-2\sigma_1^2\sigma_2^2(z-\mu_2)^2}{\sigma^2\sigma_1^2\sigma_2^2}} = \frac{1}{\sqrt{2\pi\sigma^2}} e^{-\frac{(z-(\mu_1+\mu_2))^2}{\sigma^2}} .$$

(iii) By part (i) it immediately follows that: $\frac{1}{\sqrt{2}} X_1 \pm \frac{1}{\sqrt{2}} X_2 \sim N(0, 1)$, so by (ii) we have: $\frac{1}{\sqrt{2}}(X_1 + X_2), \frac{1}{\sqrt{2}}(X_1 - X_2) \sim N(0, 1)$. To prove independence, denote the rotation of $\mathbb{R}^2$ by $\pi/4$:

$$R_{\pi/4} = \begin{bmatrix} \frac{1}{\sqrt{2}} & -\frac{1}{\sqrt{2}} \\ \frac{1}{\sqrt{2}} & \frac{1}{\sqrt{2}} \end{bmatrix}$$

and observe that: $\left(\frac{1}{\sqrt{2}}(X_1 - X_2), \frac{1}{\sqrt{2}}(X_1 + X_2)\right) = R_{\pi/4}\big(X_1, X_2\big)$. Let A_1, A_2 be two Borel subsets of $\mathbb{R}$. Then, using notation of Exercise A.20:

$$\mathbb{P}\Big(\Big\{\frac{1}{\sqrt{2}}(X_1 - X_2) \in A_1\Big\} \cap \Big\{\frac{1}{\sqrt{2}}(X_1 + X_2) \in A_2\Big\}\Big) = \mathbb{P}\Big(R_{\pi/4}(X_1, X_2) \in A_1 \times A_2\Big)$$

$$= \mathbb{P}\Big((X_1, X_2) \in R_{-\pi/4}(A_1 \times A_2)\Big) = \bar{\mathbb{P}}\Big(R_{-\pi/4}(A_1 \times A_2)\Big)$$

$$= \bar{\bar{\mathbb{P}}}\Big(R_{-\pi/4}(A_1 \times A_2)\Big) = \int_{R_{-\pi/4}(A_1 \times A_2)} \frac{1}{2\pi} e^{-\frac{x^2}{2}} \cdot e^{-\frac{x^2}{2}} \, \mathrm{d}x \, \mathrm{d}y$$

$$= \int_{A_1\times A_2} \frac{1}{2\pi} e^{-\frac{x^2+y^2}{2}} \,\mathrm{d}(x,y)$$

$$= \Big(\frac{1}{\sqrt{2\pi}} \int_{A_1} e^{-\frac{x^2}{2}} \,\mathrm{d}x\Big) \cdot \Big(\frac{1}{\sqrt{2\pi}} \int_{A_2} e^{-\frac{x^2}{2}} \,\mathrm{d}x\Big)$$

$$= \mathbb{P}\Big(\frac{1}{\sqrt{2}}(X_1 - X_2) \in A_1\Big) \cdot \mathbb{P}\Big(\frac{1}{\sqrt{2}}(X_1 + X_2) \in A_2\Big),$$

where we used the rotational invariance of the probability density:

$$\mathbb{R}^2 \ni z \mapsto \frac{1}{2\pi} e^{-\frac{|z|^2}{2}}$$

and the already noted fact that $\frac{1}{\sqrt{2}}(X_1 \pm X_2)$ are normally distributed.

Exercise C.9

(i) Consider the functions $\phi_n \in C^1(\mathbb{R})$ given, with their derivatives, by:

$$\phi_n(x) = \begin{cases} 0 & \text{for } x \le 0 \\ \frac{n}{2}x^2 & \text{for } x \in (0, \frac{1}{n}) \\ x - \frac{1}{2n} & \text{for } x \ge \frac{1}{n} \end{cases} \quad \text{so that} \quad \phi_n'(x) = \begin{cases} 0 & \text{for } x \le 0 \\ nx & \text{for } x \in (0, \frac{1}{n}) \\ 1 & \text{for } x \ge \frac{1}{n}. \end{cases}$$

Since $\phi_n(0) = 0$ and each ϕ_n' is bounded, Exercise C.8 (iv) implies that $\phi_n \circ f \in W^{1,p}(U)$ for all $n \ge 1$. We now note that the sequence $\{\phi_n \circ f\}_{n=1}^\infty$ converges to f^+ in $L^p(U)$ by the monotone convergence theorem. Further, the sequence of weak gradients $\{\nabla(\phi_n \circ f) = (\phi_n' \circ f)\nabla f\}_{n=1}^\infty$ converges in $L^p(U, \mathbb{R}^N)$ to:

$$g(x) = \begin{cases} \nabla f(x) & \text{a.e. in } \{f > 0\} \\ 0 & \text{a.e. in } \{f \le 0\} \end{cases}$$

by the dominated convergence theorem. Exercise C.8 (i) completes the proof.

(ii) The sequence $\{f_n^+\}_{n=1}^\infty$ converges to f^+ in $L^p(U)$ because: $|f_n^+ - f^+| \le |f_n - f|$ a.e. in U. Further, we have:

$$\int_U |\nabla f_n^+ - \nabla f^+|^p \,\mathrm{d}x = \int_U \big|(\mathbb{1}_{(0,\infty)} \circ f_n)\nabla f_n - (\mathbb{1}_{(0,\infty)} \circ f)\nabla f\big|^p \,\mathrm{d}x$$

$$\le C\Big(\int_U |\nabla f_n - \nabla f|^p \,\mathrm{d}x + \int_U \big|\mathbb{1}_{(0,\infty)} \circ f_n - \mathbb{1}_{(0,\infty)} \circ f\big| \cdot |\nabla f|^p \,\mathrm{d}x\Big).$$

The first term in the right hand side above converges to 0 as $n \to \infty$, by assumption. The second term converges to 0 by the dominated convergence theorem. This achieves convergence of $\{f_n^+\}_{n=1}^\infty$ to f^+ in $W^{1,p}(U)$.

Exercise C.16

(i) Fix $x \in U$. If f satisfies (C.4) and $f(x) > \lambda$, then for sufficiently small $r > 0$ we have: $\inf_{B_r(x)\cap U} f > \lambda$, proving that the set $\{f > \lambda\}$ contains an open neighbourhood of x. Conversely, if the sets $\{f > \lambda\}$ are open for all $\lambda \in \mathbb{R}$, then taking $\lambda = f(x) - \epsilon$, we obtain: $u(y) > \lambda$ for all $y \in B_r(x) \cap U$, where $r > 0$ is an appropriate radius. Consequently:

$$u(x) - \epsilon \leq \inf_{B_r(x)\cap(U\setminus\{x\})} f \leq \liminf_{y\to x} f(y)$$

and passing to the limit with $\epsilon \to 0$ there follows (C.4).

(ii) Let $\{x_n \in K\}_{n=1}^\infty$ be a sequence converging to some $x \in K$ and such that $\lim_{n\to\infty} f(x_n) = \inf_K f \in \bar{\mathbb{R}}$. Then, either $\{x_n\}_{n=1}^\infty$ has a constant subsequence so that $f(x) = \inf_K f$, or we may assume that $x_n \neq x$ for all n, in which case:

$$\inf_K f \leq f(x) \leq \inf_{n\to\infty} f(x_n) = \inf_K f$$

by (C.4) so that, again: $f(x) = \inf_K f$. This proves the main claim of the exercise. Finally, the following function $f : \mathbb{R} \to (-\infty, +\infty]$ is lower-semicontinuous but it does not attain its supremum on $K = [-1, 1]$:

$$f(x) = \begin{cases} x^2 & \text{if } x \in (-1, 1) \\ 0 & \text{if } |x| \geq 1. \end{cases}$$

(iii) If all functions in the family $\{f_i : U \to (-\infty, +\infty]\}_{i\in I}$ are lower-semicontinuous, then for every $\lambda \in \mathbb{R}$ the following set is open:

$$\big\{(\sup_{i\in I} f_i) > \lambda\big\} = \bigcup_{i\in I}\{f_i > \lambda\},$$

proving lower-semicontinuity of $(\sup_{i\in I} f_i) : U \to (-\infty, +\infty]$ by (i). If the index set I has finitely many elements, then each set $\big\{(\min_{i\in I} f_i) > \lambda\big\} = \bigcap_{i\in I}\{f_i > \lambda\}$ is also open, hence follows the second assertion of the result.

(iv) Let $\{K_n\}_{n=1}^\infty$ be an increasing sequence of compact subsets of U, such that $U = \bigcup_{n=1}^\infty K_n$. Since f is bounded from below on each K_n in view of (ii), one can construct a continuous $g : U \to \mathbb{R}$ satisfying $g \leq f$ in U (by modifying the constant lower bounds on each K_n close to the boundary). A similar reasoning yields:

$$f(x) = \sup\big\{g(x);\;\; g \in C(U) \text{ with } g \leq f \text{ in } U\big\}.$$

We now refine this construction. Let a sequence $\{B_n\}_{n=1}^\infty$ be an enumeration of the countable family of all open balls with rational centres and rational radii, compactly contained in U. For each $i, n \in \mathbb{N}$ choose a function $g_{i,n} \in C(U)$

such that $g_{i,n} \leq f$ in U and that on the ball $\frac{1}{2}B_n$ with the same centre as B_n but half its radius:

$$g_{i,n} = \begin{cases} \left(\inf_{B_n} f\right) - \frac{1}{i} & \text{if } \inf_{B_n} f < +\infty \\ i & \text{if } \inf_{B_n} f = +\infty. \end{cases}$$

Since $f = \sup_{i,n} f_{i,n}$, then taking $f_n \doteq \max_{1\leq i,k\leq n} g_{i,k}$ we obtain a nondecreasing sequence $\{f_n \in C(U)\}_{n=1}^{\infty}$ that converges pointwise to f. It is easy to modify $\{f_n\}_{n=1}^{\infty}$ to the final sequence of smooth functions, strictly increasing to f.

(v) By the local boundedness property we get $g(x) \neq -\infty$ for all $x \in U$. It further directly follows:

$$\liminf_{y\to x} g(y) = \liminf_{y\to x} \left(\liminf_{z\to y} f(z)\right) = \liminf_{y\to x} f(y) = g(x).$$

Exercise C.17

(i) By the essential local boundedness property, we get that $g : U \to (-\infty, +\infty]$. In order to show that g is lower-semicontinuous, we observe that:

$$\liminf_{y\to x} g(y) = \liminf_{y\to x} \left(\operatorname{ess}\liminf_{z\to y} f(z)\right) = \operatorname{ess}\liminf_{y\to x} f(y) = g(x).$$

(ii) Recall that $f(x) = \lim_{r\to 0} \fint_{B_r(x)} f$ for almost all $x \in U$, by Theorem C.3. On the other hand: $\operatorname{ess\,inf}_{B_r(x)} \leq \fint_{B_r(x)} f \leq \operatorname{ess\,sup}_{B_r(x)} f$, so passing to the limit with $r \to 0$ and using (C.5) we obtain that the function:

$$g(x) \doteq \lim_{r\to 0} \fint_{B_r(x)} f = \operatorname{ess}\liminf_{y\to x} f(y) = \operatorname{ess}\limsup_{y\to x} f(y)$$

is well defined and that it coincides with f almost everywhere in U. By (i) it follows that g must be continuous.

Exercise C.30

It suffices to show that if $u \in W^{1,p}_{loc}(\mathcal{D})$ satisfies:

$$\int_{\mathcal{D}} \left\langle |\nabla u|^{p-2}\nabla u, \nabla\eta \right\rangle \,\mathrm{d}x = 0 \qquad \text{for all } \eta \in C_c^\infty(\mathcal{D}) \text{ with } \eta \geq 0, \tag{D.3}$$

then the above equality holds in fact for all test functions $\eta \in C_c^\infty(\mathcal{D})$. Writing $\eta = \eta^+ - \eta^-$ where $\eta^+, \eta^- \in C_c(\mathcal{D})$ we approximate: $\eta^\pm = \lim_{n\to\infty} \eta_n^\pm$ in $W_0^{1,p}(\mathcal{D})$ by smooth compactly supported functions $\{\eta_n^\pm\}_{n=1}^\infty$. By (D.3) we have: $\int_{\mathcal{D}} \langle |\nabla u|^{p-2}\nabla u, \nabla\eta_n\rangle \,\mathrm{d}x \geq 0$ for all $\eta_n = \eta_n^+ - \eta_n^- \in C_c^\infty(\mathcal{D})$. Since $\{\eta_n\}_{n=1}^\infty$ converges to η in $W^{1,p}(\mathcal{D})$ and $|\nabla u|^{p-2}\nabla u \in L^{p'}_{loc}(\mathcal{D})$, the result follows by passing to the limit with $n \to \infty$ in view of (C.2).

Exercise C.37

(i) The claimed formula follows by direct inspection. We also note that both terms in its right hand side are nonnegative. Thus:

$$\left\langle |b|^{p-2}b - |a|^{p-2}a, b-a \right\rangle \leq 0$$

if and only if both terms equal zero: $|b-a|^2 = 0$ or equivalently: $a = b$.

(ii) Since every nonnegative function in $W_0^{1,p}(\mathcal{D})$ may be approximated by a sequence of nonnegative test functions in $C_c^\infty(\mathcal{D})$, we may apply Definition C.29 with $\eta = (u_1 - u_2)^+ \in W_0^{1,p}(\mathcal{D})$ in view of Exercise C.9. This yields:

$$\begin{aligned} 0 &\geq \int_{\mathcal{D}} \left\langle |\nabla u_1|^{p-2}\nabla u_1 - |\nabla u_2|^{p-2}\nabla u_2, \nabla(u_1 - u_2)^+ \right\rangle \,\mathrm{d}x \\ &= \int_{\{u_1 > u_2\}} \left\langle |\nabla u_1|^{p-2}\nabla u_1 - |\nabla u_2|^{p-2}\nabla u_2, \nabla u_1 - \nabla u_2 \right\rangle \,\mathrm{d}x. \end{aligned}$$

Since by (i) the last integrand above is nonnegative and it equals zero if and only if $\nabla u_1 = \nabla u_2$, there must be $\nabla u_1 = \nabla u_2$ a.e. in $\{u_1 > u_2\}$. Consequently, $\nabla(u_1 - u_2)^+ = 0$ so that $(u_1 - u_2)^+ = 0$, in view of $(u_1 - u_2)^+ \in W_0^{1,p}(\mathcal{D})$. This yields the claimed inequality: $u_1 \leq u_2$ a.e. in $\mathcal{D}$.

Exercise C.43

(i) We argue by contradiction. If $u_2(x_n) + \epsilon \leq u_1(x_n)$ along some sequence $\{x_n \in \mathcal{D}\}_{n=1}^\infty$ converging to $x \in \partial\mathcal{D}$, then:

$$\begin{aligned} \liminf_{y\to x} u_2(y) + \epsilon &\leq \liminf_{n\to\infty} u_1(x_n) + \epsilon \\ &\leq \limsup_{n\to\infty} u_1(x_n) \leq \limsup_{y\to x} u_1(y). \end{aligned} \tag{D.4}$$

In particular, $\liminf_{y\to x} u_2(y) \neq +\infty$ as $\limsup_{y\to x} u_1(y) \neq +\infty$ by assumption. Likewise, $\liminf_{y\to x} u_2(x) \neq -\infty$ by assumption that both quantities in (C.20) cannot equal $-\infty$. From (D.4) and (C.20) we now get: $\liminf_{y\to x} u_2(y) + \epsilon \leq \liminf_{y\to x} u_2(y)$ which is a contradiction.

(ii) Assume that there is a sequence $\{x_n \in \partial\mathcal{D}\}_{n=1}^\infty$ converging to some $x \in \partial\mathcal{D}$, such that $\phi_n(x_n) + \epsilon < u_1(x_n)$ for all $n \geq 1$. Let $m \geq 1$ be such that $\phi_m(x) + \epsilon > u_1(x)$. Since the function $u_1 - \phi_m$ is upper-semicontinuous, it follows that: $\phi_n > u_1 - \epsilon$ holds in some open neighbourhood of x. Thus, for all $n > m$ sufficiently large:

$$\phi_m(x_n) + \epsilon \geq \phi_m(x_n) + \epsilon > u_1(x_n),$$

which is a contradiction.

(iii) The inequality $u_1 \leq v_n + \epsilon \leq u_2 + \epsilon$ in U follows by Definition C.39. Since one may exhaust $\mathcal{D}$ by admissible domains U such that $\partial U \subset V_\delta$, it follows that $u_1 \leq u_2 + \epsilon$ in $\mathcal{D}$. Passing to the limit with $\epsilon \to 0$, we finally obtain: $u_1 \leq u_2$ in $\mathcal{D}$.

Exercise C.55
Since each term of the series is harmonic and the series converges absolutely uniformly in $\mathcal{D}$, hence u is harmonic in $\mathcal{D}$ and continuous in $\bar{D}$. On the other hand:

$$\int_{\mathcal{D}} |\nabla u(x)|^2 \, dx \geq \int_0^{2\pi} \int_0^{\rho} |\partial_r u(r,\theta)|^2 r \, dr d\theta = \sum_{n=1}^{\infty} \frac{\pi n!}{2n^4} \rho^{2n!} \geq \sum_{n=1}^{m} \frac{\pi n!}{2n^4} \rho^{2n!},$$

for any $\rho < 1$ and $m \geq 1$. This implies that $I_p(u) = +\infty$.

Exercise C.57
Since $z \mapsto -\frac{1}{\log(z-x)}$ is analytic in $\tilde{B}$, its real part v is 2-harmonic. Writing: $\log(z - x) = \log|z-x| + i\mathrm{Arg}(z-x)$, we get:

$$v(z) = -\frac{\log|z-x|}{(\log|z-x|)^2 + (\mathrm{Arg}(z-x))^2}$$

which implies the limit properties of v. Let $c \doteq \inf\big\{v(z);\ z \in B_r(x) \setminus \big(B_{r/2}(x) \cup S\big)\big\} > 0$. Then the following function $u : \mathcal{D} \to \mathbb{R}$ is a barrier at x:

$$u(z) = \begin{cases} v(z) & \text{if } z \in \mathcal{D} \cap B_{r/2}(x) \\ c & \text{if } z \in \mathcal{D} \setminus B_{r/2}(x). \end{cases}$$

Indeed, u still satisfies the required limit properties and it is p-superharmonic in $\mathcal{D}$, in view of Lemma C.42.

Exercise C.60
Assume that u is convex, fix $x_0 \in (a,b)$ and let ϕ satisfy (C.24). For every small parameter $h > 0$, we Taylor expand ϕ at $(x_0 + h)$ and $(x_0 - h)$ to get:

$$\frac{1}{2}\big(\phi(x_0+h) + \phi(x_0-h)\big) = \phi(x_0) + \frac{1}{2}h^2\phi''(x_0) + O(h^3).$$

On the other hand, convexity of u implies:

$$\frac{1}{2}(\phi(x_0+h) + \phi(x_0-h)) > \frac{1}{2}(u(x_0+h) + u(x_0-h)) \geq u(x_0) = \phi(x_0).$$

This yields: $\frac{1}{2}h^2\phi''(x_0) + O(h^3) > 0$. Dividing now by h^2 and passing to the limit with $h \to 0$ we conclude that $\phi''(x_0) \geq 0$, as claimed.

For the converse implication, we argue by contradiction. If u is not convex, then $\frac{1}{2}\big(u(x_1) + u(x_2)\big) < u\big(\frac{x_1+x_2}{2}\big)$ for some $x_1, x_2 \in (a,b)$ with $x_1 < x_2$. Without loss of generality, by adding to u a linear function if necessary, we may assume that $u(x_1) = u(x_2) = 0$. Also, by possibly shrinking the interval (x_1, x_2), we may assume that $u(x) > 0$ for all $x \in (x_1, x_2)$. For simplicity, let $x_2 = -x_1 > 0$ and fix $h > 0$ such that $x_2 + h, x_1 - h \in (a, b)$. For each $c > 0$ consider now the concave function:

$$\phi_c(x) \doteq c(x_2 + h)^2 - cx^2.$$

Since the family $\{\phi_c\}_{c>0}$ converges uniformly to 0 as $c \to 0$ on $[a, b]$, and since it strictly increases to $+\infty$ as $c \to \infty$, we may define:

$$c_0 \doteq \min\left\{c > 0;\ \ \phi_c \geq u \ \text{ on } \ [a, b]\right\} > 0.$$

Clearly, there must be $\phi_{c_0}(x_0) = u(x_0)$ for some $x_0 \in (a, b)$. Finally, define:

$$\phi(x) \doteq \phi_{c_0}(x) + \frac{1}{2}c_0(x - x_0)^2.$$

We easily see that $\phi(x_0) = u(x_0)$ and $\phi(x) > \phi_{c_0}(x) \geq u(x)$ for all $x \in [a, b]$. By modifying ϕ outside of a neighbourhood of x_0 we may actually assume that ϕ satisfies (C.24). However:

$$\phi''(x_0) = \phi''_{c_0}(x_0) + c_0 = -c_0 < 0,$$

contradicting $\phi''(x_0) \geq 0$.

References

J. Adams and J.F. Fournier. Sobolev spaces. volume 140 of *Pure and Applied Mathematics*. Academic Press, 2003.

T. Antunovic, Y. Peres, S. Sheffield, and S. Somersille. Tug-of-war and infinity Laplace equation with vanishing Neumann boundary condition. *Communications in Partial Differential Equations*, 37(10):1839–1869, 2012.

S. Armstrong and Ch. Smart. A finite difference approach to the infinity Laplace equation and tug-of-war games. *Trans AMS*, 364:595–636, 2012.

A. Arroyo and M. Parviainen. Asymptotic holder regularity for ellipsoid process. 2019.

A. Arroyo, J. Heino, and M. Parviainen. Tug-of-war games with varying probabilities and the normalized p(x)-Laplacian. *Commun. Pure Appl. Anal.*, 16(3):915–944, 2017.

A. Arroyo, H. Luiro, M. Parviainen, and Ruosteenoja. Asymptotic lipschitz regularity for tug-of-war games with varying probabilities. 2018.

A. Attouchi, H. Luiro, and M. Parviainen. Gradient and lipschitz estimates for tug-of-war type games. 2019.

Ch. Bishop and Y. Peres. Fractals in probability and analysis. Cambridge University Press, 2017.

C. Bjorland, L. Caffarelli, and A. Figalli. Nonlocal tug-of-war and the infinity fractional Laplacian. *Comm. Pure Appl. Math.*, 65(3):337–380, 2012.

P. Blanc and J.D. Rossi. Games for eigenvalues of the Hessian and concave/convex envelopes. 2018.

P. Blanc and J.D. Rossi. Game theory and partial differential equations. volume 31 of *Nonlinear Analysis and Applications*. De Gruyter Series, 2019.

P. Blanc, J. Manfredi, and J.D. Rossi. Games for Pucci's maximal operators. 2018.

H. Brezis. Functional analysis, Sobolev spaces and partial differential equations. volume 2011 Edition of *Universitext*. Springer, 2011.

R. Buckdahn, P. Cardaliaguet, and M. Quincampoix. A representation formula for the mean curvature motion. *SIAM Journal on Mathematical Analysis*, 33(4):827–846, 2001.

J.R. Casas and L. Torres. Strong edge features for image coding. pages 443–450, 1996.

F. Charro, J. Garcia Azorero, and J.D. Rossi. A mixed problem for the infinity Laplacian via tug-of-war games. *Calculus of Variations and Partial Differential Equations*, 34(3):307–320, 2009.

J. Christensen. On some measures analogous to Haar measure. *Mathematica Scandinavica*, 26: 103–103, 1970.

M. Lewicka, *A Course on Tug-of-War Games with Random Noise*, Universitext, https://doi.org/10.1007/978-3-030-46209-3

L. Codenotti, M. Lewicka, and J. Manfredi. Discrete approximations to the double-obstacle problem, and optimal stopping of tug-of-war games. *Trans. Amer. Math. Soc.*, 369:7387–7403, 2017.

M. Crandall, H. Ishii, and P.-L. Lions. User's guide to viscosity solutions of second order partial differential equations. *Bull. AMS*, 27:1–67, 1992.

F. del Teso, Manfredi J., and M. Parviainen. Convergence of dynamic programming principles for the p-Laplacian. 2018.

E. DiBenedetto. $C^{1+\alpha}$ local regularity of weak solutions of degenerate elliptic equations. *Nonlinear Anal.*, 7:827–850, 1983.

K. Does. An evolution equation involving the normalized p-Laplacian. *Comm. Pure Appl. Anal.*, 10:361–369, 2011.

J.L. Doob. Classical potential theory and its probabilistic counterpart. Springer-Verlag New York, 1984.

R.M. Dudley. Real analysis and probability. Cambridge Studies in Advanced Mathematics. Cambridge University Press, 2004.

R. Durrett. Probability: theory and examples. volume 4th edition of *Cambridge Series in Statistical and Probabilistic Mathematics*. Cambridge University Press, 2010.

P. Erdos and A.H. Stone. On the sum of two Borel sets. *Proc. Amer. Math. Soc.*, 25:304–306, 1970.

L. Evans. A new proof of local $C^{1,\alpha}$ regularity for solutions of certain degenerate elliptic P.D.E. *JDE*, 45:365–373, 1982.

L. Evans. Partial differential equations. volume 2nd edition of *Graduate Studies in Mathematics*. American Mathematical Society, 2010.

M. Falcone, S. Finzi Vita, T. Giorgi, and R.G. Smits. A semi-Lagrangian scheme for the game p-Laplacian via p-averaging. *Applied Numerical Mathematics*, 73:63–80, 2013.

D Gilbarg and N. Trudinger. Elliptic partial differential equations of second order. volume 3rd edition of *Classics in Mathematics*. Springer, 2001.

I Gomez and J.D. Rossi. Tug-of-war games and the infinity Laplacian with spatial dependence. *Communications on Pure and Applied Analysis*, 12(5):1959–1983, 2013.

W. Hansen and N. Nadirashvili. A converse to the mean value theorem for harmonic functions. *Acta Math.*, 171 (2):136–163, 1993.

W. Hansen and N. Nadirashvili. Littlewood's one circle problem. *J. London Math. Soc.*, 50 (2): 349–360, 1994.

H. Hartikainen. A dynamic programming principle with continuous solutions related to the p-Laplacian, $1 < p < \infty$. *Differential Integral Equations*, 29(5–6):583–600, 2016.

J. Heinonen and T. Kilpeläinen. On the Wiener criterion and quasilinear obstacle problems. *Trans. Amer. Math. Soc.*, 310 (1):239–255, 1988a.

J. Heinonen and T. Kilpeläinen. $\mathcal{A}$-superharmonic functions and supersolutions of degenerate elliptic equations. *Ark. Mat*, 26:87–105, 1988b.

J. Heinonen, T. Kilpelainen, and O. Martio. Nonlinear potential theory of degenerate elliptic equations. Dover Publications, Inc., Mineola, NY, 2006.

L. Helms. Potential theory. Universitext. Springer, 2014.

R. Jensen. Uniqueness of Lipschitz extensions: Minimizing the sup norm of the gradient. *Archive for Rational Mechanics and Analysis*, 123(1):51–74, 1993.

V. Julin and P. Juutinen. A new proof of the equivalence of weak and viscosity solutions for the p-Laplace equation. *Commun. Part. Diff. Eq*, 37:934–946, 2012.

P. Juutinen, P. Lindqvist, and J.J. Manfredi. On the equivalence of viscosity solutions and weak solutions for a quasi-linear elliptic equation. *SIAM J. Math. Anal.*, 33:699–717, 2001.

P. Juutinen, T. Lukkari, and Parviainen M. Equivalence of viscosity and weak solutions for the p(x)-Laplacian. *Ann. Inst. H. Poincarè Anal. Non Linèaire*, 27(6):1471–1487, 2010.

O. Kallenberg. Foundations of modern probability. volume 2nd edition of *Probability and Its Applications*. Springer, 2002.

I. Karatzas and S. Shreve. Brownian motion and stochastic calculus. Graduate Texts in Mathematics. Springer, 1991.

B. Kawohl. Variational versus PDE-based approaches in mathematical image processing. *CRM Proceedings and Lecture Notes*, 44:113–126, 2008.

B. Kawohl, J. Manfredi, and M. Parviainen. Solutions of nonlinear PDEs in the sense of averages. *J. Math. Pures Appl.*, 97(2):173–188, 2012.

O.D Kellogg. Converses of Gauss' theorem on the arithmetic mean. *Trans. Amer. Math. Soc.*, 36: 227–242, 1934.

T. Kilpeläinen. Potential theory for supersolutions of degenerate elliptic equations. *Indiana Univ. Math. J*, 38:253–275, 1989.

T. Kilpeläinen and J. Malý. The Wiener test and potential estimates for quasilinear elliptic equations. *Acta Math.*, 172:137–161, 1994.

R.V. Kohn and S. Serfaty. A deterministic-control-based approach to motion by curvature. *Comm. Pure Appl. Math*, 59:344–407, 2006.

R.V. Kohn and S. Serfaty. A deterministic-control-based approach to fully nonlinear parabolic and elliptic equations. *Comm. Pure Appl. Math*, 63:1298–1350, 2010.

S. Koike. A beginner's guide to the theory of viscosity solutions. volume 13 of *MSJ Memoirs*. Mathematical Society of Japan, 2004.

O.A. Ladyzhenskaya and N.N. Ural'tseva. Linear and quasilinear elliptic equations. volume 46 of *Mathematics in Science and Engineering*. Academic Press, 1968.

E. Le Gruyer. On absolutely minimizing Lipschitz extensions and PDE $\delta_\infty u = 0$. *NoDEA*, 14: 29–55, 2007.

J.C. Le Gruyer, E.and Archer. Harmonious extensions. *SIAM J.Math. Anal.*, 29(1):279–292, 1998.

M. Lewicka. Random tug of war games for the p-Laplacian, $1 < p < \infty$. 2018.

M. Lewicka and J. Manfredi. Game theoretical methods in PDEs. *Bollettino dell'Unione Matematica Italiana*, 7(3):211–216, 2014.

M. Lewicka and Y. Peres. The Robin mean value equation ii: Asymptotic hölder regularity. 2019a.

M. Lewicka and Y. Peres. The Robin mean value equation i: A random walk approach to the third boundary value problem. 2019b.

M. Lewicka, J. Manfredi, and D. Ricciotti. Random walks and random tug of war in the Heisenberg group. *Mathematische Annalen*, 2019.

J. Lewis. Smoothness of certain degenerate elliptic equations. *Proc. Am. Math. Soc*, 80:259–265, 1980.

J. Lewis. Regularity of the derivatives of solutions to certain degenerate elliptic equations. *Indiana Univ. Math. J*, 32:849–858, 1983.

P. Lindqvist. Notes on the stationary p-Laplace equation. SpringerBriefs in Mathematics. Springer, 2019.

P. Lindqvist and T. Lukkari. A curious equation involving the ∞-Laplacian. *Adv. Calc. Var.*, 3(4): 409–421, 2010.

P. Lindqvist and O. Martio. Two theorems of N. Wiener for solutions of quasilinear elliptic equations. *Acta Math.*, 155:153–171, 1985.

W. Littman, G. Stampacchia, and H.F. Weinberger. Regular points for elliptic equations with discontinuous coefficients. *Ann. Scuola Norm. Sup. Pisa Sci*, 17 (3):43–77, 1963.

Q. Liu and A. Schikorra. General existence of solutions to dynamic programming equations. *Communications on Pure and Applied Analysis*, 14(1):167–184, 2015.

H. Luiro and M. Parviainen. Gradient walk and p-harmonic functions. *Proc. Amer. Math. Soc.*, 145:4313–4324, 2017.

H. Luiro and M. Parviainen. Regularity for nonlinear stochastic games. *Ann. Inst. H. Poincarè Anal. Non Linèaire*, 35(6):1435–1456, 2018.

H. Luiro, M. Parviainen, and E. Saksman. Harnack's inequality for p-harmonic functions via stochastic games. *Differential and Integral Equations*, 38(12):1985–2003, 2013.

H. Luiro, M. Parviainen, and E. Saksman. On the existence and uniqueness of p-harmonious functions. *Differential and Integral Equations*, 27(3/4):201–216, 2014.

J. Manfredi, M. Parviainen, and J. Rossi. An asymptotic mean value characterization for p-harmonic functions. *Proc. Amer. Math. Soc.*, 138(3):881–889, 2010.

J. Manfredi, M. Parviainen, and J. Rossi. Dynamic programming principle for tug-of-war games with noise. *ESAIM Control Optim. Calc. Var.*, 18:81–90, 2012a.

J. Manfredi, J.D. Rossi, and S. Sommersille. An obstacle problem for tug-of-war games. *Communications on Pure and Applied Analysis*, 14(1):217–228, 2015.

J.J. Manfredi, M. Parviainen, and J.D. Rossi. On the definition and properties of p-harmonious functions. *Ann. Sc. Norm. Super. Pisa Cl. Sci.*, 11(2):215–241, 2012b.

V.G. Maz'ya. On the continuity at a boundary point of solutions of quasi-linear elliptic equations. *Vestnik Leningrad Univ. Math.*, 3:225–242, 1976.

P. Mörters and Y. Peres. Brownian motion. Cambridge University Press, 2010.

J. Moser. On Harnack's theorem for elliptic differential equations. *CPAM*, 14:577–591, 1961.

M.E. Muller. Some continuous Monte Carlo methods for the Dirichlet problem. *Ann. Math. Statist.*, 27:569–589, 1956.

K. Nyström and M. Parviainen. Tug-of-war, market manipulation and option pricing. *Math. Finance*, 27(2):279–312, 2017.

A.M. Oberman. A convergent difference scheme for infinity Laplacian: construction of absolutely minimizing lipschitz extensions. *Math. Comp.*, 74:1217–1230, 2005.

K.R. Parthasarathy. Probability measures on metric spaces. Cambridge Mathematical Textbooks. Academic Press, 1967.

M. Parviainen and E. Ruosteenoja. Local regularity for time-dependent tug-of-war games with varying probabilities. *J. Differential Equations*, 261(2):1357–1398, 2016.

Y. Peres and S. Sheffield. Tug-of-war with noise: a game-theoretic view of the p-Laplacian. *Duke Math J.*, 145:91–120, 2008.

Y. Peres, O. Schramm, S. Sheffield, and D.B. Wilson. Tug-of-war and the inifnity Laplacian. *J. Amer. Math. Soc*, 22:167–210, 2009.

Y. Peres, G. Pete, and S. Somersille. Biased tug-of-war, the biased infinity Laplacian, and comparison with exponential cones. *Calculus of Variations and Partial Differential Equations*, 38(3–4):541–564, 2010.

E. Ruosteenoja. Local regularity results for value functions of tug-of-war with noise and running payoff. *Advances in Calculus of Variations*, 9(1):1–17, 2014.

J. Serrin. A Harnack's inequality for nonlinear equations. *Bull. Amer. Math.*, 69:481–486, 1963.

J. Serrin. Local behaviour of solutions of quasi-linear equations. *Acta Math.*, 111:247–302, 1964.

P. Tolksdorf. Regularity for a more general class of quasilinear elliptic equations. *JDE*, 51:126–150, 1984.

K. Uhlenbeck. Regularity for a class of nonlinear elliptic systems. *Acta Math.*, 138:219–240, 1977.

N. Uralceva. Degenerate quasilinear elliptic systems. *Zap. Naucn. Sem. Leningrad. Otdel. Mat. Inst. Steklov*, 7:184–192, 1968.

V. Volterra. Alcune osservazioni sopra proprietáatte individuare una funzione. *Rend. Acadd. d. Lincei Roma*, 5:263–266, 1909.

N. Wiener. Certain notions in potential theory. *J. Math. Phys.*, 3:24–51, 1924.

N. Wiener. Note on a paper of o. perron. *Journal of Math. and Phys.*, 4:21–32, 1925.

D. Williams. Probability with martingales. Cambridge Mathematical Textbooks. Cambridge University Press, 1991.

Index

M. Lewicka, *A Course on Tug-of-War Games with Random Noise*, Universitext,
https://doi.org/10.1007/978-3-030-46209-3

Printed in the United States
By Bookmasters